Real World Ecology

Real World Ecology

ShiLi Miao · Susan Carstenn
Martha Nungesser

Editors

Real World Ecology

Large-Scale and Long-Term Case Studies and Methods

Foreword by Stephen R. Carpenter

 Springer

Editors

ShiLi Miao
South Florida Water
   Management District
West Palm Beach
FL, USA
smiao@sfwmd.gov

Susan Carstenn
Hawaii Pacific University
Kaneohe, HI
USA
scarstenn@hpu.edu

Martha Nungesser
South Florida Water
   Management District
West Palm Beach
FL, USA
mnunges@sfwmd.gov

ISBN 978-0-387-77916-4   e-ISBN 978-0-387-77942-3   (Softcover)
ISBN 978-0-387-77941-6   e-ISBN 978-0-387-77942-3   (Hardcover)
DOI 10.1007/978-0-387-77942-3

Library of Congress Control Number: 2008941656

Printed on acid-free paper

springer.com

To our mothers, whose love and support we appreciate.

Foreword

Ecology is not rocket science – it is far more difficult (Hilborn and Ludwig 1993). The most intellectually exciting ecological questions, and the ones most important to sustaining humans on the planet, address the dynamics of large, spatially heterogeneous systems over long periods of time. Moreover, the relevant systems are self-organizing, so simple notions of cause and effect do not apply (Levin 1998). Learning about such systems is among the hardest problems in science, and perhaps the most important problem for sustaining civilization. Ecologists have addressed this challenge by synthesis of information flowing from multiple sources or approaches (Pickett et al. 2007). Some major approaches in ecology are theoretical concepts expressed in models, long-term observations, comparisons across contrasting systems, and experiments (Carpenter 1998). These approaches have complementary strengths and limitations, so findings that are consistent among all of these approaches are likely to be most robust.

Ecosystem data are noisy. There are multiple sources of variability, such as external forcing, endogenous dynamics, and our imperfect observations. Thus it is not surprising that statistics have played a central role in ecological inference. However, with few exceptions the statistical approaches available to ecologists have been imported from other disciplines and were designed for problems that are simpler than the ones that ecologists face routinely. If you need to cut a board and all you have is a hammer, you might try pounding on the board until it breaks. Such a misapplication of force resembles some uses of statistics in ecology. But the metaphor is not quite right. It would be more accurate to say that ecosystem and landscape ecologists need to create and compare multifaceted models for large-scale processes, whereas the readily available tools were designed for testing null models that are usually trivial or irrelevant for this family of ecological questions.

The mismatch between the needs of scientists and the availability of statistical tools is acute in the analysis of ecosystem experiments. Ecosystem experiments have been an important contributor to ecological science for more than 50 years (Likens 1985, Carpenter et al. 1995). While humans have manipulated ecosystems since at least the beginnings of agriculture, if not longer, deliberate experiments for learning about ecosystems are traced to limnology in the 1940s (Likens 1985). The earliest whole-lake manipulations lacked reference systems, and so sometimes it was difficult to determine whether changes in the ecosystems were caused by the manipulations or by other environmental factors. In 1951, Arthur Hasler and his students divided an hour-glass shaped lake with an earthen dam, thereby creating two basins, Paul and Peter lakes. Peter Lake was manipulated, while Paul Lake served as an unmanipulated reference ecosystem (Johnson and Hasler 1954). The use of a reference or "control" ecosystem was a pathbreaking innovation (Likens 1985). It allowed Hasler and his students to separate the effects of the manipulations of Peter Lake from those of the environmental variability that affected both lakes (Stross and Hasler 1960, Stross et al. 1961). As a result of their experiences as students of Hasler, Gene Likens and Waldo Johnson were inspired to create two of the most influential centers of ecosystem experimentation in the world, the Hubbard Brook Ecosystem Study (Likens 2004) and the Experimental Lakes Area (Johnson and Vallentyne 1971).

Most ecosystem experiments involve spatially extensive systems (often observed at several spatial extents) over long time spans. Such experiments pose statistical challenges that cannot be handled by the methods of laboratory science or small agricultural plots (Carpenter 1998). It is not possible to substitute small-scale experiments run for short periods of time, because results of such experiments do not predict dynamics at spatial and temporal scales relevant to ecosystem science or to management (Carpenter 1996, Schindler 1998, Pace 2001). Instead, we must perform our studies at the appropriate scales – possibly at multiple scales. Then, we must learn how to learn from noisy observations of transient, heterogeneous, and non-replicable systems. This is a daunting challenge.

Thus many ecologists have broken free of the constraints of older statistical methods in order to explore new alternatives that seem better-adapted to the world of large-scale ecological change. The method of multiple working hypotheses (Chamberlain 1890) is now explicit in many ecological papers. Multiple hypotheses are expressed as quantitative models and confronted with data (Burnham and Anderson 1998, Hilborn and Mangel 1997). New approaches are explored for long-term monitoring data (Stow et al. 1998). Experiments are designed for critical tests of multiple alternative models to address fundamental questions about ecological dynamics (Dennis et al. 2001, Wootton 2004). Comparisons of multiple models are providing new insights about long-term field observations of big systems (Ives et al. 2008). These are but a few selections from a diverse and rapidly growing literature. This new phase of ecological research is turbulent and subject to rapid intellectual progress. Some of the emerging practices are nonstandard and are themselves objects of inquiry. Some approaches are tried, found wanting, and abandoned. New approaches are introduced frequently. It is an era of creativity, innovation, discarding of mistakes, and selection among alternatives – in a nutshell, a time of rapid evolution by the discipline.

The volume before you presents a sampling of case studies and syntheses from this fertile field of research. The authors and editors aim to improve our tools for ecological inference at scales that are relevant for fundamental understanding, as well as for management of ecosystems and landscapes. The book conveys the excitement and novelty of emerging approaches for learning about large-scale ecological changes.

Madison, WI   Stephen R. Carpenter

References

Burnham, K.P. and D.R. Anderson. 1998. *Model Selection and Inference*. Springer-Verlag, N.Y., USA.

Carpenter, S.R. 1996. Microcosm experiments have limited relevance for community and ecosystem ecology. Ecology 77: 677–680.

Carpenter, S.R. 1998. The need for large-scale experiments to assess and predict the response of ecosystems to perturbation. pp. 287–312 in M.L. Pace and P.M. Groffman (eds.), *Successes, Limitations and Frontiers in Ecosystem Science*. Springer-Verlag, N.Y., USA.

Carpenter, S.R., S.W. Chisholm, C.J. Krebs, D.W. Schindler, and R.F. Wright. 1995. Ecosystem experiments. Science 269: 324–327.

Chamberlain, T.C. 1890. The method of multiple working hypotheses. Science 15: 92–96.

Dennis, B., R.A. Desharnais, J.M. Cushing, S.M. Henson, and R.F. Costantino. 2001. Estimating chaos and complex dynamics in an insect population. Ecological Monographs 71: 277–303.

Hilborn, R. and D. Ludwig. 1993. The limits of applied ecological research. Ecological Applications 3: 550–552.

Hilborn, R. and M. Mangel. 1997. *The Ecological Detective*. Princeton University Press, Princeton, N.J., USA.

Ives, A.R., A. Einarsson, V.A.A. Jansen, and A. Gardarson. 2008. High-amplitude fluctuations and alternative dynamical states of midges in Lake Myvatn. Nature 452: 84–87.

Johnson, W.E. and A.D. Hasler. 1954. Rainbow trout population dynamics in dystrophic lakes. Journal of Wildlife Management 18: 113–134.

Johnson, W.E. and J.R. Vallentyne. 1971. Rationale, background, and development of experimental lake studies in northwestern Ontario. Journal of the Fisheries Research Board of Canada 28: 123–128.

Levin, S.A. 1998. Ecosystems and the biosphere as complex adaptive systems. Ecosystems 1: 431–436.

Likens, G.E. 1985. An experimental approach for the study of ecosystems. Journal of Ecology 73: 381–396.

Likens, G.E. 2004. Some perspectives on long-term biochemical research from the Hubbard Brook Ecosystem Study. Ecology 85: 2355–2362.

Pace, M.L. 2001. Getting it right and wrong: extrapolations across experimental scales. pp. 157–177 in R.H. Gardner, W.M. Kemp, V.S. Kennedy and J.E. Peterson (eds.). *Scaling Relations in Experimental Ecology*. Columbia University Press, NY, USA.

Pickett, S.T.A., J. Kolasa, and C. Jones. 2007. *Ecological Understanding*. Academic Press, Burlington, Massachusetts, USA.

Schindler, D.W. 1998. Replication versus realism: The need for ecosystem-scale experiments. Ecosystems 1: 323–334.

Stow, C.A., S.R. Carpenter, K.E. Webster, and T.M. Frost. 1998. Long-term environmental monitoring: some perspectives from lakes. Ecological Applications 8: 269–276.

Stross, R.G. and A.D. Hasler. 1960. Some lime-induced changes in lake metabolism. Limnology and Oceanography 5: 265–272.
Stross, R.G., J.C. Neess and A.D. Hasler. 1961. Turnover time and production of the
planktonic crustacean in limed and reference portions of a bog lake. Ecology 42:
237–245.
Wootton, T. 2004. Markov chain models predict the consequences of ecological extinctions.
Ecology Letters 7: 653–660.

Contents

Contributors

Brandon T. Bestelmeyer U.S. Department of Agriculture, Jornada Experimental Range, Jornada Basin Long Term Ecological Research Program, Las Cruces, NM 88003-0003, USA, bbestelmeyer@nmsu.edu

Charles D. Canham Cary Institute of Ecosystem Studies, Millbrook, NY 12545 USA, ccanham@ecostudies.org

Steve R. Carpenter University of Wisconsin, Center for Limnology, Madison, WI 53706 USA, srcarpen@wisc.edu

Susan M. Carstenn Hawai'i Pacific University, College of Natural Sciences, Kaneohe, Hawaii 96744 USA, scarstenn@hpu.edu

Guangsheng Chen Auburn University, School of Forestry and Wildlife Sciences, Ecosystem Science and Regional Analysis Laboratory, Auburn, AL 36849, USA chengu1@auburn.edu

Chris Edelstein City of Griffin, Stormwater Department, 100 S Hill St., Griffin, GA 30224 USA, cedelstein@cityofgriffin.com

Marie-Josée Fortin University of Toronto, Department of Ecology and Evolutionary Biology, Toronto, Ontario, Canada M5S 3G5, mariejosee.fortin@utoronto.ca

James B. Grace U.S. Geological Survey, National Wetlands Research Center, Lafayette, LA 70506 USA, gracej@usgs.gov

Binhe Gu South Florida Water Management District, Everglades Division, West Palm Beach, FL 33406 USA, bgu@sfwmd.gov

Kris M. Havstad U.S. Department of Agriculture, Jornada Experimental Range, Jornada Basin Long Term Ecological Research Program, Las Cruces, NM 88003-0003, USA, khavstad@nmsu.edu

Jeffrey E. Herrick U.S. Department of Agriculture, Jornada Experimental Range, Jornada Basin Long Term Ecological Research Program, Las Cruces, NM 88003-0003, USA, jherrick@nmsu.edu

Dafeng Hui Tennessee State University, Department of Biological Sciences, Nashville, TN 37209 USA dhui@tnstate.edu

Alan K. Knapp Colorado State University, Graduate Degree Program in Ecology and Department of Biology, Fort Collins, CO 80524, USA, alan.knapp@ColoState.edu

E. Conrad Lamon Duke University, Nicholas School of the Environment and Earth Sciences, Levine Science Research Center, Durham, NC 27708 USA, conrad@duke.edu

Peter R. Leavitt University of Regina, Department of Biology, Regina, SK, Canada S4S 0A2, Peter.Leavitt@uregina.ca

Mingliang Liu Auburn University, School of Forestry and Wildlife Sciences, Ecosystem Science and Regional Analysis Laboratory, Auburn, AL 36849, USA, liuming@auburn.edu

Dengsheng Lu Auburn University, School of Forestry and Wildlife Sciences, Ecosystem Science and Regional Analysis Laboratory, Auburn, AL 36849, USA, luds@auburn.edu

Yiqi Luo University of Oklahoma, Department of Botany/Microbiology, Norman, OK 73019 USA, ylou@ou.edu

Suzanne McGowan University of Nottingham, School of Geography, University Park, Nottingham, NG7 2RD, UK, suzanne.mcgowan@nottingham.ac.uk

Stephanie J. Melles University of Toronto, Department of Ecology and Evolutionary Biology, Toronto, Ontario, Canada M5S 3G5, stephajm@zoo.utoronto.ca

ShiLi Miao South Florida Water Management District, STA Management Division, 3301 Gun Club Road, West Palm Beach, FL 33406 USA, smiao@sfwmd.gov

H. Curtis Monger New Mexico State University, Department of Plant and Environmental Sciences, Las Cruces, NM 88003-8003, USA, cmonger@nmsu.edu

Martha Nungesser South Florida Water Management District, Everglades Division, 3301 Gun Club Road, West Palm Beach, FL 33406 USA, mnunges@sfwmd.gov

Michael L. Pace Department of Environmental Sciences, University of Virginia, Charlottesville, VA 22904-4123, USA, pacem@virginia.edu

Shufen Pan Auburn University, School of Forestry and Wildlife Sciences, Ecosystem Science and Regional Analysis Laboratory, Auburn, AL 36849, USA, panshuf@auburn.edu

Debra P. C. Peters USDA-ARS, Jornada Experimental Range, New Mexico State University, Las Cruces, NM 88003-0003 USA, debpeter@nmsu.edu

Song S. Qian Duke University, Nicholas School of the Environment and Earth Sciences, Durham, NC 27708 USA, song@duke.edu

Kenneth H. Reckhow Duke University, Nicholas School of the Environment and Earth Sciences, A317 Levine Science Research Center, Durham, NC 27708 USA, reckhow@duke.edu

Wei Ren Auburn University, School of Forestry and Wildlife Sciences, Ecosystem Science and Regional Analysis Laboratory, Auburn, AL 36849, USA, renwei1@auburn.edu

Samule M. Scheiner National Science Foundation, Division of Environmental Biology, Arlington, VA 22230, USA, sscheine@nsf.gov

Erik Sindhøj Swedish University of Agricultural Sciences (SLU), Faculty of Natural Resources and Agricultural Sciences, SE-750 07 Uppsala, Sweden, Erik.Sindhoj@mv.slu.se

Particia A. Soranno Department of Fisheries and wildlife, Michigan state University, East Lansing, MI 48824, USA, soranno@msu.edu

Craig A. Stow NOAA Great Lakes Environmental Research Laboratory, Aquatic Ecosystem Modeling, Ann Arbor, MI 48105-2945 USA, craig.stow@noaa.gov

Cassondra R. Thomas TBE Group, West Palm Beach, FL 33411, USA, crthomas@tbegroup.com

Hanqin Tian Auburn University, School of Forestry and Wildlife Sciences, Auburn, AL 36849, USA, tianhan@auburn.edu

Xiaofeng Xu Auburn University, School of Forestry and Wildlife Sciences, Auburn, AL 36849, USA, xuxiaof@auburn.edu

Andrew Youngblood La Grande Forestry and Range Sciences Laboratory, La Grande, Oregon 97850-3368 USA, ayoungblood@fs.fed.us

Chi Zhang Auburn University, School of Forestry and Wildlife Sciences, Auburn, AL 36849, USA, zhangch@auburn.edu

Chapter 1
Introduction – Unprecedented Challenges in Ecological Research: Past and Present

ShiLi Miao, Susan Carstenn and Martha Nungesser

1.1 Unprecedented Challenges in Ecological Research

The focus of ecological research has been changing in fundamental ways as the need for humanity to address large-scale environmental perturbations and global crises increasingly places ecologists in the limelight. Ecologists are asked to explain and help mitigate effects from local to global scale issues, such as climate change, wetlands loss, hurricane devastation, deforestation, and land degradation. The traditional focus of ecology as "the study of the causes of patterns in nature" (e.g., Tilman 1987) has shifted to a new era in which ecological science must play a greatly expanded role in improving the human condition by addressing the sustainability and resilience of socio-ecological systems (Millennium Ecosystem Assessment 2003, Palmer et al. 2004). In the twenty-first century, scientists studying ecological science are required not only to understand mechanisms of ecosystem change and develop new ecological theories but also to contribute to a future in which natural and human systems can coexist sustainably on the Earth (Carpenter and Turner 1998, Hassett et al. 2005). This unprecedented challenge demands that ecologists link science to planning, decision- and policy-making, forecasting ecosystem states, and evaluating ecosystem services and natural capital (Carpenter et al. 1998, Clark et al. 2001b). To realize these goals, ecologists must expand temporal and spatial scales of research, develop novel design approaches and analytical tools that meet the demands of this increasingly complex milieu, and provide education and training in using these tools.

Ecological research began with observational field studies, then moved to experimentation, at which time the difficulty of isolating and controlling the variables that influence ecosystems became apparent (McIntosh 1985). In response, ecologists tried to reproduce systems on a smaller spatial scale using microcosms and mesocosms, where the influence of variables could be systematically isolated, controlled, and tested (Forbes 1887, Beyers 1963, Hutchinson

S. Miao (✉)
South Florida Water Management District, STA Management Division, 3301 Gun Club Road, West Palm Beach, FL 33406, USA
e-mail: smiao@sfwmd.gov

S. Miao et al. (eds.), *Real World Ecology*, DOI 10.1007/978-0-387-77942-3_1,
© Springer Science+Business Media, LLC 2009

1964, Abbott 1966, 1967). Emphases on whole ecosystem studies followed Tansley (1935) and H.T. Odum and E. P. Odum's ecosystem concepts (Odum and Odum 1955, Odum 1955) have been pursued for over half a century by ecologists working in a wide array of ecosystems including forests (Edmisten 1970, Likens et al. 1970, Beier and Rasmussen 1994), lakes (Schindler 1971, 1973, Carpenter 1996, Vitousek et al. 1997, Carpenter 1998, Lamon III et al. 1998), deserts (Schlesinger 1990, Havstad et al. 2006), grasslands (Risser and Parton 1982, McNaughton and Chapin 1985), estuaries (Martin et al. 1990, Martin et al. 1994), and wetlands (Odum et al. 1977, Woodwell 1979, Likens 1985, Niswander and Mitsch 1995, Mitsch et al. 1998). From the persistent efforts of ecologists, including ecosystem and landscape ecologists, ecosystem science has developed into a well-established and diverse discipline that bridges the gap between fundamental research and applied problem solving (Carpenter and Turner 1998, Schindler 1998, Turner 2005). However, these and other studies collectively revealed critical issues for the advancement of ecology: multiple scales of spatial and temporal extent and variability, complex interactions, and system feedbacks.

Ecologists must now reach beyond focusing on simple systems with few variables to addressing complex ecosystems and landscapes with many uncontrolled variables operating across multiple spatial and temporal scales (Carpenter 1996, Peters et al. 2008). Conceptually, these ideas are illustrated in Fig. 1.1. Traditional ecological studies have focused largely on small-scale, short-term questions that can be addressed by replicated designs and statistical null hypothesis testing. Statistical tools for these questions are well understood and widely applied. These studies have tried to define causal relationships and develop ecological theories that lead to greater understanding of natural systems. The applicability of these approaches to provide greater understanding is limited by space, time, complexity, and the ability to replicate the study sites. However, many of the statistical techniques (in the toolbox on the left, Fig. 1.1) used to design and analyze these studies are not transferable to large-scale ecological research, as it generally encompasses large spatial scales, long temporal scales, and high complexity including feedbacks and nonlinear dynamics resulting in statistical uncertainty. New tools have been and continue to be developed to address the demands for analytical procedures that support predictions of future ecological conditions, policy development, environmental management and sustainability, and assessing environmental impacts, but refinement is needed for both experimental design and analysis.

To some ecologists, it has become apparent that many classical statistical approaches developed in other fields, such as randomized block design and analysis of variance, no longer fit the scope and objectives of large-scale and long-term ecosystem and landscape studies (Carpenter 1998, Grace et al. Chapters 2 and 11, Canham and Pace Chapter 8). Increasingly, ecologists question the appropriateness of null hypothesis testing in impact assessment and ecosystem restoration (Carpenter 1998, Green et al. 2005, Stephens et al. 2006). There is an

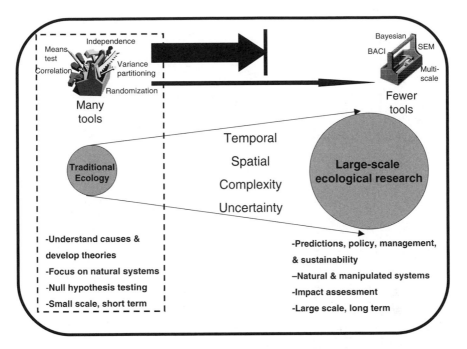

Fig. 1.1 We propose a modified paradigm for ecology that encompasses a much broader physical and temporal scale than traditionally taught ecology, requiring a very different approach to analysis and design of experiments. The approach is similar to that of the Gordian knot, known in mythology as a seemingly intractable problem that can be solved by a bold stroke. The problems addressed are those involving management, sustainability, policy, impact assessment, and predictions based upon issues that encompass large spatial scales, long temporal scales, complexity, and uncertainty that render traditional, null hypothesis based studies irrelevant. New experimental designs and analytical tools are required to address these questions. Traditional ecology, in contrast, focuses on smaller-scale, short-term questions that can be addressed by replicated designs and statistics that test a null hypothesis. Statistical tools for these questions are well understood and widely addressed, and while some may be applicable to these non-traditional approaches, most are generally inappropriate for these complex and difficult ecological questions

increasing call for a paradigm shift in statistical methodology (McBride et al. 1993, Maurer 1998, Germano 1999, Johnson 1999, Anderson et al. 2000, McBride 2002), and leadership is urgently needed to ask appropriate questions, design better studies, and develop analytical practices for ecological research in today's world.

In this volume, we offer our experience in design and analysis to meet the challenges and find solutions to large-scale and long-term ecological research issues. In each case study, alternative designs and/or analytical approaches are applied to research where replication was not practical, incorporating temporal and spatial scaling, and other challenges of non-traditional research (Table 1.1).

Table 1.1 Highlights of nine case-study chapters including ecological issues, systems, scales, and their design and analytical features

Chapter	Authors	Ecological issues	Systems	Scales	Response variables or Parameters	Disturbance	Design or Statistical issue	Analytical approach		
								Statistical modeling	Conceptual or Empirical modeling	
2	Grace & Youngblood	Fire & forest management	Forest	Multiple spatial	Vegetation & pine bark beetle	Pulse	Partitioning variance vs ecological understanding	Structural equation modelling	
3	Peters et al.	Regime shift grassland to shrublands	Grassland	Cross-scale	Native plant cover, density, & spatial distribution	Press	Cross-scale design	Quantile regression	Simulation, Cellular automata
4	Miao et al.	Fire & ecosystem restoration	Wetland	Multiple temporal & spatial	Water, soil, & vegetation total phosphorus	Pulse; Press	Multiple-scale design & asymmetric sampling scheme	Moving regression vs. ANOVA	
5	Stow et al.	Ecosystem management	Lake	Multiple temporal & spatial	Water column chlorophylla & total phosphorus	Press	Multi-level, cross-system inference & prediction	Bayesian hierarchical models	
6	Fortin & Melles	Avian response to forest loss	Forest	Large-scale pattern	Ovenbird	Press	Interplay of data acquisition, resolution & spatial structure scales	Univariate	

Table 1.1 (continued)

Chapter	Authors	Ecological issues	Systems	Scales	Response variables or Parameters	Disturbance	Design or Statistical issue	Analytical approach	
								Statistical modeling	Conceptual or Empirical modeling
7	McGowan & Leavitt	Climate change, fisheries, & lake management	Lake	Multiple temporal & spatial	Sediments, isotopes, pigments, water, quality, & salmon	Multiple	Retrospective analysis	Synchrony, variance partitioning, time series, & correlations	explicit spatial contrasts
8	Canham & Pace	Watershed nutrient loading	Watersheds & Lake	Linkages between scales	Chemical constituents	Press	Identifying model parameters	Spatial regression	Empirical
9	Tian et al.	Human-induced changes & Scaling	Forest, grassland, cropland	Regional (US & China)	Carbon storage & flux	Press	Extrapolation; assessment; prediction		Integrated regional modelling
10	Luo & Hui	Climate Change	Telexstrial Ecosystem	Multiple	Photosynthesis; C partitioning & respiration	Step v. Gradual	Prediction		Inverse analysis

1.2 Major Developments of Alternative Experimental Designs

The evolution of alternative design approaches started in the middle of the twentieth century and continues today. Non-replicated experimental design, such as paired treatment and control or reference, dates back to 1948, when Hasler and his colleagues (Hasler et al. 1951) experimented with two lakes in Chippewa County, Wisconsin; one lake was experimentally manipulated (limed) and the other served as a control or reference. The paired treatment–control design was applied by Likens and his colleagues to study forested ecosystem processes and associated aquatic ecosystems on a watershed scale at the Hubbard Brook Experimental Forest in New Hampshire (Bormann et al. 1968, Likens 1985 and references therein). Later, Box and Tiao (1965, 1975b) developed the Before–After (BA) design to assess the effects of new environmental laws and a new freeway on Los Angeles air-pollution levels, after which it was applied to other air pollution studies (Hilborn and Walters 1981, Morrisey 1993). The BA design has no "control," therefore it cannot eliminate the possibility that an effect may have resulted from something other than the studied impacts or treatments. To address this shortcoming, Green (1979) recommended sampling both an impact and a control site before and after a disturbance, i.e., Before–After-Control-Impact (BACI), as an appropriate design for environmental assessment, emphasizing the necessity of the control site.

Some approaches to BACI were criticized by Hurlbert (1984) because of potential problems in statistical inference arising from the lack of independent replicates, both spatial and temporal, which was termed "pseudoreplication." Though Hurlbert's argument was countered for the specific issue of "pseudoreplication in time" (Stewart-Oaten et al. 1986), the central premise of the paper stimulated a discussion among ecologists, statisticians, and editors of ecological journals that has highlighted limitations in both classical statistical inference and ecological experimentation (Carpenter 1990, Hargrove and Pickering 1992, Carpenter 1996).

Furthermore, some scientists developed the Before–After-Control-Impact Paired Series design (BACIPS), which estimates not only the spatial variability of data collected from a treatment and control but also the temporal variability of the data (Bernstein and Zalinski 1983, Stewart-Oaten et al. 1986). The BACIPS design was further developed through theoretical and practical applications and summarized in a book edited by Osenberg and Schmitt (1996) and in a monograph by Stewart-Oaten and Bence (2001). In spite of continued improvements, the BACIPS design has not been enthusiastically received. One reason hindering BACIPS wide-scale acceptance and application is that it requires extensive sampling both before and after the treatment or impact, and often these data are not available. Nonetheless, various modified non-replicated designs have continued to emerge including Beyond BACI (Underwood 1992, 1993, 1994) and multiple BACI (MBACI) (Keough and Quinn 2000). In an attempt to increase statistical rigor, the Beyond BACI design employs multiple randomly selected control locations, and the MBACI design includes both multiple controls and multiple treatment sites. Non-replicated designs have received limited acceptance for environmental monitoring and are virtually ignored by scientists in experimental ecological fields.

The BACI design and its more recent modifications have considerably improved the power and sensitivity of statistical procedures to detect impacts by minimizing the confounding effects imparted by natural variation and factors other than experimental manipulation. This design has been used in both the design of environmental assessments (Stewart-Oaten et al. 1992) and ecosystem evaluation studies (Anderson and Dugger 1998). For example, the largest river restoration project in the world, Florida's Kissimmee River Restoration Project, incorporates ecological monitoring studies that use BACI- and BACIPS-like sampling designs to evaluate restoration success (Bousquin et al. 2005).

More recently, Legendre et al. (2002) considered whether spatial autocorrelation effects could be eliminated by varying the design of field surveys and by conducting stochastic simulations to evaluate which design provides the greatest statistical power. In addition, multi-scale experimental designs have emerged as a powerful tool for identifying the mechanisms underlying ecosystem change. Ellis and Schneider (1997) integrated Control/Impact (CI) and BACI designs along an environmental gradient to detect the spatial extent and varying magnitude of environmental impacts. Petersen et al. (2003) proposed multi-scale experiments in coastal ecosystems. Peters et al. (Chapter 3) applied a design incorporating multiple interacting scales to assess pattern and mechanisms of woody species encroachment into grasslands, and Miao et al. (Chapter 4) applied MBACI designs to assess ecological impacts of repeated fires on wetland ecosystem restoration. Moreover, Barnett and Stohlgren (2003) and Hewitt et al. (2007) demonstrated the effectiveness and required spatial extent of a monitoring program by applying a multi-scale nested sampling design to assess local and landscape-scale heterogeneity of plant species richness. They argued that spatial and temporal nesting increased cost-effectiveness of assessing cumulative effects of diffuse impacts and numerous point sources. These studies demonstrate that ecologists have gradually realized that rather than struggle with controlling or minimizing spatial and temporal variations, they should incorporate and account for variation as well as natural history and other prior knowledge, including long-term monitoring data (Peters et al. 2006, Hewitt et al. 2007, Miao et al. Chapter 4).

1.3 Major Developments of Alternative Analytical Approaches

The development of non-replicated experimental designs has paralleled the development of alternative statistical analyses in the 1970s and 1980s, with advances in one providing impetus to the other. Alternative experimental designs such as BA, BACI, or BACIPS require analyzing time–series data within one site or comparing unreplicated impact and control sites. Intervention Analysis (Box and Tiao 1975a) based on a BA design used a mixed autoregressive moving average model and maximum likelihood estimates for model parameters to detect an effect resulting from a disturbance. Furthermore, Randomized Intervention Analysis (RIA), based on Monte Carlo simulations, was applied to detect whether an impacted ecosystem changed relative to a control ecosystem, while considering serial correlation within the time-series data (Carpenter et al. 1989). These methods were employed because the data did not meet classical statistical assumptions of an ordinary t-test: normality, constant variance, and independence. BACI designs, t-tests, or ANOVA models, with or without modifications for variance allocation, were applied after designing a sample scheme that would avoid serial correlation among the data and ascertaining whether the assumption of independence was met (Smith 2002). A t-test was proposed by Stewart-Oaten and Bence (2001) for the BACIPS design, while an asymmetric ANOVA was recommended by Underwood (1993, 1994) for the Beyond BACI design. However, an ecologist's statistical tools must move beyond the ANOVA paradigm (Grace et al. Chapters 2 and 11) to maximize our understanding of ecological systems and processes and identify the mechanisms that underlie ecosystem change, rather than simply accepting or rejecting a null hypothesis.

Contemporary statistical tools such as maximum likelihood, meta-analysis, information theory, Bayesian statistics, structural equation modeling (SEM), and inverse analysis, readily applied in other scientific fields including conservation biology, wildlife management, meteorology, and paleoecology are increasingly applied to ecology (Clark et al. 2001a, Holl et al. 2003, Pugesek et al. 2003, Grace et al. 2005, Green et al. 2005, Hilty et al. 2006, Hobbs and Hilborn 2006). Likelihood methods are extremely flexible when identifying best fit parameters, including strongly skewed and non-normal data (Aguiar and Sala 1999, Pawitan 2001, Hobbs and Hilborn 2006), allowing both linear and nonlinear models to fit to data. The likelihood approach also provides a basis for meta-analysis, information theory, and Bayesian analyses. Meta-analysis incorporates disparate, albeit carefully selected experimental data including pseudoreplicated studies, into a statistical statement of cumulative knowledge (Hedges and Olkin 1985, Hunt and Cornelissen 1997, Gurevitch et al. 2001). Bayesian statistics have received more attention than the others as a result of persistent efforts by a group of leading ecologists including Reckhow (1990), Ellison (1996, 2004), Lamon and Stow (2004), Clark (2005), and McCarthy (2007). Bayesian statistical models are designed to incorporate information from multiple sources to explicitly use results of previous studies as well as current experiments, observations, or manipulations. This multi-source feature allows relatively wide application for resource management and policy decisions. Bayesian models offer distinct advantages over classical null hypothesis testing. They provide a posterior probability distribution for the model parameters which potentially can be used to support a wide range of decisions that apply multiple decision criteria and prediction function testing, while approaches used in classical null hypothesis testing are much more constrained.

Structure Equation Model (SEM) is essentially a multivariate extension of regression and correlation analyses derived from the original concept of path analysis (Grace et al. Chapter 2). It is a powerful tool for inferring cause and effect relationships in the absence of experimental manipulation (Pugesek et al. 2003, Grace 2006) and therefore offers a more comprehensive, efficient, and effective framework than the traditional ANOVA-based experimental approaches for learning about processes from data (Grace et al. Chapter 2). Inverse analysis is an approach that focuses on data analysis to estimate parameters and their variability in order to evaluate model structure and information content of data. Overall, novel experimental design and analytical approaches such as those mentioned above are capable of addressing the complexity and uncertainty of large-scale ecosystem studies.

1.4 Ongoing Issues

These developments in design and analysis are not yet mainstream in ecological studies. In 1990, Carpenter and others contributed to a special edition of the journal *Ecology* in which they called for developing non-replicated experimental design and novel statistical analyses. In the 10 years (1990–2000) following Carpenter's appeal for development of new statistical methods, relatively few papers used BACI, BACIPS, MBACI, and similar approaches, with that number increasing slightly from 2000 to 2006. Since 1990, over 140 papers appeared in refereed journals that used a version of these methods: 42% in the USA, 19% in Australia–New Zealand, and 18% in Europe. Studies using BACI or one of its variants were conducted most frequently in aquatic ecosystems (marine 34%, freshwater 29%) while only 12% were conducted in forests. Fifty-three percent included animals (19% fisheries and 14% birds). Use of these analytical techniques was most common for impact assessment (53%) and management issues (11%), though more typical research questions (14%) were also reported. Surprisingly, only 13% of the articles addressed restoration and habitat improvement. Because this survey was conducted for scientific papers searchable online, this review is likely incomplete, and there may be a bias against some of the earlier research.

Numerous reasons exist for the lag in acceptance of alternative design and analytical approaches. First of all, ecologists traditionally have been trained to design field experiments using randomized complete block design or orthogonal designs with systematic or random sampling, particularly when prior knowledge about spatial structure and patterning of the system does not exist. Additionally, ecologists have long relied on a relatively narrow set of statistical techniques to ask questions that can be answered using existing statistical frameworks (Grace et al. Chapter 2). Hobbs and Hilborn (2006) pointed out that "There is danger that questions are chosen for investigation by ecologists to fit widely sanctioned statistical methods rather than statistical methods being chosen to meet the needs of ecological questions." It is clear that students need experience conducting both traditional and novel analyses, but their professors, well versed in traditional analytical methods, are often not sufficiently experienced to engage their students in the use of novel approaches. Finally, established journals and their editors may shy away from reporting non-replicated designs and their associated data analyses because they, too, must obtain reviews from scientists who are most comfortable with classical approaches.

As part of our efforts to design a large-scale ecosystem study in the Florida Everglades, we (Miao and Carstenn 2005) attempted to integrate ecological research and management needs four years ago to advance the field of ecosystem ecology. In the process, we were confronted by many, if not most, of the design and analytical challenges addressed in this book. Echoing the concerns of the 1990 Ecology Special Feature, we organized a symposium for the 2006 Ecological Society of America (ESA) Annual Conference to share our and others' experiences with integrating new statistical approaches into the design and analysis of large-scale and long-term ecosystem and landscape studies. Following the ESA Symposium and a *Frontiers in Ecology and the Environment* editorial (Miao and Carstenn 2006), we heard repeatedly from eminent and junior scientists alike that there is not only a need to change techniques but also a need for guidance and examples of "how to change." In this book, we have united a group of scientists who have been working in the field of ecology for decades to present the ecological issues, challenges, novel solutions, and implications of their research.

1.5 Major Features of the Book

This book fills a unique niche in ecological methodology. It is neither a statistical book nor an experimental design book. Instead, it is a "how-to" book integrating design, analysis, and interpretation of large-scale and long-term case studies based on real-world ecological issues. Authors have emphasized their thought processes, communicating why they applied particular experimental designs and/or analytical approaches to answer their research questions. In doing so, each case study begins with issue identification; includes experimental design, analysis, and interpretation; and concludes with appropriate management recommendations. Each chapter emphasizes the reasoning behind the approach rather than simply the results of an experiment, giving each chapter a flavor very different from scientific journals. It offers a unique and rich "behind the scenes" learning experience to readers that they usually do not gain from scientific journal articles covering the same topics. This educational aspect encourages multiple readings of chapters where the approach may not be familiar.

Overall, the structure of the book is broken into design, analysis, and modeling. The book offers an array of alternative perspectives and options for the design and analysis of large-scale and long-term ecological studies (Table 1.1). Grace et al. (Chapter 2) critique conventional experimental practices that use ANOVA-based experimental approaches and strongly recommend rethinking the dominant role of ANOVA in ecological studies. ANOVA models (including their derivatives ANCOVA and MANOVA) have dominated ecological analyses and are often considered to be the preferred model for analyses. Overemphasized in the biological sciences, they are poorly suited to the analysis of systems. For example, one striking characteristic of the ANOVA approach is its use of "replications." In this book, authors from diverse backgrounds and ecological fields have shown that for many large-scale and long-term studies including watersheds, wetlands, fire, global climate change, landscape regime shifts, and paleoecology (Table 1.1), replication of ecosystem and landscape disturbances or treatments is neither possible nor desirable under real-world circumstances (Schindler 1998). Alternatives better suited to the study of multi-process system models deserve more attention (Grace et al. Chapter 2). It is time for large-scale ecological studies to develop alternatives rather than just applying replication and randomization to cope with system variation (Carpenter 1990, Hewitt et al. 2001, Hewitt et al. 2007, Miao et al. Chapter 4). For example, Canham and Pace (Chapter 8) employed an inverse approach to asking research questions about processes and answered their questions using an alternative modeling approach. They argue that instead of focusing on "statistical significance" of an effect of a manipulative experiment, ecosystem ecologists and/or resource managers should address the questions of where, when, and most importantly, how much a system was affected by the manipulation. Traditionally, ecologists and hydrologists have devoted enormous efforts to the intensive *direct* measurement of one or a few variables, yet these data provide little insight into predicting whole system performance. An inverse approach which asks "what would the rate of the process have to be given the data available" uses readily measured variables (e.g., lake chemistry) to model processes, then predicts process responses to changing variables.

Another unique feature of this book is that it not only stimulates scientific aspirations for alternative novel approaches but also provides diverse *solutions* to individual problems in research design, statistical analysis, and modeling approaches to assess ecological responses to natural and anthropogenic disturbances at ecosystem and landscape levels. Several chapters present readers with a clear picture of steps taken by the authors to move beyond the dilemmas they faced and overcame obstacles by linking design and analytical techniques. For example, Peters et al. (Chapter 3) outlined a multi-scale experimental design with relevant analytical techniques to examine the key processes influencing woody plant encroachment from fine to broad scales. Miao et al. (Chapter 4) applied multi-scale spatial controls to contend with variation arising from system spatial structure and asymmetric temporal sampling schemes for response variables which operate at different biological organization levels, thereby revealing both short- and long-term fire effects on a wetland ecosystem.

In addition to design, several chapters provide solid arguments and examples of alternative statistical methods to solve real-world problems, particularly those related to spatial and temporal variation. For example, Grace and colleagues (Chapter 2) applied SEM to two field experimental studies, plant diversity in coastal wetlands and the effects of thinning and burning on delayed mortality in Ponderosa pine forests. They illustrated how the application of SEM to ecological problems, especially large-scale studies, can contribute to the scientific understanding of natural systems. Stow and his colleagues (Chapter 5) advocate a popular cross-system approach for large-scale ecological inference in limnological studies (Cole et al. 1991). They developed several alternative multilevel Bayesian models for chlorophyll a concentrations and total phosphorus concentrations, and suggested that working in a Bayesian framework provides measures of uncertainty that can be used to evaluate the probability that management objectives will be achieved under differing strategies. Fortin and Melles (Chapter 6) analyzed spatial responses of avian bird species to forest spatial heterogeneity at the landscape level. They addressed data acquisition, resolution, spatial structure, and statistical analyses; identified statistical challenges that emerged while analyzing spatially autocorrelated data; and proposed a series of widely applicable analytical steps to help determine which spatial and numerical methods best estimated species' responses to changes in forest cover at the regional scale. McGowan and Leavitt (Chapter 7) highlighted the role of paleoecology in ecosystem science by demonstrating how the modes and causes of ecological variation can be identified by analysis of long time series (100–1000s year) using numerous statistical approaches, including ecosystem synchrony, variance partitioning analysis, and explicit spatial contrasts among lakes. These retrospective studies were used to generate clear management options for pressing environmental issues such as sustainable fisheries, management, and climate change.

Modeling efforts have been widely recognized and accepted for scaling-up traditional experiments and solving management problems. However, most current modeling approaches are still constrained when ecological and management issues are addressed on regional spatial scales and/or long temporal scales (King 1991, Tian et al. 1998, Tian et al. Chapter 9). Canham and Pace (Chapter 8) present a new approach to analyzing the linkages between watersheds and lakes based on a simple, spatially explicit, watershed-scale model of lake chemistry. Their modeling approach provides a means to test questions on regional scales using the power of data from large numbers of watersheds that produce robust parameter estimates and comparisons of models. Tian and others (Chapter 9) attempted to predict the growth of plants, animals, and ecosystems in the future when climate, CO_2, and other factors will likely differ greatly from today. They employed an integrated regional modeling approach to effectively reorganize data collected on multiple scales to make them consistent with the study scale while preventing information loss and distortion.

Luo and Hui (Chapter 10) applied inverse analysis to Duke Forest Free Air CO_2 Enrichment (FACE) experimental data demonstrating that uncertainty in both parameter estimations and carbon sequestration in forest ecosystems can be quantified. They argued that inverse analysis will play a more important role in global change ecology. The combination of forward and inverse approaches allows us to probe mechanisms underlying ecosystem responses to global change. Finally, Grace et al. (Chapter 11) presents a framework to describe how different types of analyses depend on the amount of data available and the amount of knowledge about mechanisms. The flexibility of model analysis procedures proposed permits a greater integration of process with data than up to this point, suggesting at least one way forward for the study of large-scale systems.

A further innovation of this book is that the authors present a comprehensive framework for ecological problem solving using new and recently published data (e.g., Chapters 4 and 6) rather than summaries of previously published research. Each chapter addresses the development of one or more new methodologies and their underlying philosophies to an extent that cannot usually be addressed adequately in a typical journal article. For each chapter, the methods are the primary message while the case study is the context in which the authors present their methods. This approach is intended to help researchers design and analyze their own work using similar methods by clearly connecting the challenges of ecological research, the limitations of traditional statistical paradigms, and the goals and purposes of scientific investigations.

Moreover, all chapters of the book were subjected to rigorous anonymous peer review. The chapters were first reviewed by the three editors, revised, and then submitted to two or three external reviewers to assure an extensive peer-review process. These reviews ensure more extensive critiques and editing than many journal articles receive.

Overall, from our unique perspectives, the authors of this book illustrate how we, as ecologists, can effectively address ecological questions under spatial, temporal, and budgetary constraints while using defensible quantitative but non-traditional techniques. The authors highlight successful case studies that use novel approaches to address large-scale or long-term environmental investigations. This collection of case studies showcases innovative experimental designs, analytical options, and interpretations currently available to theoretical and applied ecologists, practitioners, and biostatisticians. These case studies begin to address the challenges that ecologists increasingly face in understanding and explaining large-scale, long-term environmental change.

Acknowledgement We thank the South Florida Water Management District for supporting the Fire project that stimulated our investigation of the issues that eventually inspired us to pursue collaboration on this book, and for support for the book project. We also thank Hawaii Pacific University for providing release time for S. Carstenn. We appreciate C. Stow's generous contribution of his time for reviewing and raising important questions, J. Grace's enthusiastic support and discussion of statistical issues, critical comments from S. Carpenter and P. Leavitt on earlier drafts of this chapter, and S. Bousquin, D. Peters, B. Bestelmeyer, and A. Knapp for providing some references.

References

Abbott, S. 1966. Microcosm studies of estuarine waters: The replicability of microcosms. Journal of Water Pollution Control Federation **1**:258–270.

Abbott, S. 1967. Microcosm studies of estuarine waters: The effects of single doses of nitrate and phosphate. Journal of Water Pollution Control Federation **2**:113–122.

Aguiar, M. R. and O. E. Sala. 1999. Patch structure, dynamics and implications for the functioning of arid ecosystems. Trends in Ecology and Evolution **14**:273–277.

Anderson, D. H. and B. D. Dugger. 1998. A Conceptual Basis for Evaluating Restoration Success. American Wildlife and Natural Resource Conference, Lake Placid, FL.

Anderson, D. R., K. P. Burnham, and W. L. Thompson. 2000. Null hypothesis testing: problems, prevalence, and an alternative. The Journal of Wildlife Management **64**:912–923.

Barnett, D. T. and T. J. Stohlgren. 2003. A nested-intensity design for surveying plant diversity. Biodiversity and Conservation **12**:255–278.

Beier, C. and L. Rasmussen. 1994. Effects of whole-ecosystem manipulations on ecosystem internal processes. Trends in Ecology and Evolution **9**:218–223.

Bernstein, B. B. and J. Zalinski. 1983. An optimum sampling design and power tests for environmental biologists. Journal of Environmental Management **16**:35–43.

Beyers, R. J. 1963. The metabolism of twelve aquatic laboratory microcosms. Ecological Monographs **33**:255–306.

Bormann, F. H., G. E. Likens, D. W. Fisher, and R. S. Pierce. 1968. Nutrient loss accelerated by clear cutting of a forest ecosystem. Science **159**:882–884.

Bousquin, S. G., D. H. Anderson, D. J. Colangelo, and G. E. Williams. 2005. Introduction to baseline studies of the channelized Kissimmee River. Establishing a Baseline: Pre-Restoration Studies of the Channelized Kissimmee River **1**:1.1–1.19.

Box, G. E. P. and G. C. Tiao. 1965. A change in level of a nonstationary time series. Biometrika **52**:181–192.

Box, G. E. P. and G. C. Tiao. 1975a. Intervention analysis with applications to economic and environmental problems. Journal of the American Statistical Association **70**:70–79.

Box, G. E. P. and G. C. Tiao. 1975b. Intervention analysis with applications to economic and environmental problems. Journal of the American Statistical Association **70**:70–79.

Carpenter, S., N. F. Caraco, D. L. Correll, R. W. Howarth, A. N. Sharpley, and V. H. Smith. 1998. Nonpoint pollution of surface waters with phosphorus and nitrogen. Ecological Applications **8**:559–568.

Carpenter, S. R. 1990. Large-scale perturbations: Opportunities for innovation. Ecology **71**:2038–2043.

Carpenter, S. R. 1996. Microcosm experiments have limited relevance for community and ecosystem ecology. Ecology **77**:677.

Carpenter, S. R. 1998. The need for large-scale experiments to assess and predict the response of ecosystems to perturbation. Pages 287–312 *in* M. L. Pace and P. M. Groffman, editors. Limitations and frontiers in ecosystem science. Springer-Verlag.

Carpenter, S. R., T. M. Frost, D. Heisey, and T. K. Kratz. 1989. Randomized intervention analysis and the interpretation of whole-ecosystem experiments. Ecology **70**:1142–1152.

Carpenter, S. R. and M. G. Turner. 1998. At last: A journal devoted to ecosystem science. Ecosystems **1**:1–5.

Clark, D. A., S. Brown, D. W. Kicklighter, J. Q. Chambers, J. R. Thomlinson, J. Ni, and E. A. Holland. 2001a. Net primary production in tropical forests: an evaluation and synthesis of existing field data. Ecological Applications **11**:371–384.

Clark, J. S. 2005. Why environmental scientists are becoming Bayesians. Ecology Letters **8**:2–14.

Clark, J. S., S. R. Carpenter, M. Barber, S. Collins, A. Dobson, J. A. Foley, D. M. Lodge, M. Pascual, R. Pielke Jr., W. Pizer, C. Pringle, W. V. Reid, K. A. Rose, O. Sala, W. H. Schlesinger, D. H. Wall, and D. Wear. 2001b. Ecological forecasts: An emerging imperative. Science **293**:657–660.

Cole, J., G. Lovett, and S. Findlay, editors. 1991. Comparative analyses of ecosystems: patterns, mechanisms and theories. Springer-Verlag.

Edmisten, J. 1970. Studies of *Phytolacca icosandra*. Pages D183–188 *in* H. T. Odum and R. F. Pigeon, editors. A Tropical Rainforest. U.S. Atomic Energy Commission, Oak Ridge, Tennessee.

Ellis, J. I. and D. C. Schneider. 1997. Evaluation of a gradient sampling design for environmental impact assessment. Environmental Monitoring and Assessment **48**:157–172.

Ellison, A. M. 1996. An introduction to Bayesian inference for ecological research and environmental decision-making. Ecological Applications **6**:1036–1046.

Ellison, A. M. 2004. Bayesian inference in Ecology. Ecology Letters **7**:509–520.

Forbes, S. A. 1887. The lake as a microcosm. Bulletin of the Science Association of Peoria **15**:537–550.

Germano, J. D. 1999. Ecology, statistics, and the art of misdiagnosis: The need for a paradigm shift. Environmental Reviews **7**:167–190.

Grace, J. B., editor. 2006. Structural equation modeling and natural systems. Cambridge University Press, Cambridge.

Grace, J. B., L. K. Allain, H. Q. Baldwin, A. G. Billock, W. R. Eddleman, A. M. Given, C. W. Jeske, and R. Moss. 2005. Effects of prescribed fire in the coastal prairies of Texas. USGS Open File Report 2005–1287.

Green, J. L., A. Hastings, P. Arzberger, F. J. Ayala, K. L. Cottingham, K. Cuddington, F. Davis, J. A. Dunne, M.-J. Fortin, L. Gerber, and M. Neubert. 2005. Complexity in ecology and conservation: Mathematical, statistical, and computational challenges. BioScience **55**:501–510.

Green, R. H. 1979. Sampling Design and Statistical Methods for Environmental Biologists. John Wiley & Sons, University of Western Ontario.

Gurevitch, J., P. S. Curtis, and M. H. Jones. 2001. Meta-analysis in ecology. Advanced Ecological Research **32**:199–247.

Hargrove, W. W. and J. Pickering. 1992. Pseudoreplication: A *sine qua non* for regional ecology. Landscape Ecology **6**:251–258.

Hasler, A. D., O. M. Brynildson, and W. T. Helm. 1951. Improving conditions for fish in brown-water bog lakes by alkalization. Journal of Wildlife Management **15**:347–352.

Hassett, B., M. Palmer, E. Bernhardt, S. Smith, J. Carr, and D. Hart. 2005. Restoring watersheds project by project: trends in Chesapeake Bay tributary restoration. Frontiers in Ecology and the Environment **3**:259–267.

Havstad, K. M., L. F. Huenneke, and W. H. Schlesinger, editors. 2006. Structure and function of a Chihuahuan Desert ecosystem: the Jornada Basin Long Term Ecological Research site. Oxford University Press, Oxford.

Hedges, L. V. and I. Olkin. 1985. Statistical methods for meta-analysis. Academic Press, Orlando, Florida, USA.

Hewitt, J. E., S. E. Thrush, and V. J. Cummings. 2001. Assessing environmental impacts: Effects of spatial and temporal variability at likely impact scales. Ecological Applications **11**:1502–1516.

Hewitt, J. E., S. F. Thrush, P. K. Dayton, and E. Bonsdorff. 2007. The effect of spatial and temporal heterogeneity on the design and analysis of empirical studies of scale-dependent systems. The American Naturalist **169**:398–408.

Hilborn, R. and C. J. Walters. 1981. Pitfalls of environmental baseline and process studies. Environmental Impact Assessment Review **2**:265–278.

Hilty, L. M., P. Arnfalkb, L. Erdmannc, J. Goodmand, M. Lehmanna, and P. A. Wägera. 2006. The relevance of information and communication technologies for environmental sustainability – A prospective simulation study. Environmental Modelling and Software **21**:1618–1629.

Hobbs, N. T. and R. Hilborn. 2006. Alternatives to statistical hypothesis testing in ecology: A guide to self teaching. Ecological Applications **16**:5–19.

Holl, K. D., E. E. Crone, and C. B. Schultz. 2003. Landscape restoration: Moving from generalities to methodologies. BioScience **53**:491–502.

Hunt, R. and J. H. C. Cornelissen. 1997. Components of relative growth rate and their interrelations in 59 temperate plant species. New Phytologist **135**:395–417.

Hurlbert, S. H. 1984. Pseudoreplication and the design of ecological field experiments. Ecological Monographs **54**:187–211.

Hutchinson, B. E. 1964. The lacustrine microcosm reconsidered. American Science **52**:334–341.

Johnson, D. H. 1999. The insignificance of statistical significance testing. Journal of Wildlife Management **63**:763–772.

Keough, J. M. and G. P. Quinn. 2000. Legislative vs. practical protection of an intertidal shoreline in Southeastern Australia. Ecological Applications **10**:871–881.

King, D. A. 1991. Correlations between biomass allocation, relative growth rate and light environment in tropical forest saplings. Functional Ecology **5**:485–492.

Lamon, E. C. and C. A. Stow. 2004. Bayesian methods for regional-scale lake eutrophication models. Water Research **38**:2764–2774.

Lamon III, E. C., S. R. Carpenter, and C. A. Stow. 1998. Forecasting PCB concentrations in Lake Michigan Salmonids: A dynamic linear model approach. Ecological Applications **8**:659–668.

Legendre, P., M. R. T. Dale, M.-J. Fortin, J. Gurevitch, M. Hohn, and D. Meyers. 2002. The consequences of spatial structure for the design and analysis of ecological field surveys. Ecography **25**:601–615.

Likens, G. E. 1985. An experimental approach for the study of ecosystems: The Fifth Tansley Lecture. Journal of Ecology **73**:381–396.

Likens, G. E., F. H. Bormann, N. M. Johnson, D. W. Fisher, and R. S. Pierce. 1970. Effects of forest cutting and herbicide treatment on nutrient budgets in the Hubbard Brook watershed-ecosystem. Ecological Monographs **40**:23–47.

Martin, J. H., K. H. Coale, K. S. Johnson, S. E. Fitzwater, R. M. Gordon, S. J. Tanner, C. N. Hunter, V. A. Elrod, J. L. Nowicki, T. L. Coley, R. T. Barber, S. Lindley, A. J. Watson, K. Van Scoy, C. S. Law, M. I. Liddicoat, R. Ling, T. Stanton, J. Stockel, C. Collins, A. Anderson, R. Bidigare, M. Ondrusek, M. Latasa, F. J. Millero, K. Lee, W. Yao, J. Z. Zhang, G. Friederich, C. Sakamoto, F. Chavez, K. Buck, Z. Kolber, R. Greene, P. Falkowski, S. W. Chisholm, F. Hoge, R. Swift, J. Yungel, S. Turner, P. Nightingale, A. Hatton, P. Liss, and N. W. Tindale. 1994. Testing the iron hypothesis in ecosystems of the equatorial Pacific Ocean. Nature **371**:123–129.

Martin, J. H., M. Gordon, and S. Fitzwater. 1990. Iron in Antarctic waters. Nature **345**:156–158.

Maurer, B. A. 1998. Research: Ecology-Ecological science and statistical paradigms at the threshold. Science **279**:502–503.

McBride, G. B. 2002. Statistical methods helping and hindering environmental science and management. Journal of Agricultural Biological and Environmental Statistics **7**:300–305.

McBride, G. B., J. C. Loftis, and N. C. Adkins. 1993. What do significance tests really tell us about the environment? Environmental Management **17**:423–432.

McCarthy, M. A., editor. 2007. Bayesian Methods for Ecology, Cambridge.

McIntosh, R. P. 1985. The background of ecology: Concept and Theory. Cambridge University Press, Cambridge.

McNaughton, S. J. and F. S. Chapin, III. 1985. Effects of phosphorus nutrition and defoliation on C_4 graminoids from the Serengeti Plains. Ecology **66**:1617–1629.

Miao, S. and S. Carstenn. 2006. A new direction for large-scale experimental design and analysis. Frontiers in Ecology 4:227.

Miao, S. L. and S. Carstenn. 2005. Assessing long-term ecological effects of fire and natural recovery in a phosphorus enriched Everglades wetlands: cattail expansion phosphorus biogeochemistry and native vegetation recovery. . West Palm Beach, Florida.

Millennium Ecosystem Assessment. 2003. Ecosystems and human well-being. Millennium Ecosystem Assessment.

Mitsch, W. J., X. Wu, R. W. Nairn, P. E. Weihe, N. Wang, R. Deal, and C. E. Boucher. 1998. Creating and restoring wetlands: a whole-ecosystem experiment in self-design. BioScience 48:1019–1030.

Morrisey, D. J. 1993. Environmental impact assessment—a review of its aims and recent developments. Marine Pollution Bulletin 26:540–545.

Niswander, S. F. and W. J. Mitsch. 1995. Functional analysis of a two-year-old created in-stream wetland: Hydrology, phosphorus retention, and vegetation survival and growth. Wetlands 15:212–225.

Odum, E. P. and H. T. Odum. 1955. Trophic structure and productivity of a windward coral reef community on Eniwetok Atoll. Ecological Monographs 35:291–320.

Odum, H. T. 1955. Trophic structure and productivity of Silver Springs, Florida. Ecological Monographs 27:55–112.

Odum, H. T., K. C. Ewel, W. J. Mitsch, and J. W. Ordway. 1977. Recycling treated sewage through cypress wetlands. Pages 35–67 in F. M. D'Itri, editor. Wastewater Renovation and Reuse. Marcel Dekker, New York.

Osenberg, C. W. and R. J. Schmitt, editors. 1996. Detecting ecological impacts caused by human activities. Academic Press, Inc., San Diego, CA.

Palmer, M., E. Bernhardt, E. Chornesky, S. Collins, A. Dobson, C. Duke, B. Gold, R. Jacobson, S. Kingsland, R. Kranz, M. Mappin, M. L. Martinez, F. Micheli, J. Morse, M. Pace, M. Pascual, S. Palumbi, O. J. Reichman, A. Simons, A. Townsend, and M. Turner. 2004. Ecology for a crowded planet. Science 304:1251–1252.

Pawitan, Y. 2001. In all likelihood: Statistical modeling and inference using likelihood. Oxford Scientific Publications, Oxford, UK.

Peters, D. P. C., B. T. Bestelmeyer, J. E. Herrick, H. C. Monger, E. Fredrickson, and K. M. Havstad. 2006. Disentangling complex landscapes: new insights to forecasting arid and semiarid system dynamics. BioScience 56:491–501.

Peters, D. P. C., P. M. Groffman, K. J. Nadelhoffer, N. B. Grimm, S. L. Collins, W. K. Michener, and M. A. Huston. 2008. Living in an increasingly connected world: a framework for continental-scale environmental science. Frontiers in Ecology and the Environment 6: 229–237.

Petersen, J. E., W. M. Kemp, R. Bartleson, W. R. Boynton, C.-C. Chen, J. C. Cornwell, R. H. Gardner, D. C. Hinkle, E. D. Houde, T. C. Malone, W. P. Mowitt, L. Murray, L. P. Sanford, J. C. Stevenson, K. L. Sunderburg, and S. E. Suttles. 2003. Multiscale experiments in coastal ecology: Improving realism and advancing theory. BioScience 53:1181–1197.

Pugesek, B. H., A. von Eye, and A. Tomer, editors. 2003. Structural Equation Modeling: Applications in Ecological and Evolutionary Biology. Cambridge University Press, Cambridge.

Reckhow, K. H. 1990. Bayesian inference in non-replicated ecological studies. Ecology 71:2053–2059.

Risser, P. and W. J. Parton. 1982. Ecological analysis of a tallgrass prairie: Nitrogen cycle. Ecology 63:1342–1351.

Schindler, D. W. 1971. Carbon, nitrogen, phosphorus and the eutrophication of freshwater lakes. Journal of Phycology 7:321–329.

Schindler, D. W. 1973. Experimental approaches to liminology – an overview. Journal of the Fisheries Research Board of Canada 30:1409–1413.

Schindler, D. W. 1998. Replication versus realism: The need for ecosystem-scale experiments. Ecosystems 1:323–334.

Schlesinger, W. L. 1990. Evidence from chronosequence studies for a low carbon-storage potential of soils. Nature 348:232–234.

Smith, E. P. 2002. BACI design. Pages 141–148 in A. H. El-Shaarawi and W. W. Piegorsch, editors. Encyclopedia of Environmetrics.

Stephens, P. A., S. W. Buskirk, and C. Martínez del Rio. 2006. Inference in ecology and evolution. Trends in Ecology and Evolution 22:192–197.

Stewart-Oaten, A. and J. R. Bence. 2001. Temporal and spatial variation in environmental impact assessment. Ecological Monographs 71:305–339.

Stewart-Oaten, A., J. R. Bence, and C. W. Osenberg. 1992. Assessing effects of unreplicated perturbations: no simple solutions. Ecology 73:1396–1404.

Stewart-Oaten, A., W. Murdoch, and K. R. Parker. 1986. Environmental impact assessment: "Pseudoreplication" in time? Ecology 67:929–940.

Tansley, A. G. 1935. The use and abuse of vegetational concepts and terms. Ecology 16:284–307.

Tian, H., C. A. Hall, and Y. Qi. 1998. Modeling primary productivity of the terrestrial biosphere in changing environments: Toward a dynamic biosphere model. Critical Reviews in Plant Sciences 15:541–557.

Tilman, D. 1987. Secondary succession and the pattern of plant dominance along experimental nitrogen gradients. Ecological Monographs 57:189–214.

Turner, M. G. 2005. Landscape ecology in North America: Past, present, and future. Ecology 86:1967–1974.

Underwood, A. J. 1992. Beyond BACI: The detection of environmental impacts on populations in the real, but variable, world. Journal of Experimental Marine Biology and Ecology 161:145–178.

Underwood, A. J. 1993. The Mechanics of Spatially replicated sampling programs to detect environmental impacts in a variable world. Australian Journal of Ecology 18:99–116.

Underwood, A. J. 1994. On Beyond BACI: Sampling designs that might reliably detect environmental disturbances. Ecological Applications 4:3–15.

Vitousek, P. M., J. Aber, R. W. Howarth, G. E. Likens, P. A. Matson, D. W. Schindler, W. H. Schlesinger, and G. D. Tilman. 1997. Human alteration of the global nitrogen cycle: Causes and consequences. Issues in Ecology 1:1–16.

Woodwell, G. M. 1979. Leaky ecosystems: Nutrient fluxes and succession in the pine barrens vegetation. Pages 333–343 in R. T. T. Forman, editor. Pine Barrens: Ecosystem and Landscape. Academic Press, Inc., New York, NY.

Chapter 2
Structural Equation Modeling and Ecological Experiments

James B. Grace, Andrew Youngblood and Samule M. Scheiner

2.1 Introduction

Practicing ecologists are generally aware that the conventional approaches to experimental design and analysis presented in standard textbooks fall short of satisfying their scientific aspirations (Carpenter 1990, Miao and Carstenn 2006). It is the thesis of this chapter that an alternative framework, structural equation modeling (SEM), provides both a perspective for seeing some of the limitations of conventional procedures and also suggests expanded possibilities for the design and analysis of experiments. In our presentation, we will first provide a brief description of SEM. Our emphasis shall be on a few key distinctions that help us to describe the essential features of SEM that separate it from other methods. Second, we will examine the analysis of variance (ANOVA) model from the perspective of SEM. We believe that SEM provides a good point of contrast for better understanding the strengths and weaknesses of ANOVA. Following this material, we describe two experimental studies: one conducted in coastal wetlands and another conducted in low elevation interior coniferous forests. In these examples, we first consider the challenges that large-scale ecological experiments can pose to traditional analysis methods such as ANOVA. We then derive approaches to analyzing the data in these studies using SEM. Finally, we end the chapter by discussing the potential for SEM to contribute to the design and analysis of ecological experiments.

Before launching into a discussion of SEM and ANOVA, we would like to make an important distinction that generally applies to our topic. Above we used the phrase "scientific aspirations." We believe that it is important to recognize that there is a difference between our scientific aspirations and the statistical procedures that may be used in pursuit of those aspirations. Text-books dealing with statistics in general and experimental statistics in particular seem to imply that we are interested in a very limited set of scientific goals and

J.B. Grace (✉)
US Geological Survey, National Wetlands Research Center, Lafayette, Louisiana
70506, USA
e-mail: gracej@usgs.gov

S. Miao et al. (eds.), *Real World Ecology*, DOI 10.1007/978-0-387-77942-3_2,
© Springer Science+Business Media, LLC 2009

that these goals can be achieved through the procedures for data analysis they present. Further, statistical procedures for the design and analysis of experimental studies are often presented in the form of protocols or prescriptions. Such presentations imply that somehow the process of scientific inquiry is satisfied through the application of, say, a factorial random block design. We believe that this codification of science through statistical protocols has been so influential in ecology that its limitations are invisible to many people. We agree with Abelson (1994) that the idea that fixed statistical protocols somehow lead automatically to scientific learning is misguided. Our presentation of methods is designed to illustrate the scientific structure of statistical models and, thereby, clarify the roles that univariate and structural equation models can play in experimental studies.

2.2 What Is Structural Equation Modeling?

SEM in its modern form represents a scientific framework that uses statistical procedures to further our learning about causal processes (Bollen 1989, Grace 2006). Importantly, SEM is not a fixed procedure, but instead involves a highly flexible modeling methodology. A great variety of statistical techniques and tools have been and can be used in the process of estimating parameters and evaluating models. What has been constant since its roots in path analysis (Wright 1921) has been SEM's emphasis on addressing questions about causal processes (as opposed to being focused on selecting a parsimonious set of predictors). SEM techniques have evolved substantially over time, and users of this methodology have been quick to incorporate new statistical techniques as they become available. For the past several decades, SEM applications have relied heavily on maximum likelihood procedures. Recent innovations in Markov Chain Monte Carlo procedures and Bayesian estimation are available to SEM practitioners and have been incorporated into some SEM software packages (Arbuckle 2007, Lee 2007).

What essential features distinguish SEM from other analysis frameworks and modeling approaches? In this chapter, we will specifically contrast SEM with univariate modeling methods such as ANOVA. In general, the implications of SEM for science are made clear by considering some of its characteristic features (Table 2.1). One feature is that at a fundamental level, the "structural" part of SEM refers to the objective of learning about causal processes from data. This ambitious goal has contributed to some of the distinctive features of SEM practice, which include (a) tests of mediation (defined and illustrated below) and the study of indirect effects, (b) sequential learning, and (c) the use of latent (unmeasured) variables. The first of these, testing for mediation and the study of indirect effects, is one of the most distinctive features of SEM. The second of these, sequential learning, is characteristic of SEM in the degree to which it plays a central role in the modeling process. Third, the ability of

Table 2.1 Some attributes of structural equation modeling

Is a scientific framework, not a specific statistical technique
Flexible method for representing a wide range of models
Requires support from theoretical knowledge
A graphical modeling method
Involves the concept of mediation
Emphasizes the study of pathways to learn about causal processes
Relies on sequential learning (hypothesize -> test -> refine hypothesis -> retest) to build
  confidence in interpretation
Evaluates theoretically specified models rather than null-hypotheses
Often involves the inclusion of latent variables
Sheds light into the simultaneous functioning of multiple processes, contributing to a system-
  level understanding

SEM to incorporate and evaluate latent variables is another major capability of the methodology.

An additional characteristic of SEM is that it is a member of the family of techniques known as graphical models. The use of diagrams to graphically represent the structure of models was invented by mathematicians interested in representing causal relationships (Borgelt and Kruse 2002). In graphical models, the flow of causation is more intuitively represented than can be achieved simply using sets of simultaneous equations (Pearl 2000). It should become readily apparent in this chapter how graphical representation of models helps to convey the scientific logic associated with SEMs.

A simple example can be used to illustrate the role of theory and the concepts of mediation and sequential learning. In a study of chaparral vegetation recovery after wildfires, Grace and Keeley (2006) observed a negative relationship between the age of the shrubland stand that burned and the degree of post-fire recovery (Fig. 2.1A). An association between the ages of a shrubland stands prior to burning and the degree of vegetative recovery after fire contains the temporal elements needed to plausibly support an interpretation of causal influence. But, what, we may ask, is the basis for such an influence? In SEM practice, we seek to understand relationships through intervening variables. The authors hypothesized that the basis for the relationship in Fig. 2.1A was because older stands had more fuel to burn and, therefore, sustained more severe fires. This hypothesis implies a second model that is shown in Fig. 2.1B. A test of mediation involves solving both models A and B, and then evaluating the equation

$$\gamma_{net} = \gamma_{11} \times \beta_{21} \tag{2.1}$$

where γ_{net} is the net effect of stand age on recovery, γ_{11} is the effect of stand age on fire severity, and β_{21} is the effect of fire severity on recovery. Analysis of these relationships by Grace and Keeley (2006) showed the condition in

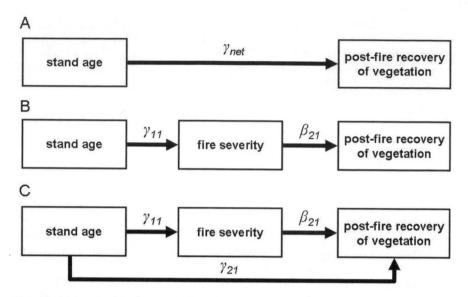

Fig. 2.1 Models illustrating the concept of mediation. (**a**) Model representing the net effect of stand age on post-fire recovery. (**b**) Model testing whether fire severity mediates (or explains) relationship in Model A. (**c**) Alternative model in which fire severity does not completely explain effects of stand age on recovery

equation (2.1) to be satisfied, which supports the interpretation that older stands affected recovery through their more severe fires.

It should be pointed out that alternative outcomes might have been found. If either γ_{11} or β_{21} had been found to not represent significant effects, it would have been concluded that fire severity did not mediate the effect in Fig. 2.1A. If both γ_{11} and β_{21} were deemed significant, but the condition in equation (2.1) was not met, then that would imply an additional, separate process whereby recovery was influenced by stand age (as shown in Fig. 2.1C). Grace and Keeley (2006) considered that older stands might have also possessed seed banks that were depleted by the passage of time (compared to younger stands). Such a process would imply an additional effect γ_{21} that would be expected to have a negative sign. Since the condition in equation (2.1) held, the conclusion was drawn that the operation of an additional process was not detectable in this case.

The above example shows how theoretical knowledge can be combined with evaluations of mediation to evaluate possible causal interpretations based on the relationships in the data. Through this process, net relationships are partitioned into a network of direct and indirect pathways, each representing separate processes. Additional characteristics of SEM listed in Table 2.1 are illustrated in later sections where examples are presented in more detail.

2.3 The Univariate Model and Analysis of Variance

The univariate model is a logical point of comparison for structural equation models. For the purposes of this chapter, the collection of univariate models associated with ANOVA procedures will be of special interest. It is fair to say that ANOVA models (including ANCOVA, analysis of covariance) currently play a dominant role in ecological studies. One of our objectives in this chapter is to put ANOVA models into perspective through comparison with structural equation models.

Our emphasis on ANOVA in this chapter is based in large part on the influence this model continues to have on the practice of ecology. At the most fundamental level, ANOVA has been of importance to ecologists, as with other natural scientists, because of an interest in drawing confident inferences. Through the efforts of Sir Ronald Fisher (1956) and the suggestions of Jerzy Neyman and Egan Pearson (Neyman 1976), the concept of controlled, manipulative experiments in combination with a focus on null hypothesis testing has come to represent to many the gold standard of statistical procedures in the natural sciences.

In addition to providing a means of analyzing the data from experimental studies, ANOVA models are closely associated with the topic of experimental design. Strictly speaking, the issues of design and analysis are somewhat separate. However, in practice, they are often linked, primarily because they are often taught together as a set of protocols. For most practicing ecologists who are not statistical specialists, what they are taught (or advised to do) is to design their studies in such a fashion that the data are appropriate for ANOVA, ANCOVA, or MANOVA (multivariate analysis of variance). This typically means (1) establishing replicate experimental units with high homogeneity and similarity, (2) manipulating a few factors with a small number of levels for each factor to permit replication of each treatment combination, and (3) using covariates, blocking, or physical control to remove the effects of environmental context from the analysis. This is, of course, an oversimplification, and advanced biometricians will see the above list as ignoring the important case of response-surface, regression-based designs. Still, we believe the above to be a fair description of typical training and applications by many practitioners, particularly when emphasis is placed on understanding differences between means.

For the sake of illustration, let us consider a single-factor experiment with a single covariate. We can represent such a situation in the univariate framework as

$$y_{ij} = \mu + \beta_1 x_i + \beta_2 c_{ij} + \beta_3 x_i c_{ij} + \varepsilon_{ij}, \tag{2.2}$$

where y is a response of interest across i treatments and j replicates of each treatment, μ is the intercept, x is a treatment variable, β_1 is the effect that the treatment has on y, c is a covariate, β_2 is the effect of the covariate on y, β_3 is the interactive effect of the treatment and covariate, and ε_{ij} represents the error of prediction for y. There are several variants of this equation possible, depending on the number of treatments, whether treatments have fixed or random effects, and how we choose to model treatment effects (e.g., if the treatment is nominal or ordered categorical).

There is, of course, nothing inherently wrong with the ANOVA approach to experimental studies. However, we believe it is important for ecologists to realize that there are opportunities to address additional questions beyond those easily evaluated using ANOVA. Oversimplifying a bit, the ANOVA framework is well suited for the study of net effects of select factors, but limited in its capacity to accommodate certain complexities that we will discuss in this chapter. In this chapter, we will use several different approaches to explain and illustrate why we make these contentions. Our main points, however, are summarized in Table 2.2.

The concept of variance partitioning (the analysis of variances) is based on the premise that treatments and covariates are orthogonal or uncorrelated (Fig. 2.2A). The assumption of orthogonality between treatments and covariates is required if we are to partition variance so as to uniquely ascribe effects to individual factors (it is perhaps worth noting that the interaction terms in such models are not orthogonal to the main factors, however). When treatments and covariates are correlated, it means that a portion of the variance explanation for our y variable is shared between x_1 and c (see Grace and Bollen 2005 for a discussion of this concept). The burden of achieving orthogonality, as required for ANOVA and ANCOVA, falls on the study design, and as we shall discuss, a special set of circumstances in which treatment manipulations express themselves consistently across replicates. If factors affecting y (including treatments and covariates) are intercorrelated, the concept of variance partitioning becomes inappropriate because variance explanation cannot be partitioned cleanly. How is one to analyze experimental data if not through variance partitioning? The answer lies in the realization that regression is a more general framework for statistical modeling and one that can more readily accommodate correlated factors (Cohen 1968). In order to accommodate correlated factors, a regression approach replaces an emphasis on comparing means and on variance partitioning with an emphasis on regression/path coefficients and the estimation of model parameters. Regression/path coefficients can be seen as parameters that describe processes rather than particular outcomes (slopes rather

Table 2.2 Limitations associated with ANOVA univariate models commonly used in the natural sciences

Historically has emphasized means rather than process parameters
Built on premise of uncorrelated (orthogonal) factors
Best suited for studying isolated effects, one or a few at a time
Encourages limited representation of sample space and particularly narrow environmental conditions
Historically has been linked to null hypothesis testing

Fig. 2.2 (**a**) Graphical presentation of an analysis of covariance model, where y is a response of interest, x_1 is a treatment factor, and c is a covariate that is uncorrelated with the treatment factor. The interactive effect of x_1 and c is signified by the joint arrows. (**b**) Structural equation model, in which mediating conditions generated by the treatments can be included in the model

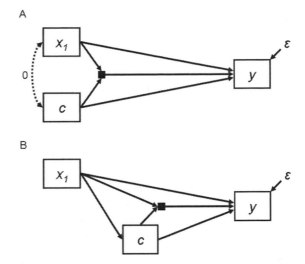

than means) (Pedhazur 1997). Regression, and SEM by extension, involves a parameter-based approach to analyze data that more naturally accommodate situations where factors are correlated.

It is useful for us to consider under what circumstances we might expect experimental studies to involve important covariates that correlate with treatments. There is always an unlucky situation that can occur where an important covariate happens to correlate with a particular treatment. However, there is an important category of circumstances in which covariates are generated by the treatments themselves. One such circumstance is when treatment applications are not uniformly expressed within experimental units. Let us take the example of a study involving prescribed fire as a case where treatment manipulations might not be expressed consistently across experimental units. Imagine a number of experimental units such as field plots (in forest, grassland, or shrubland) that are subjected to three prescribed fire treatments (1) no fire, (2) dormant season fire, and (3) growing season fire. It is nearly a universal feature of burning across landscapes that fire behavior is variable. In particular, it can vary temporally as weather conditions vary throughout a day or from day to day. We can also expect it to vary from season to season, and this can further be influenced by differing fuel conditions, which can vary significantly over short time periods and even more dramatically among seasons. It would be very natural in such circumstances for the investigators to adjust for the variation among plots in actual fire effects, by measuring things such as the completeness or severity of burns that took place. We might find, for example, that dormant season burns were consistently hot and complete, while growing season burns were more variable because they occur when fuels have greater variability in fuel moisture. In such a case, the covariate "completeness of burn" would correlate with the season of burn and, therefore, violate the assumptions of ANCOVA. Figure 2.2B represents such a case.

Another major feature associated with conventional experimental designs is the focus on one or a few factors of interest and a desire to keep other environmental factors as constant as possible. Indeed, it seems that the more the environmental conditions can be forcibly controlled, for example through the use of laboratory or other artificial conditions, the more the assumptions and purpose of ANOVA can be met. It is, in essence, this gradation from artificial, controlled conditions to natural, less controlled conditions that is the focus of this book. Many practicing ecologists view the emphasis that conventional experimental procedures and the assumptions of ANOVA place on orthogonality of factors and homogeneity of experimental units as something to be overcome and not something to be quietly accepted.

One can ask what it is about conventional designs and analyses that tend to cause an emphasis on the study of a few processes? Actually, there are a number of contributing factors. An emphasis on understanding mean differences among treatments, for example, promotes designs that require numerous replicates in order to lend precision to these point estimates. Such replicated designs make multi-factor studies replicate-hungry, placing logistic demands on experiments. Another important issue is that variance partitioning leads to analytical results in the form of ANOVA tables in which multi-factor studies can yield significant multi-way interactions. Below we shall compare the presentation of results from ANOVAs to those from SEM to illustrate the differences in perspective resulting from the two approaches. For now, we simply note that finding multi-way interactions using ANOVA is often less instructive than the researcher might wish. We should also point out that the design of multi-factor experiments commonly creates physical conditions (and a structuring of the data) in which interactions are conspicuous. Collectively, these things mean that often ANOVA results tend to feel very specific and less comprehensive than often wished by scientists desiring to understand the mechanistic processes controlling systems.

Overall, the act of comparing mean responses to manipulations through the partitioning of variances (characteristic of ANOVA) involves a different mindset than does the estimation of treatment effects on regression or path coefficients (characteristic of regression or SEM approaches). The former tends to encourage the researcher to cover a narrower range of environmental conditions and to replicate intensively rather than extensively. Unreplicated, regression-based designs, in which experimental factors are treated as continuous variables, provide a greater opportunity to represent a larger volume of sample space and are encouraged by the SEM approach. By adopting such an approach, one not only comes closer to studying the system as a whole, but also positions themselves to accomplish what ecologists often say they wish to accomplish – to provide information that will help us to predict the future behavior of systems.

There is at least one additional characteristic of conventional applications of ANOVA procedures that stand in contrast to SEM. Historically, ANOVA has been framed in terms of null hypothesis testing. There is currently an active discussion on the merits of null hypothesis testing across many fields of study (Harlow, Mulaik and Steiger 1997, Guthery et al. 2005). This topic is both interesting and important; however, it is beyond our purpose to consider it here in any depth. As will be demonstrated in the examples that follow, SEM practice is based on the study of a priori, theoretically justified models, not the rejection of null models of no effect. This perspective is consistent with its foundations in likelihood statistics.

Ultimately, we believe that consideration of the SEM perspective leads one to the conclusion that typical ANOVA approaches and their influence on experimental studies has resulted in the codification of a methodology that is not well suited for the study of systems, but instead, more appropriate for isolating the effects of individual processes. In the following sections, we will illustrate the alternatives provided by SEM.

2.4 SEM Example #1: The Factors Controlling Plant Diversity in Coastal Wetlands

This first example will serve the purpose of allowing a fairly simple comparison of SEM to traditional ANOVA results. In this example, we will describe a study that was designed with SEM in mind and for which the experimental design matches the analyses in a straightforward way. The results of this study, which addressed the question of what factors control patterns of plant diversity in coastal wetlands, have been summarized in several papers (Gough and Grace 1998a, 1998b, 1999). The presentation will be brief and the reader is referred to the references mentioned if interested in more details.

2.4.1 Background

This study was inspired by the desire to understand the multi-process control of plant species diversity. As characteristic of most ecological topics, the study of diversity has emphasized the derivation and evaluation of a whole host of simple theories based on one or a few processes. As Palmer (1994) has pointed out, there are over 100 theories/models/hypotheses that have been proposed to explain patterns of diversity, and there has been little consolidation and resolution of ideas until recently (Grace 1999, Scheiner and Willig 2006). The reason that ecologists have placed so much emphasis on studies of individual mechanisms can be traced to the dominating influence of the univariate paradigm (and the above-described features of ANOVA) in biology (Grace 2006, Chapter 12). In this study of coastal wetlands, the overall goal was the identification of the system of factors controlling plant diversity. This involved the manipulation of four major factors based on insights obtained from a previous nonexperimental SEM study of the same geographic area (Grace and Pugesek 1997).

2.4.2 Methods

The experimental design for this study involved four factors of interest: (1) salinity, (2) herbivory, (3) flooding, and (4) soil nutrient enrichment. The experimental unit was a 0.33 m-diameter plot. Salinity was manipulated by planting or transplanting sods in fresh versus brackish marshes. Herbivory was manipulated using exclosures that excluded mammalian herbivores (nutria, muskrats, hogs, rabbits, and deer). Flooding was manipulated by lowering or elevating sods by 10 cm. Soil fertility was manipulated by applying nutrients or native sediments. Each individual sod was exposed to some combination of salinity, herbivory, flooding, and soil fertility treatment. There were a total of 254 plots and the study was allowed to proceed for three growing seasons before final measurement of above-ground biomass. Measurements of environmental conditions including soil salinity, soil elevation, soil carbon and nitrogen, and visually apparent disturbance were also taken at the end of the study. For further details regarding the methodology, see Gough and Grace (1999).

2.4.3 Results

The analysis of such an experiment can be approached from either an ANOVA or a SEM framework and the results from both analyses were published by the authors. Here we present a sample of results from the two approaches for illustrative purposes. Table 2.3 reproduces the results from the repeated measures ANOVA published in Gough and Grace (1998a). The main effects conspicuous in this table are (1) enrichment, fencing, and salinity effects (loss of species) intensified over time, (2) there was an overall decline in species richness due to enrichment, and (3) richness was lower in brackish marsh than in fresh marsh. Certainly there was much gleaned by examining graphs of means and their differences in conjunction with this table of ANOVA results beyond this brief summary of statistical findings.

SEM was used by Gough and Grace (1999) to represent the combined influences of manipulated and covarying factors (Fig. 2.3). In that paper, they also compared the values of species richness predicted from a prior nonexperimental SEM analysis with those observed from the experimental manipulations. Here we focus briefly on just a few of the main interpretations from the SEM results and again refer those interested in more details to the original paper.

As seen in the SEM diagram (Fig. 2.3), latent variables were used to represent certain conceptual entities. "Richness" or species richness was estimated from a single indicator (sden), the number of species observed in a plot at the end of the study. "Disturbed" refers to the disturbance regime of a plot, which was indicated by a single measure of visually apparent disturbance at the end of the study. The direction of arrows (from latent to indicator) follows the

Table 2.3 Source table for repeated measures ANOVA of species richness from Gough and Grace (1998a). Note that marsh represents the salinity effect, fencing the herbivory effect, and enrichment the soil fertility effect

Source	df	MS	F	P
Marsh	1	1805.03	124.38	0.0001
Fence	1	0.17	0.01	0.91
Marsh × fence	1	1.78	0.12	0.73
Replicate (marsh × fence)	28	14.51	–	–
Enrichment	2	20.49	3.98	0.02
Marsh × enrichment	2	2.24	0.43	0.65
Fence × enrichment	2	15.66	3.04	0.06
Marsh × fence × enrichment	2	5.04	0.98	0.38
Error	55	5.15	–	–
Time	6	66.89	51.56	0.0001
Time × marsh	6	5.36	2.46	0.03
Time × fence	6	4.05	1.86	0.09
Time × marsh × fence	6	3.76	1.72	0.12
Time × replicate (marsh × fence)	168	2.18	–	–
Time × enrichment	12	3.53	2.72	0.002
Time × marsh × enrichment	12	1.13	0.87	0.58
Time × fence × enrichment	12	1.73	1.33	0.20
Time × marsh × fence × enrichment	12	0.73	0.56	0.87
Error (time)	330	1.30	–	–

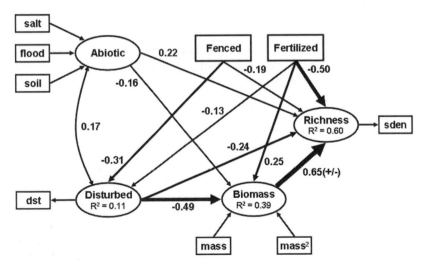

Fig. 2.3 Structural equation model results presented by Gough and Grace (1999) with minor modification (kind permission obtained from Ecology)

convention of a L→M (latent to manifest) relationship (Grace and Bollen 2008), where the observed variable reflects the causal influence of an underlying (latent) process. In Fig. 2.3, "Biomass" refers to the standing community biomass (live and dead) that was harvested at the end of the study. Two indicators (measured biomass in grams per meter squared and the square of the biomass value) were used to represent biomass effects on richness in a polynomial equation, because the observed relationship was found to be unimodal. In this case, the diagram follows the convention of a M→L (manifest to latent) relationship because the abstract concept (Biomass) arises from a collection of measured entities (mass and mass squared). "Abiotic" in the model represents the collective effects of abiotic factors (salinity, flooding, and other soil properties such as spatial variations in soil organic content) on richness and biomass production.

It should be pointed out that there are certain approximations in this model. All pathways are approximated as linear effects, except the one from biomass to species richness, which is a second-order modal relationship. Also, while the path coefficients are unbiased, the standard errors associated with those parameters may be slightly underestimated because certain features of the sample design (e.g., nesting) were ignored. From a likelihood perspective (in contrast to a null hypothesis perspective), this poses little problem for our interpretations. Also, it should be noted that the use of standardized coefficients, as was done in this paper, is an approximation that should be interpreted carefully (see Grace and Bollen 2005 for a succinct discussion of this issue).

2.4.4 Discussion

The SEM results presented in Fig. 2.3 are presumed to summarize the effects of various mechanisms. The focus in this model is on explaining spatial variation in species richness, and each direct or indirect pathway represents a distinct process. For example, the indirect path Disturbed→Biomass→Richness represents the effect of disturbance on species richness mediated by influences on standing biomass. In contrast to this, the pathway Disturbed→Richness represents an effect of disturbance on richness that is independent from biomass. As discussed previously with reference to Fig. 2.1, the interpretations of direct pathways are unevaluated, since no test of mediation has been performed. In this example, the authors suspected that the direct path from disturbed to richness represented a morphological influence of disturbance on plant form, but certainly other processes could be involved.

There is a great deal we could infer from Fig. 2.3 and it is beyond our scope here to give a complete presentation. However, for illustrative purposes, we consider the various components behind the effects of fencing on richness (Table 2.4). There are three pathways in our model that lead from fencing to richness. Perhaps the most conceptually obvious is the pathway

Fenced→Disturbed→Biomass→Richness. Slightly less obvious is the pathway Fenced→Disturbed→Richness, which may involve a selective loss of species from herbivory. The third pathway is the direct one, Fenced→ Richness. The mechanism behind this path is a bit of a mystery and there are several possible candidates. None of these three pathways is terribly strong in this case, though all contain links deemed to be statistically significant. Of further interest is that these three pathways, whereby fencing influences richness, are largely offsetting when taken all together. The concept of *total effect* in structural equation models refers to the net influence of several pathways. In this case, we can see (Table 2.4) that the three pathways from Fenced to Richness add up to $-0.19 + 0.074 + 0.100 = -0.016$, which we can see from the ANOVA results in Table 2.3, is indistinguishable from a value of zero (i.e., nonsignificant). While the ANOVA results only provide an assessment of the net effect, the SEM results yield both the net effect and the various components. In this case, as is typical in our experience, ANOVA results fail to reveal the wealth of information about underlying processes that can be extracted using SEM. Further, the form of information provided by the SEM analysis seems much more naturally suited to an understanding of system function.

Table 2.4 List of pathways and inferred processes involved in the relationship between fencing and species richness in Fig. 2.3. Pathway strength is calculated from the product of the individual path coefficients along the pathway

Pathway	Inferred process	Strength
Fenced→Richness	Effect of fencing on richness independent of influences mediated through disturbance or biomass. Perhaps a selective loss of species via herbivory.	−0.19
Fenced→Disturbed→Richness	Fencing effect on richness caused by reduced disturbance yet unrelated to biomass reduction.	0.074
Fenced→Disturbed→ Biomass→Richness	Fencing effect on richness caused by reduced disturbance and subsequent reduction in biomass.	0.100
Net effect of Fenced on Richness	Averages across offsetting processes.	−0.016
Fenced→Disturbed	Reduction in plot-level disturbance resulting from fencing.	−0.31
Disturbed→Biomass	Degree of reduction in biomass resulting from disturbance.	−0.49
Biomass→Richness	Degree to which richness varied with biomass. Since relationship is first increasing and then decreasing (unimodal), this is a nonlinear effect.	0.65 (+/−), net effect positive
Disturbed→Richness	Reduction in richness from disturbance that is not proportional to reduction in biomass. Could be selective mortality of herbivory-intolerant species.	−0.24

2.5 SEM Example #2: The Effects of Forest Treatments on Post-fire Mortality in Ponderosa Pine

The next example presented in this chapter is one that confronts some of the bigger challenges associated with large-scale ecological experiments. In particular, we deal here with the situation where experimental units are very large, and treatment manipulations have heterogeneous effects. The study was part of a national Fire and Fire Surrogate (FFS) study (see http://frames.nbii.gov/ffs). The FFS study involved a network of 13 study sites in forests of the contiguous U.S. and sought to evaluate thinning and prescribed burning treatments designed to reduce fuels and restore historical forest structures. The work reported here is from the northeastern Oregon site, which is dominated by ponderosa pine (*Pinus ponderosa*) and Douglas-fir (*Pseudotsuga menziesii*). The information for the following presentation was taken from Youngblood et al. (in press). Additional information about this study and resulting changes in stand structure, understory vegetation, and fuelbed characteristics can be found in Youngblood et al. (2006) and Youngblood et al. (2008), respectively.

2.5.1 Background

It is generally believed that prior to the 20th century, dry conifer forests in the western US were subjected to frequent, low to mixed-severity fires (Agee 1993, Youngblood et al. 2004). These fires, which were likely most prevalent during seasonal dry periods, selected against fire-intolerant species, prevented heavy fuel buildup and maintained an open forest structure with a widely spaced overstory of large trees, a sparse woody understory, and an abundant herbaceous fine-fuel layer (Wickman 1992, Arabas, et al. 2006). Following decades of altered fire regimes, these forests now commonly possess quite different characteristics, including greater accumulation of woody fuels, more ladder fuels, a substantially higher understory tree density, and as a result, conditions conducive to a greater risk of catastrophic fires. Contributors to these changes are thought to include fire exclusion, livestock grazing, selective timber harvests of fire-intolerant species, and changes in climate (Bergoffen 1976, Steele et al. 1986, Dolph et al. 1995, Arno et al. 1997).

One component of this system that contributes to both forest dynamics and fire conditions is a guild of bark beetles that attack both large and small trees. Fires have been shown to attract at least some species of bark beetles and to increase tree susceptibility to beetle damage (de Groot and Turgeon 1998, McCullough et al. 1998, Parker et al. 2006). Mortality from bark beetles that occurs during outbreaks can also contribute to fire-promoting conditions by increasing woody fuel mass (Youngblood et al. in press). As a consequence, delayed tree mortality following fire from bark beetles is an integral part of the picture when considering forest management in these systems.

In recent decades, awareness of the importance of altered forest structure to fire risk has led to ambitious national programs that seek to apply management treatments so as to reduce fire risk. Federal legislation and initiatives relating to these interests include the National Fire Plan, the 10-Year Comprehensive Strategy, the Healthy Forests Initiative, and the Healthy Forests Restoration Act of 2003. These activities promote large-scale fuel reduction efforts and the restoration of forest structure and ecological characteristics. Most commonly, selective removal of smaller trees, either by mechanical thinning or prescribed burning (or both), has been recommended as an approach to the problem. This is to be accomplished, though, with a minimal impact on the large trees in the overstory. The main objective of the FFS study was to evaluate the effectiveness of these kinds of treatments and also to consider any potentially adverse side effects, such as mortality of large trees from post-treatment beetle invasions.

2.5.2 Field Methods

This study was conducted in the Blue Mountains of northeastern Oregon, USA. The study area, broadly defined, covers approximately 10,000 ha and occurs within an elevation range of 1,000–1,500 m. Because of the dissection of forest by ravines and valleys, most contiguous stands in this area are less than 200 ha in size. Previous timber harvests have resulted in the forest overstory being composed primarily of 70 to 100-year-old trees of ponderosa pine and Douglas fir, with scattered survivors from the previous populations. Significant recruitment of trees into the understory had occurred in the decades preceding the study. More information about site history and study methodology can be found in Youngblood et al. (2006).

The experimental units in this study were forest stands 10 – 20 ha in size. The reason that units this large were used was because of a desire to examine operational-scale activities representative of those that would occur in routine management applications. The experimental units included in the study were randomly selected from a population of possible candidate sites with similar topographic features and forest stand structure. The four treatments applied were (1) no manipulation, (2) thin, which was a single entry thinning of understory trees, (3) burn, which involved a single prescribed burn, and (4) thin + burn, with thinning taking place shortly before burning. Treatments were allocated to experimental units using a completely random design, with four replicate units per treatment and a total of 16 units in the study. The study was established in 1998, with mechanical thinning treatments being applied that same year. Burning treatments were applied in September 2000 and sampling continued through 2004.

In 1998, prior to treatment applications, permanent 0.04 ha sampling plots were established within the units, since a complete sampling of the 180 ha contained within the units was deemed infeasible. Plots were located across a systematic grid of sample points spaced 50 m apart and $\geq$ 50 m from the unit boundaries. Because units varied in size, the number of plots per unit varied from 10 to 30. Within all sample plots, all trees $\geq$ 1.37 m in height were inventoried by species, noted as live or dead, and their height and diameter at breast height were measured. Individual trees were examined for signs of bark beetle attack. Trees with external indications of bark beetle attack were tagged and numbered. Dead trees were autopsied by removing portions of the bark to identify any signs of characteristic beetle damage. The most common bark beetles and wood borers found in the study included the following: the western pine beetle (*Dendroctonus brevicomis*), mountain pine beetle (*D. ponderosae*), red turpentine beetle (*D. valens*), pine engraver (*Ips pini*), Douglas-fir beetle (*D. pseudotsugae*), Douglas-fir engraver (*Scolytus unispinosus*), flatheaded wood borer (Buprestidae), and roundheaded borer (Cerambycidae). There were numerous other bark beetles found during the inventory, but none that contributed substantially to tree death. New mortality found post-treatment that was associated with scorched trees having no sign of fresh beetle damage was attributed to the fire. Post-treatment tree inventories were conducted following treatment applications in 2001 and 2004.

As part of the study, a large number of additional variables were measured. These included measures of (1) fuelbed characteristics and mass, (2) fire temperatures, (3) fire severity, and (4) physical damage to trees. Additional details relating to these measurement procedures can be found in Youngblood et al. (in press).

2.5.3 Univariate Analyses

ANOVA was considered for analyzing unit-level tree mortality. As illustrated in Fig. 2.4,, mortality rates ascribed to specific causes were highly variable at the unit level, indicative of a binomial process with a low probability of mortality. The observed high level of variability for mortality responses at the unit level was examined, and it was determined that the low probability of beetle-caused mortality combined with the large amount of within-unit heterogeneity in fire effects made standard, unit-level analysis inappropriate/ineffective for this study. For this reason, data were summarized at the plot level for analysis. Again, ANOVA was considered for analysis of mortality as a function of treatment types using the plot-level data ($n = 380$ plots). In all treatment types, mortality at the plot level was evidenced as a low probability binomial process, characterized by a high percentage of plots having zero mortality values. To simplify the problem, the authors collapsed tree size classes into two categories, small trees ($<$ 25 cm dbh) and larger trees ($\geq$ 25 cm dbh), and

Fig. 2.4 Illustration of the variation in the unit-level tree mortality observed for one of the most common bark beetle species, the wood borer (Youngblood et al. in press)

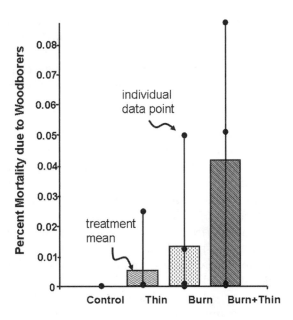

then performed a frequency table analysis. Fisher's exact test (Zar 1984) was used in the frequency table analysis because of the small numbers of non-zero observations for some causes of mortality.

2.5.4 Structural Equation Modeling

This study is a dramatic example of a case where the experimental manipulations of "thinning" and "burning" represent broad dichotomous terms belying a much greater range of actual variation in the intervening causal factors potentially influencing mortality. Conventional ANOVA analyses in this case ascribe net results to the particular thinning treatment and the particular burn event described in the methods of Youngblood et al. (in press). Based on only such analyses, we have little capacity to extrapolate to any thinning treatment other than the one performed (set to achieve a particular final density) and no set of burn conditions other than those that occurred in that brief window of time. The designers of the study had the foresight to measure in some detail causal agents that might mediate the effects of thinning and burning on tree mortality (despite the fact that these covariates do not qualify as suitable for use in an ANCOVA to isolate the effect of the manipulations, as seen in Fig. 2.2).

Youngblood et al. (in press) chose to focus their SEM on trees exposed to fire and to deal only with ponderosa pine. This decision was based on the finding that virtually all of the beetle-caused mortality occurred in burned or thinned and burned units (as did, of course, all fire-related tree mortality) and because there was little mortality in Douglas-fir. The authors also decided that the plot-level data were more appropriate for capturing the process-level events. The authors began the SEM process by developing a conceptual model representing the processes hypothesized to connect treatment manipulations to tree mortality (Fig. 2.5). The logic expressed in the conceptual model was that the combination of thinning with burning may influence fuel loads, which in turn would potentially affect fire temperatures (which measure fire intensity) and fire severities (which measure the amount of impact). Of great interest was the possibility that the degree of fire severity might influence the incidence of beetle-caused and other mortality. Trees were grouped into large and small size classes to correspond with the management objectives of reducing densities of small-diameter trees while preserving large trees. In addition to the above-described cascade of effects from treatment → fuels → fire intensity → fire severity → mortality, it was recognized that additional processes could be at work. For example, there could be treatment effects on fire intensity and fire severity not related to fuel loads, such as from the piling of downed fuels associated with the thinning treatment. Also, it was considered possible that tree mortality could relate to fire intensity in ways not entirely explained by fire severity. For example, intense fires could generate more smoke, which might attract beetles from further away, even if individual trees were only lightly scorched.

For the specification of the actual structural equation model(s) to be examined, a first consideration was how to model the binomial response of mortality. Mortality itself was observed to be an occasional, stochastic event, and as a

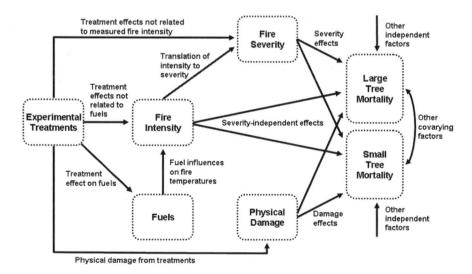

Fig. 2.5 Conceptual model representing hypothesized relationships among factor contributing to delayed conifer mortality following burning (from Youngblood et al. in press)

result, inherently unpredictable. It was decided that what was of primary interest was to model the probability of mortality. To accomplish this, mortality variables were specified using a probit transformation (Muthén and Muthén 2007). This approach essentially presumes that beneath the dichotomous events of dying or surviving lies a continuous probability of mortality operating as a latent process. In such a formulation, the model essentially represents how the probabilities of mortality vary with treatment and mediating conditions. To accommodate the high degree of nonnormality of the responses, the software used (Mplus in this case, Muthén and Muthén 2007) employs a weighted least squares estimation procedure.

A second statistical issue that the authors addressed was to adjust for the fact that plots were nested within units. The nesting of plots within units does not affect the parameter estimates (e.g., path coefficients) themselves. However, it can bias the sizes of the standard errors of the parameters. This was handled within Mplus by accounting for the hierarchical nature of the data using the methods of Hox (2002). Additional details regarding the statistical procedures can be found in Youngblood et al. (in press).

2.5.5 SEM Results

Results indicated that mortality of both large and small trees related to fire severity (Fig. 2.6). For small trees, the estimated probabilities of death from both pine engraver beetles and wood borers were substantially higher where fires were severe. Mortality from other causes included both deaths from fire itself and also an increased chance of death where trees were physically damaged prior to burning. The curved, double-headed arrows on the right side of the mortality boxes in Fig. 2.6 represent the influences of unmeasured agents that caused mortalities to be positively correlated. For large trees, there was little death ascribed solely to fire (the mortality from other causes) and it was unrelated to any of the predictors. Mortality from wood borers and from western pine beetles was strongly increased by severe fires. Note that a full understanding of the results requires reference to the magnitudes of responses and not just the relationships presented in Fig. 2.6. For our purposes, however, we ignore those additional considerations (although they are addressed in Youngblood et al. in press).

Two forms of fire severity were found to contribute to mortality in this study: severe scorching of the ground surface and direct charring of the tree boles (Fig. 2.6). In this model, the authors chose to present the combined influences of fire severity from these two different causes so as to reduce the complexity of the presentation. The technique used to represent the combined influences of fire severity was compositing (Grace and Bollen 2008), which is a handy capability of SEM when working with concepts such as fire severity (and fuel type) that are multifaceted. As for the rest of the model, it was found that fire severity was

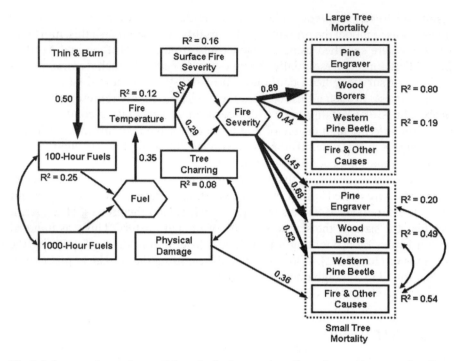

Fig. 2.6 Structural equation model results for large and small ponderosa pine mortality (from Youngblood et al. in press). Rectangles represent observed variables while hexagons represent composites summarizing the effects of multiple predictors on individual responses. Path coefficients presented are standardized values. R^2 values represent the estimated variance explained in underlying continuous mortality probabilities based on model relationships

related to measured fire temperatures, which was in turn related to the combined effects of 100-hour fuels (mass of woody fuels between 2.54 and 7.62 cm in diameter) and 1000-hour fuels (mass of woody fuels greater than 7.62 cm in diameter). Finally, the 100-hour fuel load was higher in units where mechanical thinning preceded burning. Overall, mortality was found to be more likely where fires were most severe and fires tended to be more severe in units where thinning increased fuel loads prior to burning.

2.5.6 Implications of Findings

Examination of unit-level mortality patterns revealed very clearly that a simplistic consideration of treatment effects on mortality was inadequate (Fig. 2.4). For example, while some thin + burn units showed high levels of mortality from wood borers, other similarly treated units did not. Also, there were units subjected to burning without thinning that experienced more mortality than some of the thin + burn units. We can certainly expect that if we were to repeat the experiment many times, we would find variation in tree response at the level of the sample unit to be significant. It would take a huge number of such experiments, we imagine, to obtain a broad enough sample of thinning treatments (with different degrees of thinning under different forest conditions) and burns (under different conditions) to achieve much generality at the level of the treatment type. Using the plot-level data, we were able to take advantage of the heterogeneity in actual conditions within units to shed some light on the processes of importance.

In this particular example, there are some important implications of these findings for science applications. Of particular value is the fact that there are no important, major pathways directly from the treatment variable to responses other than the 100-hour fuels. This means that the treatment variable "drops out of the model" in the sense that we have an explanation for what the thin + burn treatment did (increase 100-hour fuels) and we have a basis for discussing our findings that does not deal in terms of the specific thinning and burning treatments that were conducted. For example, since we recognize the predominant role of fire severity in influencing mortality, management practices can be designed around fire severity targets. In conditions where fuels indicate the potential for high severity fires, burning techniques and the judicious selection of burn times might be used to lower burn intensity near large trees. In any event, we feel that this application of SEM improved both the ability to extract information about the system and the ability to obtain information that can be used to predict future outcomes.

2.6 The Potential Utility of SEM in Experimental Studies

2.6.1 The Challenge of Exploring the Processes Behind Net Effects

Modern principles of experimental design and analysis developed from a desire to obtain unambiguous interpretations about causal factors (Salsburg 2001). Manipulation of treatment factors, physical control of nontreatment factors, randomization, replication, and null hypothesis testing have all been carefully considered by a generation of statisticians. The purpose of these procedures is to convince even a skeptical person that a specific causal process operates (Mayo 2004). To accomplish this, factors must be isolated and assumed to be uncorrelated to one another.

The SEM tradition shares with the biometric tradition of using ANOVA to analyze results from controlled experimental studies a desire to understand causes. SEM practice does not reject the above-mentioned elements of experimental procedures, but rather, views the ANOVA-based experimental philosophy as a special case that lives within a broader body of possible objectives and emphases. In virtually any analysis of experimental data, there is a potential role for ANOVA in establishing the net effects. However, we think that following such analyses with the evaluation of SE models can advance our evaluation of mechanisms behind the responses.

Within SEM practice, there are additional alternatives for addressing causation that are not a part of the ANOVA approach. One is the critical role played by path models, which permit causal interpretations to be evaluated and multiple causal pathways to be partitioned. All these originate from the test of mediation (see Fig. 2.1), which is absent from the ANOVA framework (Fig. 2.2A versus B). The ANOVA-based approach precludes testing causal interpretations because mediating causes are not evaluated.

A second basis for complementarity between the SEM and ANOVA perspectives is the desirability of working with fully orthogonal models. There is substantial agreement about the value of isolating treatments from confounding background conditions. Part of the tension that exists here between the aspirations of field ecologists and ANOVA-based approaches comes from the lack of recognition that ANOVA is best suited for studying individual processes when we do not wish to extrapolate to the behavior of whole systems. Take for example microcosm studies where small containers of organisms are used as surrogates for ecological communities. Studies conducted under such conditions (which are often viewed to be ideal for ANOVA designs and analyses) are capable of establishing the validity of processes *in principle*. That does not mean that their outcomes extrapolate directly to natural systems (though they may extrapolate in concept). Understanding the behavior of whole systems requires a consideration of multiple, interacting processes. As we have shown in this chapter, univariate models are not designed for evaluating hypotheses about systems, and even when experiments are conducted under field conditions, ANOVA-based studies generally fail to reflect the natural importance of processes (Grace et al. 2007). The microcosm study based on an ANOVA approach suffers doubly from both the limits of the univariate model and from the artificiality of the conditions. From the SEM perspective, it is highly desirable to allow many parts of the system to freely intercorrelate during experiments so as to retain some degree of naturalness for the results. In our second example, we argued that the authors gained understanding by including measures of the cascading set of mediating responses (such as fuels, fire temperatures, and fire severity) despite the fact that these covariates correlated strongly with the treatments. It is the flexibility of the SEM framework that permits such intercorrelated factors to be included in models. We feel this enhances the scientist's ability to discern the effects and importance of underlying causal processes compared to what can be achieved within the ANOVA framework.

A third characteristic of SEM is the emphasis on estimating path coefficients rather than partitioning variances and expressing results in terms of differences among group means. This is not to say that the study of fixed (versus random) factors and the analyses of mean differences are not possible within SEM, because they are (Grace 2006, Chapter 5). However, the user of SEM aspires to permit interpolation and even extrapolation where possible. Path coefficients support this aspiration. Among the many merits derived from estimating path coefficients instead of means and their standard errors is that all the samples in a study contribute to the estimate of each coefficient. Estimation of path coefficients encourages a broad sampling of conditions that expands sample space and reveals relationships.

A fourth characteristic of SEM is the reliance on sequential hypothesis testing. While sequential hypothesis testing may occur within an ANOVA environment, it receives greater emphasis with SEM. As mentioned earlier, beyond the most exploratory applications, SEM practice involves (1) development of competing a priori hypotheses expressed as models, (2) evaluation of those models, (3) refinement of hypotheses and models, and (4) retesting of models with new data. In SEM, we place limited emphasis on the conclusions drawn from a single study; rather, the emphasis is on consistency at large. Also, progressively refined models that seek to understand direct paths by including hypothesized mediating variables is another common element of SEM practice. Taken together, we can see that one of our goals in SEM is to build understanding through sequential studies. Again, this is not unique to SEM practice, though its emphasis is quite a bit stronger because SE models are able to bring along the knowledge gained from previous studies in the form of the structure of the prior models used in the next study.

2.6.2 Technical Issues Associated with Using SEM in Experimental Studies

In this treatment, we have ignored many important points relating to technical issues. This has been deliberate because (1) the objective of the book in which this chapter resides is to speak to ecologists, not statisticians and (2) SEM is a scientific framework and from that perspective technical issues can be viewed as an ongoing process of invention and refinement. Omitted from our discussion has been the multiple-group methodology routinely implemented in SEM software that is very well suited to evaluating experimental data. In multi-group analysis, SE models are evaluated across groups (e.g., across males and females or across populations). This framework permits both an easy approach to comparing model parameters across groups, and it automates the evaluation of interactions involving grouping variables. Several examples of multi-group SEM have now been developed in the ecological sciences (e.g., Grace and Jutila 1999, Tonsor and Scheiner 2007).

While SEM provides a framework for specifying and evaluating models, it does not guarantee that all the important statistical issues are addressed (nor does it inhibit the proper formulations either). A main strength of SEM is its flexibility. At the same time, this flexibility means that it is not feasible to present to users some limited set of protocols for its use (as is often done with ANOVA). Experiments frequently impose a whole set of characteristics on the data when they involve nested designs, split-plot arrangements, blocking, and the like. Most such features actually do not influence the parameter estimates, but they do influence the standard errors for the parameters and require some considerations. The estimation tools and techniques included in specialized SEM software frequently permit the accommodation of many technical issues (Grace 2006). Recently, the development of Markov Chain Monte Carlo (MCMC) methods in conjunction with Bayesian estimation procedures has opened the door to substantial refinements in statistical modeling (Congdon 2001). This allows users to specify hierarchical models more easily, to accommodate nonnormal processes more readily, and to relax many of the assumptions that have restricted analyses in SE models (Lee 2007). From a SEM perspective, many discussions of such things as null hypothesis testing versus model selection and Bayesian versus frequentist inference are important statistical details to be addressed. From the viewpoint of using data to enhance scientific understanding, we emphasize the differences between univariate models and structural equation models.

2.7 Summary

As Kuhn (1970) pointed out, paradigm shifts from one view of the world to another are not easily accomplished. Two events are required before a shift occurs: (1) a demonstration that the new paradigm has certain advantages and (2) a realization of the limitations of the existing paradigm. While SEM has become accepted as a means of learning from nonexperimental data in the natural sciences (e.g., Grace 2008), we believe its acceptance as a framework for experimental studies will require a bit of a paradigm shift because the ANOVA perspective is so entrenched. In this paper, we attempt to demonstrate some of the advantages of using SEM in experimental studies. To emphasize why we think a change in practice is needed, we also point out what we believe to be weaknesses of the ANOVA-based approach. In practice, both ANOVA and SEM can contribute to our analysis of experimental data.

While a growing number of ecologists are coming to recognize the value of SEM for experimental studies, there remain many (especially biometricians) who seem resistant to recognizing the limitations of variance partitioning and univariate models such as ANOVA for the study of systems. Since the professional biometricians tend to be the ones who teach most University courses in applied statistics, we appear at the moment to be deeply entrenched in a limited paradigm. We view this as a transitional condition. In fact, SEM is widely accepted for use in experimental studies in the human sciences while approaches such as ANOVA are viewed as simplistic approximations (Hair et al. 1995). It is our hope that explicit presentations of the issues, as attempted in this chapter, along with persistence in applications and illustrations, will permit SEM to become sufficiently established so that it can contribute to the study of large-scale and long-term studies to its full potential.

Acknowledgments We thank ShiLi Miao, Craig Stow, Beth Middleton, Jack Waide, and two anonymous reviewers for helpful comments and suggestions on an earlier version of the manuscript. The use of trade names is for descriptive purposes only an does not imply endorsement by the US Government. This manuscript is based in part on work done by SMS while serving at the National Science Foundation. The views expressed in this paper do not necessarily reflect those of the National Science Foundation or the United States Government.

References

Abelson, R.P. 1995. Statistics as Principled Argument. Hillsdale, New Jersey: Lawrence Erlbaum Publishers

Agee, J.K. 1993. Fire Ecology of Pacific Northwest Forests. Washington, D.C.: Island Press

Arabas, K.B., K.S. Hadley, and E.R. Larson. 2006. Fire history of a naturally fragmented landscape in central Oregon. Canadian Journal of Forest Research **36**:1108–1120.

Arbuckle, J.L. 2007. Amos 7.0 User's Guide. Chicago, Illinois: SPSS Inc.

Arno, S.F., H.Y. Smith, and K.A. Krebs. 1997. Old growth ponderosa pine and western larch stand structures: influences of pre-1900 fires and fire exclusion. USDA Forest Service Research Paper INT-RP-495.

Bergoffen, W.W. 1976. 100 years of federal forestry. USDA Forest Service Agriculture Information Bulletin 402.

Bollen, K.A. 1989. Structural equations with latent variables. New York: John Wiley & Sons

Borgelt, C. and R. Kruse. 2002. Graphical Models. New York: John Wiley & Sons

Carpenter, S.R. 1990. Statistical analysis of ecological response to large-scale perturbations. Ecology **71**:2037.

Cohen, J. 1968. Multiple regression as a general data-analytic system. Psychological Bulletin **70**:426–443.

Congdon, P. 2001. Baycsian statistical modeling. New York: John Wiley & Sons

de Groot, P., and J.J. Turgeon. 1998. Insect-pine interactions. In Ecology and Biogeography of *Pinus*. (ed.) D.M. Richardson, pp. 354–380. Cambridge: Cambridge University Press

Dolph, K.L., S.R. Mori, and W.W. Oliver. 1995. Long-term response of old-growth stands to varying levels of partial cutting in the eastside pine type. Western Journal of Applied Forestry **10**:101–108.

Fisher, R.A. 1956. Statistical Methods and Scientific Inference. Edinburgh: Oliver and Boyd

Gough, L. and J.B. Grace. 1998a. Herbivore effects on plant species density at varying productivity levels. Ecology **79**:1586–1594.

Gough, L. and J.B. Grace. 1998b. Effects of flooding, salinity, and herbivory on coastal plant communities, Louisiana, United States. Oecologia **117**:527–535.

Gough, L. and J.B. Grace. 1999. Predicting effects of environmental change on plant species density: experimental evaluations in a coastal wetland. Ecology **80**:882–890.

Grace, J.B. 1999. The factors controlling species density in herbaceous plant communities: an assessment. Perspectives in Plant Ecology, Evolution and Systematics **2**:1–28.

Grace, J.B. 2006. Structural equation modeling and natural systems. Cambridge: Cambridge University Press.

Grace, J.B. 2008. Structural equation modeling for observational studies. Journal of Wildlife Management **72**:14–22.

Grace, J.B. and K.A. Bollen. 2005. Interpreting the results from multiple regression and structural equation models. Bulletin of the Ecological Society of America **86**:283–295.

Grace, J.B. and K.A. Bollen. 2008. Representing general theoretical concepts in structural equation models: the role of composite variables. Environmental and Ecological Statistics **15**:191–213.

Grace, J.B. and H. Jutila. 1999. The relationship between species density and community biomass in grazed and ungrazed coastal meadows. Oikos **85**:398–408.

Grace, J.B., and J.E. Keeley. 2006. A structural equation model analysis of postfire plant diversity in California shrublands. Ecological Applications **16**:503–515.

Grace, J.B. and B. Pugesek. 1997. A structural equation model of plant species richness and its application to a coastal wetland. American Naturalist **149**:436–460.

Grace, J.B., T.M., Anderson, M., Smith, E., Seabloom, S., Andelman, G., Meche, E., Weiher, L.K., Allain, H., Jutila, M., Sankaran, J., Knops, M.E. Ritchie, and M.R. Willig. 2007. Does species diversity limit productivity in natural grassland communities. Ecology Letters **10**:680–689.

Guthery, F.S., L.A., Brennan, M.J., Peterson, and J.J. Lusk. 2005. Information theory in wildlife science: critique and viewpoint. Journal of Wildlife Management **69**:457–465.

Hair, J.F., Jr., R.E., Anderson, R.L., Tatham, and W.C.Black. 1995. Multivariate Data Analysis. Fourth Edition, Englewood Cliffs, NJ: Prentise Hall.

Harlow, L.L., S.A., Mulaik, and J.H. Steiger (eds.). 1997. What If There Were No Significance Tests? Mahwah, NJ: Lawrence Erlbaum Associates Publishers.

Hox, J. 2002. Multilevel Analysis. Mahwah, NJ: Lawrence Erlbaum Associates Publishers.

Kuhn, T.S. 1970. The Structure of Scientific Revolutions. Second Edition, Chicago: University of Chicago Press, Chicago.

Lee, S.Y. 2007. Structural Equation Modeling: A Bayesian Approach. New York: John Wiley and Sons.

Mayo, D.G. 2004. An error-statistical philosophy of evidence. pp 79–118, In: Taper, M.L. and S.R. Lele (eds.) 2004. The nature of scientific evidence. Chicago: University of Chicago Press.

McCullough, D.G., R.A. Werner, and D. Neumann. 1998. Fire and insects in northern and boreal forest ecosystems. Annual Review of Entomology **43**:107–127.

Miao, S.L. and S.M. Carstenn, 2006. A new direction for large-scale experimental design and analysis. Frontiers in Ecology and the Environment **4**:227.

McCarthy, M.A. 2007. Bayesian methods for ecology. Cambridge: Cambridge University Press.

Muthén, L.K. and B.O. Muthén. 2007. Mplus user's guide. Third Edition. Los Angeles, CA: Muthén and Muthén.

Neyman, J. 1976. Tests of statistical hypotheses and their use in studies of natural phenomena. Communications in Statistics and Theoretical Methods **8**:737–751.

Palmer, M.W. 1994. Variation in species richness: towards a unification of hypotheses. Folia Geobotanic Phytotaxa Praha, **29**:511–530.

Parker, T.J., K.M. Clancy, and R.L. Mathiasen. 2006. Interactions among fire, insects and pathogens in coniferous forests of the interior western United States and Canada. Agricultural and Forest Entomology **8**:167–189.

Pearl, J. 2000. Causality. Cambridge: Cambridge University Press.

Pedhazur, E.J. 1997. Multiple regression in behavioral research. Third Edition, New York: Wadsworth Press.

Salsburg, D. 2001. The lady tasting tea. New York: Henry Holt & Company.

Scheiner, S.M. and M.R. Willig. 2006. Developing unified theories in ecology as exemplified with diversity gradients. American Naturalist **166**:458–469.

Steele, R., S.F. Arno, and K. Geier-Hayes. 1986. Wildfire patterns change in central Idaho's ponderosa pine-Douglas fir forest. Western Journal of Applied Forestry **1**:16–18.

Tonsor, S.J. and S.M. Scheiner. 2007. Plastic trait integration across a CO_2 gradient in Arabidopsis thaliana. American Naturalist **169**:E119–140.

Wickman, B.E. 1992. Forest health in the Blue Mountains: The influence of insects and disease. USDA Forest Service General Technical Report PNW-GTR-295.

Wright, S. 1921. Correlation and causation. Journal of Agricultural Research **10**:557–585.

Youngblood, A., T. Max, and K. Coe, 2004. Stand structure in eastside old-growth ponderosa pine forests of Oregon and northern California. Forest Ecology and Management **199**:191–217.

Youngblood, A., K.L. Metlen, and K. Coe. 2006. Changes in stand structure and composition after restoration treatments in low elevation dry forests of northeastern Oregon. Forest Ecology and Management **234**:143–163.

Youngblood, A., C.S. Wright, D. Ottmar, and J.D. McIver. 2008. Changes in fuelbed characteristics and resulting fire potentials after fuel reduction treatments in dry forests of the Blue Mountains, northeastern Oregon. Forest Ecology and Management In press.

Youngblood, A., J.B. Grace, and J.D. McIver. in press. Delayed conifer mortality after fuel reduction treatments: interactive effects of fuels, fire intensity, and bark beetles. Ecological Applications.

Zar, J.H. 1984. Biostatistical Analysis. Englewood Cliffs, New Jersey: Prentice Hall.

Chapter 3
Approaches to Predicting Broad-Scale Regime Shifts Using Changing Pattern-Process Relationships Across Scales

Debra P. C. Peters, Brandon T. Bestelmeyer, Alan K. Knapp, Jeffrey E. Herrick, H. Curtis Monger, and Kris M. Havstad

3.1 Introduction

Shifts from one ecosystem state to another with dramatic consequences for ecosystem organization and dynamics are increasing in frequency and extent as a result of anthropogenic global change (Higgens et al. 2002, Foley et al. 2003, Scheffer and Carpenter 2003). Ecosystem state changes (i.e., regime shifts) have been well-documented for a number of different systems, from lakes (Carpenter 2003) to oceans (Beaugrand 2004), coral reefs (McCook 1999), and grasslands (Rietkerk and van de Koppel 1997). Ecosystem state changes are usually characterized by a shift in dominant species that persists through time. For example, shifts in dominant fish species in lakes can result in significant changes in prey populations (Magnuson et al. 2006), and shifts to woody plant dominance in grasslands result in increased rates of erosion and land degradation (Schlesinger et al. 1990). Ecosystem state changes can also occur with changes in the production of a single species in association with modification of one or more biophysical processes (e.g. Howes et al. 1986). In many cases, ecosystem state changes are "ecological surprises" in that they are observed and confirmed after they occur. These surprises result from our inability to understand the full suite of mechanisms driving and maintaining these shifts. New approaches are needed to improve our understanding and to allow the detection and prediction of impending ecosystem state changes, in particular those that impact the delivery of goods and services to human populations.

Experimental and analytical approaches designed to detect and predict ecosystem state changes must enhance our understanding of cross-scale interactions and elucidate the role of these interactions in determining ecosystem thresholds (the level or magnitude of an ecosystem process that results in a sudden or rapid change in ecosystem state).

D.P.C. Peters (✉)
USDA ARS, Jornada Experimental Range, Jornada Basin Long Term Ecological
Research Program, Las Cruces, NM 88003-0003, USA
e-mail: debpeter@nmsu.edu

S. Miao et al. (eds.), *Real World Ecology*, DOI 10.1007/978-0-387-77942-3_3,
© Springer Science+Business Media, LLC 2009

Critical thresholds are often crossed during or following a state change such that a return to the original state is difficult or seemingly impossible (Bestelmeyer 2006). Here, we define thresholds as points in time where a change in an environmental driver results in a discontinuous increase or decrease in the rate of a process and the resultant change in a state variable (Peters et al. 2004a, Groffman et al. 2006). Thresholds can occur either in the environmental driver, the rate of a process or a state variable. Thresholds indicate that a change in a dominant process has occurred and that distinct exogenous drivers or endogenous positive feedbacks are governing rates of change (Scheffer et al. 2001, Peters et al. 2004a). Feedbacks tend to maintain a state, and it is often the change in these feedbacks and the resultant alteration in pattern–process relationships that differentiate a regime shift from a reversible ecosystem change that is not maintained through time (Carpenter 2003, Peters et al. 2007). For example, shifts from grasslands to woodlands can be maintained for hundreds of years by positive feedbacks between woody plants and soil properties. Once critical thresholds in surface soil properties are crossed, soil water availability is modified to promote woody plant growth and limit the establishment of grasses; thus promoting the maintenance of woodlands. These state change dynamics are differentiated from successional dynamics following disturbances that remove the vegetation without major changes to soil properties (e.g., wildfire). Critical thresholds in soil properties do not exist following these disturbances; thus a return to grass dominance through a succession of species is possible.

In some cases, state changes are driven by processes at one spatial or temporal scale interacting with processes at another scale (Carpenter and Turner 2000, Gunderson and Holling 2002, Peters et al. 2007). For example, the Dust Bowl of the 1930s in the US resulted from interactions among broad-scale patterns of extremely low rainfall, landscape-scale patterns in the number and spatial arrangement of abandoned agricultural fields, and fine-scale patterns in plant mortality, which resulted in shifts from vegetated to bare states throughout the Central Great Plains (Peters et al. 2004a). Observations at single or even multiple, independent scales would have been insufficient to predict such cross-scale interactions that ultimately generated the emergent behavior characterized as an ecosystem state change (Michener et al. 2001). Detailed monitoring of vegetation on small plots during the 1930s was insufficient to account for the spatial extent and rate of loss of vegetation being driven by landscape- and broader-scale processes, on a continental-scale (Weaver and Albertson 1940).

Approaches that explicitly account for thresholds and cross-scale interactions are expected to improve our understanding of the mechanisms driving ecosystem state changes and to allow more informed predictions of impending changes (Ludwig et al. 2000, Diffenbaugh et al. 2005, Bestelmeyer et al. 2006b). Alternatively, downscaling the effects of broad-scale drivers on fine-scale patterns can, under some conditions, improve understanding of key ecological processes driving these patterns. Extrapolating information about fine-scale processes to broad scales can be used, under some conditions, for predicting responses through time and space. Previous studies have described hierarchical, multi-scale approaches to improve both understanding and prediction (e.g., Wiens 1989, Stohlgren et al. 1997, Petersen et al. 2003). Our approach considers the interactions and feedbacks in patterns and processes across land areas from an ecosystems perspective that links different kinds of data across spatial and temporal scales. Most studies of cross-scale interactions have documented changing patterns in vegetation through time and across space, and then assumed changing patterns resulted from changing ecological processes (Peters et al. 2004a). However, an approach that combines pattern analyses with experimental manipulation of processes and simulation modeling of rates of ecosystem change under different drivers is needed to tease apart the role of drivers and processes in determining patterns at different scales.

We had three objectives: (1) to describe an experimental approach that would identify changing pattern–process relationships across scales and the dominant processes within a scale that drive and maintain state changes in terrestrial systems; (2) to describe statistical approaches for detecting and predicting state changes; (3) to apply these approaches to broad-scale shifts from grasslands to woody plant dominance in the Chihuahuan Desert, but they are applicable to the United States and globally. Our approach built on existing methods that examined multiple scales hierarchically, and combined observation, manipulation, and simulation modeling (e.g., Wiens 1989, Petersen et al. 2003, Stohlgren et al. 1997). However, our approach included three new components: (1) an explicit focus on stratification of the landscape by variation in drivers, in addition to the more common approach of stratifying by underlying environmental (c.g., soils, landuse) and biotic (vegetation cover, species composition) heterogeneity, (2) experimental manipulations of drivers at multiple scales, and (3) sufficient levels of manipulation to allow thresholds to be detected and examined.

3.2 The Shrinking Grasslands: Woody Plant Encroachment into Perennial Grasslands

Woody plant encroachment is a pervasive problem throughout perennial grasslands globally (Scholes and Archer 1997, Briggs et al. 2005, Knapp et al. 2008). In the United States alone, 220–330 million ha of non-forested land either has changed or is changing from a grassland to a woodland (Houghton et al. 1999, Pacala et al. 2001). The consequences of a grassland to woodland ecosystem state change are consistent among grassland types: local ecosystem properties are modified, including primary production, biodiversity, and rates and patterns of nutrient cycling (Schlesinger et al. 1990, Ricketts et al. 1999, Huenneke et al. 2002, Briggs et al. 2005, Knapp et al. 2008). Regional to global processes are altered, including transport of dust to the atmosphere, redistribution of water to the oceans and groundwater reserves, and feedbacks to weather (Jaffee et al. 2003, McKergow et al. 2005, Pielke et al. 2007).

This state change is occurring in most grassland types located along a precipitation gradient from the xeric desert grasslands in the Southwest to the mesic tallgrass prairie in the central Great Plains and the barrier islands along the Atlantic coast (Archer 1994, Briggs et al. 2005, Young et al. 2007), and along a temperature gradient from warm deserts to alpine meadows and arctic tundra (Marr 1977, Shaver et al. 2001, Epstein et al. 2004). Woody plant encroachment most often involves increases in cover and density of native or non-native shrubs or small trees and loss of perennial grasses. At the dry end of the gradient, the size and density of bare soil patches and concentration of soil resources under sparse woody plants increases to create a landscape that is more arid and desert-like (i.e., desertification). At the wet end of the gradient, woody plant encroachment results in an increase in leaf area of woody plants, but no increase in bare ground (i.e., afforestation; Briggs et al. 2005).

During an ecosystem state change from grassland to woodland, the fundamental unit of change is the individual, either a grass or woody plant. Adjacent grass- and woody plant-dominated communities have a dynamic transition zone or ecotone that consists of individual plants and patches of plants. The location of the ecotone shifts through time and space as ecological driving forces fluctuate to favor either grasses or woody plants (Peters et al. 2006b). Historical evidence of a grassland–woodland transition zone that has shifted repeatedly at large spatial and long temporal scales – a tension zone – comes from the paleorecord in the central United States (Grimm 1983) and in the Chihuahuan Desert based on packrat middens (Van Devender 1990), carbon isotopes (Monger et al. 1998) , and fossil pollen (Hall 2005). However, during an ecosystem state change, environmental drivers reach a threshold in magnitude or rate of change that favors woody plant encroachment, woody plants disperse into the grassland, germinate and become established (Peters et al. 2006d). In many cases, woody plants modify the water, light, and nutrient conditions of their microsites to create a positive feedback that promotes their persistence (Schlesinger et al. 1990, Lett and Knapp 2003, McCarron and Knapp 2003, Lett et al. 2004). Successful individuals, by producing seed or through vegetative reproduction, can increase the density of woody plants throughout the transition zone. A broad-scale state change results when woody plants coalesce into patches that then dominate the cover of the community at the landscape scale (Peters et al. 2006b).

On some soil types (e.g., sandy soils in arid systems), when a threshold density of woody plants is reached, a change from biotic (e.g., competition, dispersal) to abiotic control (i.e., wind erosion) takes place. Upon reaching the threshold, the shift from grassland to woodland occurs rapidly (years to decades) (Peters et al. 2004a). Because encroachment by woody plants often results in changes to surface soil properties, a state change can be documented by a combination of changes in vegetation dominance and soil properties (Bestelmeyer et al. 2006a).

It is the change in soil properties and the positive feedbacks with the vegetation that may determine the irreversibility of the system (Davenport et al. 1998).

Although shifts from grasslands to woodlands occur over a very large and diverse area that includes much of the continental United States, there is a common set of drivers and processes that determine these ecosystem dynamics (Fig. 3.1). The relative importance of specific drivers and processes varies in each case; however, the suite of possibilities that determine changing pattern–process relationships across scales is similar. Drivers, such as climate, provide both the spatial and temporal context for finer-scale dynamics, and can interact with finer-scale drivers to influence state change dynamics across scales (Fig. 3.1). At the grassland regional scale, drivers include climate (precipitation, temperature) and atmospheric properties (e.g., nitrogen deposition, carbon dioxide levels, solar radiation) (Bahre and Shelton 1993, Archer et al. 1995, Polley et al. 2002, Knapp et al. 2008) as well as the regional pool of species and their dispersal attributes (Ricklefs 1987). Multi-year drought is the broad-scale driver most often implicated in desertification, whereas extended wet periods can increase the rate of afforestation in more mesic environments (Humphrey 1958, Grimm 1983, Briggs et al. 2005).

As the spatial scale of resolution decreases to the landscape, additional drivers become important, such as patterns of livestock grazing, fire, land use, soils, and geomorphology, which interact with climate variability at multiple time scales (Lyford et al. 2003, Monger and Bestelmeyer 2006). In general, broad-scale drought (xeric systems) or wet periods (mesic systems) combine with landscape-scale variation in intensity and frequency of livestock grazing (xeric systems) and fire (mesic systems) to promote woody plant expansion (Schlesinger et al. 1990, Briggs et al. 2005). At the fine scale of individual plants, patterns in life history traits (e.g., lifespan, drought and grazing tolerance, water- and nitrogen-use efficiency) influence a plant's ability to respond to micro-site conditions.

Processes follow a similar hierarchy of scales. At the finest scale of individuals, recruitment, competition for resources vertically in the soil profile, and mortality of grasses and woody plants as well as plant-soil feedbacks are important to local dominance (Brown and Archer 1989). As the scale of interest increases to the landscape and region, additional processes become important that redistribute resources and propagules horizontally. Transport vectors (i.e., water, wind, animals, humans, and disturbance) redistribute water, soil, nutrients, and seeds, and act to connect adjacent or non-contiguous spatial units (Ludwig et al. 2005; 2007, Fredrickson et al. 2006, Okin et al. 2006). We define contagious processes as those processes that connect spatial units at the same or different scales via transport vectors (Peters et al. 2008).

Because of the explicit linkages among scales, the importance of contagious processes can only be understood with studies of multiple interacting scales. For example, seeds can be dispersed long distances (tens of kilometers) by birds and other vectors to initiate fine-scale woody plant invasion in non-contiguous areas (Chambers et al. 1999, Lyford et al. 2003). Thus, construction of livestock exclosures, designed to alter local scale processes of herbivory on seedlings, can

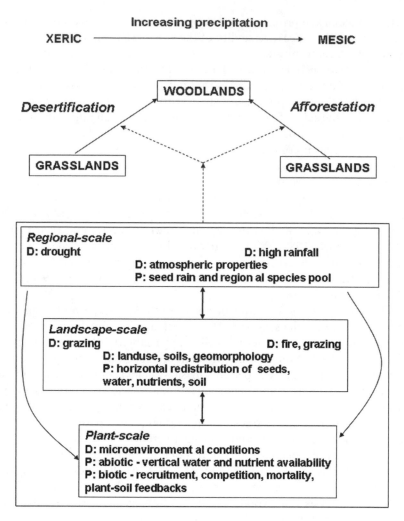

Fig. 3.1 Conceptual diagram of key patterns in drivers and processes associated with grass-land to woodland transitions along a precipitation gradient that includes desertification (xeric) and afforestation dynamics (mesic). Pattern–process relationships at three major scales are shown that influence shifts from grasslands to woodlands (*dashed arrows*): regional-, landscape-, and plant-scale. For each scale, patterns in drivers are denoted by D and processes are denoted by P. Dynamics at any given scale are hierarchical and determined by interactions with the next higher and lower scale in the hierarchy. In addition, regional-scale drivers can influence plant-scale dynamics directly. In most cases, similar drivers and processes are important across the precipitation gradient. The exceptions are: (**a**) drought is a key broad-scale driver for desertification whereas afforestation is affected by periods of high rainfall, and (**b**) fire is more important in mesic than xeric areas

be ineffective in delaying or stopping the advancement of woody plants if large numbers of seeds are dispersed by wind and small animals or if seeds were already in the soil seed bank at the time of exclosure construction (Peters et al. 2006a).

3.3 Limits of Current Approaches

In spite of decades-old recognition that scale and spatial heterogeneity are important, there is little systematic understanding of how to design cross-scale studies (Viglizzo et al. 2005, Bestelmeyer et al. in review). Traditional approaches to studying landscapes tend to blend areas of ecologically significant heterogeneity at particular spatial scales and to emphasize randomness within a scale. Fine-scale heterogeneity may be acknowledged, but only certain kinds and scales of heterogeneity are recognized as sufficiently important by the investigator to be sampled. Observer perception that a particular process is important often dictates the scale of interest and the associated sampling design (Wiens 1989). For example, broad climate zones used to examine the effects of drought on grassland to woodland conversion either ignore fine-scale variation in soil properties or stratify by these properties at the landscape or regional scale (Charney et al. 1975, Claussen 1997, Martin and Asner 2005). These designs ignore variation in the distribution of patches within a landscape that may be critical to predicting shifts from grasslands to woodlands (Kéfi et al. 2007, Scanlon ct al. 2007).

Within climate zones, random sampling is typically used and heterogeneity is ignored. If heterogeneity is important to understanding ecosystem dynamics, then random sampling can lead to under-sampling of relatively small areas that have a proportionately greater effect on the process of interest. Experimental manipulations, especially those that are intensive and long-term, focus by necessity on relatively few areas and represent selected components of spatial heterogeneity. The results of these focused studies are often over-generalized to the larger suite of conditions that are too costly to explore directly. Furthermore, spatial stratification is most often based on patterns in the environment (e.g., soils, vegetation) and ignores variability in transfer processes that redistribute resources and propagules (Peters et al. 2004b). Contagious processes require a spatially explicit approach that considers spatial heterogeneity and accounts for the transport of material and seeds among spatial units at the same and different scales (Peters et al. 2004b).

Similarly, the temporal scale of studies has often been limited. Studies of state changes have emphasized short-term (year to decade-long) studies of pattern–process relationships performed at different stages of the regime shift. The complete state change is seldom directly observed, but presumably the data collected are related to causes of the shift. Thresholds (especially grassland–woodland transitions) often take a long time to develop (e.g., 30–50 years; Peters et al. 2004a). There are few examples in which ecosystem drivers and processes have been assessed, monitored or evaluated before, during and after reaching a threshold. In most cases, thresholds are observed using analyses of historical data where key processes driving threshold behavior have been inferred (e.g., Peters et al. 2004a). Space for time substitutions have also been used to infer long term dynamics that can not be observed

directly, although applicability of space-for-time substitutions has been ques-
tioned (Hargrove and Pickering 1992).

3.4 Cross-Scale Approach

Similar challenges in design and analysis of studies exist for grassland–woodland
transitions occurring throughout the precipitation and temperature gradients in
the US which must be accounted for in any experimental approach (Fig. 3.1).
These challenges include: (1) high spatial and temporal heterogeneity in environmental drivers and conditions, and vegetation across a hierarchy of interacting
scales, (2) threshold behavior with positive feedbacks that result in a persistent
shift in dominance and change in ecosystem state, and (3) the importance of
spatial context and connections among spatial units via contagious processes,
such as seed dispersal.

We propose that a general design for a cross-scale approach to
grassland–woodland transition would include the following steps: (1) identify
patterns in broad-scale drivers, (2) determine hierarchical structure of the
spatial units, (3) stratify and map the focal region based on these spatial units
and environmental gradients, (4) sample and correlate attributes across strata
within each spatial scale, (5) perform experimental manipulations of pattern–
process relationships within key strata, and (6) use simulation modeling to
investigate possible interactions between broad- and fine-scale drivers.
Although each step has been conducted in previous studies, either individually
or in combination with one or more other steps, the synthesis of all six steps into
one approach is novel, in particular for understanding state changes associated
with woody plant expansion. With such general approaches, we can explain or
evaluate the potential for ecosystem change under a broader array of circumstances than is currently possible. We describe this approach using a case study
from the Southwestern US.

3.5 Case Study: State Changes in the Chihuahuan Desert

A multi-scale sampling approach was used to improve understanding and
prediction of grassland–woodland transitions in the northern Chihuahuan
Desert. Landscapes of southern New Mexico exemplify arid and semiarid
regions of the world where perennial grasslands have transitioned to xerophy-
tic, unpalatable shrubs over the past 150 years (Gibbens et al. 2005). In the
1980s and 1990s, most studies of ecosystem state changes focused on the
importance of plant-scale processes and the vertical and horizontal redistribution of resources among individual grasses, shrubs, and bare interspaces
(Wright and Honea 1986, Schlesinger et al. 1990). The studies revealed that as
grasslands degrade and shrubs invade, bare soil patches increase in spatial

extent. Wind and water transport soil nutrients from bare areas to shrubs, where they accumulate to form "islands of fertility" maintained by positive plant–soil feedbacks. These plant-scale results were linearly extrapolated to explain state changes at broad scales (Schlesinger et al. 1990). More recently, the importance of spatial processes (e.g., seed dispersal) and redistribution of resources (water, nutrients) within and among patches of plants and larger landscape units have been recognized as influencing spatial and temporal variation in grassland–shrubland transitions (Peters et al. 2006a). Transport vectors (wind, water, animals, humans) act to move materials (seeds, propagules) and nutrients (nitrogen, water) within and among spatial units to result in patterns that can not be explained using a nonspatial, point-based approach.

For example, recent analyses show that redistribution of water from upslope locations to playas during high rainfall events can explain both high and low values of aboveground net primary production (ANPP) that likely depend on timing of the excess water relative to plant growth (Fig. 3.2a; from Peters et al. 2006a). Excess water that precedes the growing season can result in extremely large values of ANPP, whereas large amounts of water during the growing season can result in very small values of ANPP as a result of standing water that kills plants. In another example, spatially explicit, plant-based sampling that accounted for small arroyos where water accumulates and moves downslope was required to locate remnant grass plants in a shrubland (Fig. 3.2b; Peters et al. 2006c). Random sampling or sampling stratified by dominant vegetation characteristics likely would have missed the few plants found on average (0.1 grass/m^2) in this area. Because these grasses can be used as source plants for restoration, the locally, concentrated plants related to heterogeneous redistribution of water are more important to measure than the average value obtained from random sampling.

A recognition of the complexity of landscapes led to the need for a multi-scale approach to understand how changing pattern–process relationships across scales influence the rate and pattern of these transitions. Of particular interest is how plant-scale processes associated with individual grasses and shrubs influence spatial processes, resource redistribution among spatial units, and broad-scale drivers (e.g.., rainfall, temperature) to influence state changes from grasslands to shrublands. We focus on the role of three factors in determining these transitions: (1) spatial variation in environmental drivers (e.g., soil properties, geomorphology, weather), (2) the extent to which resource distribution is modified by patch structure interacting with transport vectors and environmental drivers, and (3) the extent to which changes in patch structure overwhelm within-patch processes (e.g., competition). These factors are examined within the temporal context of known historic legacies and measurements of environmental drivers (Peters et al. 2006a). This approach is designed to test the following hypothesis: as connectivity in bare patches increases, the rate and magnitude of material transfer that connects bare soil patches via wind and water increases. A threshold level of connectivity among bare patches (based on their size and spatial arrangement) is reached where the transfer of

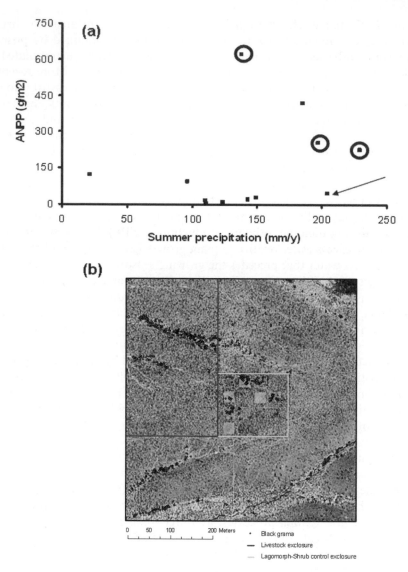

Fig. 3.2 Examples of the importance of transport vectors to landscape-scale dynamics. (a) Aboveground net primary production (ANPP) is not always related to rainfall on the site. Redistribution of water during wet years can create flooding events that either lead to higher (*circles*) or lower (*arrow*) than expected values depending on the timing of the excess water relative to plant growth. Redrawn from Peters et al. (2006a) with kind permission from BioScience. (b) Low average density (0.1 plant/m^2) of the historically dominant perennial grass, *Bouteloua eriopoda*, is a misleading value when plants are restricted to arroyos that receive additional water during rain events. These remnant plants can provide seed and vegetative propagules for revegetation in systems that are currently dominated by shrubs. Redrawn from Peters et al. (2006c) with kind permission from Rangeland Ecology and Management Journal

materials among patches increases nonlinearly. Because of positive feedbacks between bare ground connectivity and shrub dominance, the spatial extent of shrub dominance also increases nonlinearly, and resource losses from the system are escalated across increasing scales.

A pattern analysis of historical aerial photos and satellite images that started in the 1930s in southern New Mexico, USA provides support for this hypothesis (Peters et al. 2004a). Images show grasslands became dominated by woody plants through time, but the change in area dominated by woody plants increased nonlinearly with three distinct thresholds (Fig. 3.3). These thresholds likely occurred with a change in the dominant process driving shrub invasion and grass loss. The first threshold (T1) occurred at the point in time when transport of woody plant seeds into the area was the dominant process. This introduction phase was followed by a phase where patch expansion through seedling establishment was the dominant process. This shift in dominant process occurred at a second threshold (T2) where the slope of the line is smaller than during the introduction phase. The patch expansion phase lasted until the mid-1980s. At that time, the number and spatial arrangement of bare soil patches accompanying woody plant expansion was sufficient for another shift in dominant process to wind erosion and deposition (Okin et al. 2005). A third threshold (T3) was crossed with this shift in dominant process, and the rate of increase in woody plant-dominated area increased dramatically. A large area became dominated by coppice dunes centered on individual woody plants within 10 years.

Processes associated with these changes in patterns between the 1930s and present were surmised based on previous studies and past experience (e.g., Schlesinger et al. 1990). The connectivity hypothesis of changing pattern–process relationships is now being directly tested at the Jornada Basin Long Term Ecological Research (JER) site located in the northern Chihuahuan Desert of southern New Mexico, USA (32.5° N, 106.8° W, 1188 m a.s.l.) (Fig. 3.4a). The cross-scale approach to test this hypothesis is as follows. The study began in 2007, thus our results are preliminary and limited for Steps 3–5.

3.5.1 Step 1. Identify patterns in broad-scale drivers

We first identified the key broad-scale drivers for our system (wind, water) based on past experience and studies conducted at the JER (Havstad et al. 2006), and then determined the patterns in these drivers that can be used to stratify the landscape in Step 2. Landscape stratification is most often conducted using vegetation, soil, climate or topography. Here, we focus on spatial variability in drivers that influence state change dynamics in arid systems: wind that causes soil movement associated with desertification and water redistribution that influences plant success. In our initial analyses, we used satellite images and aerial photos to qualitatively determine broad-scale spatial

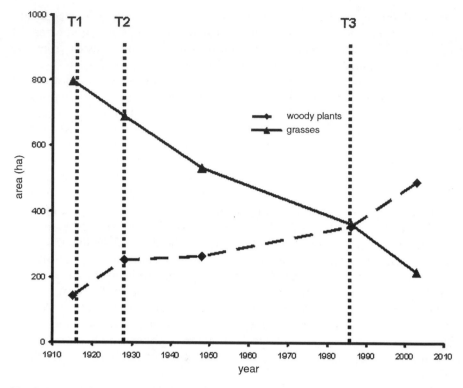

Fig. 3.3 In southern New Mexico, cover of woody plants (mesquite) has increased through time with three thresholds associated with different dominant processes: T1 was associated with the introduction and establishment of woody plants (1912–1928), T2 occurred when patch expansion processes were driving woody plant dynamics (1929–1988), and T3 occurred when wind erosion and deposition overwhelmed biotic processes to become the dominant processes determining woody plant dominance (1989–2004). Redrawn from Peters et al. (2004a)

variability in the effects of wind and water on resource redistribution (Fig. 3.4b). Data collected as part of the research project were used to confirm the quantitative characterizations of patterns in these drivers.

Wind: Patterns in sand deposition follow the prevailing winds from the major source of sand (Rio Grande) to the northeast across the JER (wind arrows; Fig. 3.4b). Sand deposition is less spatially extensive in the southern part of the JER, presumably because the Doña Ana Mountains obstruct wind flow. Thus, initial pattern analyses of sand deposition suggest that the JER can be divided into two general regions based on the effects of strong wind (north and central) and weak wind (south). Because sand deposition is determined by both wind speed and a sand source, long-term measurements of both wind speed and sand movement are needed to create a map of the effects of wind across the JER. These measurements were initiated as part of the multi-scale design.

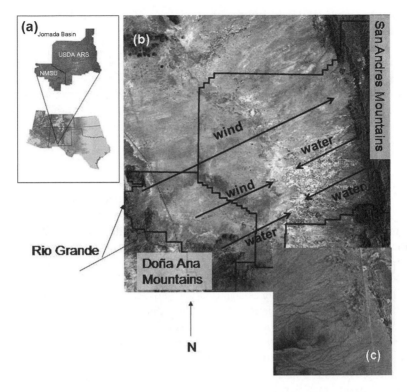

Fig. 3.4 (**a**) Location of the Jornada in southern New Mexico, USA where mean annual precipitation over the past 80 years is 24 cm, and average monthly maximum temperatures range from 13°C in January to 36°C in June. (**b**) Satellite image showing spatial variation in wind redistribution of soil particles, primarily from the southwest to the northeast (*wind arrows*). The southeastern part of the Jornada is blocked from sand movement by the Doña Ana Mountains, but has high redistribution of water from the mountains in the east (San Andres) and west (Doña Ana) to the basin (*water arrows*). (**c**) Redistribution of water from the mountains to lower elevations by arroyos is visible in close up images

Water: The spatial extent of the Jornada Basin is sufficiently large (200,000 ha) that rainfall is spatially variable across the Basin (Wainwright 2006); however, this variation is small (<50 mm/year rainfall; Peters et al. 2006e) relative to the redistribution of water along arroyos visible on high-resolution aerial photos (Fig. 3.4c). In general, the JER can be divided into areas containing arroyos that redistribute rainfall from the San Andres Mountains in the east and the Doña Ana Mountains in the southwest to lower elevations (water arrows; Fig. 3.4b), and relatively flat areas without broad-scale redistribution of rainfall. Stream gauges and small flumes distributed throughout the JER are being used to characterize the amount of water redistribution in these different areas such that a quantitative analysis can be used in the future to differentiate variability in the effects of water.

Wind and water: The resulting qualitative map of broad-scale drivers shows three main areas of the Jornada based on the relative importance of wind and water: (1) high wind, low water, (2) high wind, high water; and (3) low wind, high water (Fig. 3.4b). Interestingly, a transition zone occurs in the central to southern Jornada Basin where these two drivers either act in opposite directions (wind SW to NE; water NE to SW) or water operates from west to east or east to west. These patterns in broad-scale drivers provide an opportunity to evaluate their relative importance through landscape stratification followed by observations, experimental manipulation, and modeling in the steps below.

3.5.2 Step 2. Identify hierarchical levels of spatial units

The next step is to identify hierarchical levels of spatial units and delineate gradients or discrete unit classes (e.g., digital climate, landforms, soils, patch types/ecotones) using existing maps coupled to high-resolution remote sensing and novel analytical approaches, such as the use of object-oriented image classification, to map patterns at user-defined scales (e.g., Laliberte et al. 2007). Fine-grained patch distinctions and gradients that are difficult to detect in imagery at any scale can be complemented by field observations. Identifying hierarchical levels is a step that is often used in multi-scale studies (e.g., Wiens 1989). In our study, we identified four major hierarchical levels of spatial units based on previous research (Fig. 3.5; Peters et al. 2006a): (a) individual plants and associated bare interspaces, (b) patches or groups of plants and interspaces, (c) ecological sites or landscape units, and (d) soil-geomorphic units. The landscape unit (i.e., either a grassland or shrubland) is the spatial scale at which state changes in vegetation are often defined; thus, the landscape unit contained within a larger soil-geomorphic unit was the area of interest in the next step of stratification and mapping.

3.5.3 Step 3. Stratify and map the areas of interest

We used maps of landforms, soils, and vegetation to determine the location of our broadest hierarchical level, the soil-geomorphic unit, of which there are four on the JER (Monger et al. 2006): (1) the *basin floor sand sheet* (SS) on flat (<1% slope) loamy sands dominated by upland grasses and shrubs, primarily honey mesquite, (2) the *piedmont slope (bajada) sand sheet* on loamy sands dominated by honey mesquite, (3) the *piedmont slope (bajada)* (P) on silty and gravelly soils currently dominated by shrubs (creosotebush) with grasses in the understory, and (4) the *transition zone* (TZ) on low-gradient (<1% slope) loamy soils located between the sand sheet and piedmont slope bajada, and often characterized by banded vegetation patterns that commonly occur in arid regions on gently sloping terrain (Tongway et al. 2001) (Fig. 3.6a). Each

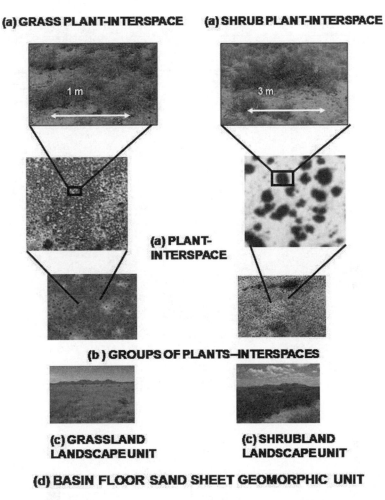

(a) GRASS PLANT-INTERSPACE   **(a) SHRUB PLANT-INTERSPACE**

1 m

3 m

(a) PLANT-INTERSPACE

(b) GROUPS OF PLANTS-INTERSPACES

(c) GRASSLAND LANDSCAPE UNIT   **(c) SHRUBLAND LANDSCAPE UNIT**

(d) BASIN FLOOR SAND SHEET GEOMORPHIC UNIT

Fig. 3.5 Spatial hierarchy of arid systems includes four major scales: (**d**) the basin floor sand sheet soil-geomorphic unit consists of (**c**) two landscape units (grasslands, shrublands), each of which contains (**b**) patches of plants and bare interspaces; (**a**) each patch consists of individual plants and an associated bare interspace shown on aerial photo and close up photo

geomorphic unit exhibits variation in vegetation and soils across a hierarchy of scales (Fig. 3.5), and contains similar landscape units defined by topography (uplands, slopes, playas) and vegetation (grasslands, shrublands). However, the geomorphic units differ with respect to: (1) the two physical transport vectors (wind, water) at broad scales and all three vectors (including animals) at multiple, finer scales, and (2) feedbacks between spatial variation in patch structure and the other system elements.

Aerial photos were used to document patch structure within each of three of the geomorphic units (SS, TZ, P) and to assess how this patch structure changes

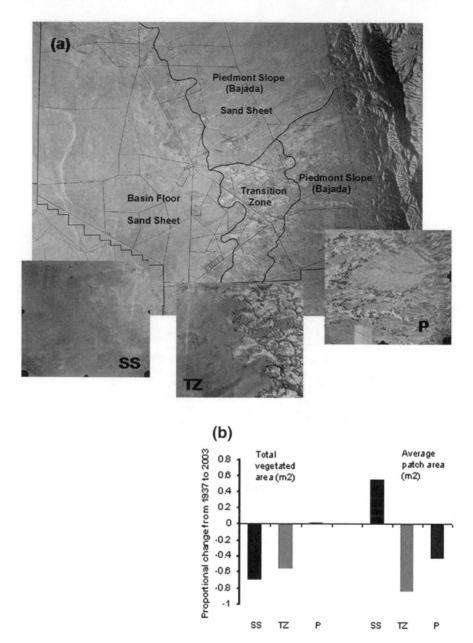

Fig. 3.6 (a) Location of four geomorphic units at the Jornada: patterns in soil and vegetation differ based on enlarged images for three of the units. (b) Changes in pattern from 1937 to 2003 in randomly selected 500 m × 500 m areas within each of three geomorphic units derived from aerial photos using Erdas Imagine and APACK software. The basin floor sand sheet (SS) decreased in vegetated area, yet increased in average patch size. The transition zone (TZ) decreased in vegetated area and patch size. The piedmont slope (bajada) (P) did not change in vegetated area, yet decreased in patch size. The piedmont slope sand sheet was not analyzed

through time (Fig. 3.6b). In general, different patterns were found through time for each geomorphic unit. Total vegetated cover decreased and average patch area increased for the SS whereas both total cover and patch area decreased for the TZ. The piedmont slope (P) decreased in average patch area with little change in total vegetated cover.

Data were collected to stratify each geomorphic unit into the next smaller level of landscape units. Patterns of vegetation and surface soil features of each landscape unit were quantified using high-resolution aerial imagery from a variety of sources, including unmanned aerial vehicles (UAVs; ground resolution of ca. 5 cm), aircraft images (ca. 25 cm ground resolution), and high- (60–70 cm QuickBird; 1 m IKONOS) and low-resolution (15–30 m resolution ASTER and Landsat) satellite data. Patches of bare ground and vegetation were extracted separately for all imagery and compared to show how landscape metrics for each patch type change with spatial scale. Field measurements were strategically employed to validate image interpretation. Each landscape unit was internally stratified based on levels of connectivity in bare soil patches determined from image analyses and spatial analysis software (Bestelmeyer et al. 2006a), and broad-scale patterns in water and wind vectors was quantified from a sensor network being developed as part of Step 1.

3.5.4 Step 4. Sample and correlate attributes

Field sampling of vegetation and soil was conducted within each landscape unit to document relationships both within and among patches. Sampling across each spatial unit (landscape or patch) in a "snapshot" fashion was used to quantify how patterns and processes underlying transitions vary between spatial units, and to ask how spatial context within the broader-scales affects these relationships. This approach is a relatively simple way to measure cross-scale effects. One example is to ask how erosion rates vary between patch types with 75% grass cover and 25% grass cover, and how this difference, in turn, is affected by the occurrence of these patch types in landscape units containing soils of high versus low erodibility, that, in turn, occur in soil-geomorphic units of high versus low wind speed.

3.5.5 Step 5. Experimental manipulations of drivers

The next step called for conducting experimental manipulations in conditions identified from step 4 as critical to state change dynamics. For example, replicate intensive study sites were selected within each landscape unit to represent a range of broad-scale drivers, in this case, high versus low wind, and a range of connectivity in resource redistribution levels (high to low) at finer scales, as determined from the pattern analyses in Step 3. Manipulations of patch structure were used to either increase or decrease bare ground connectivity. Areas of high bare ground cover were manipulated by adding structures to mimic grass/shrub effects to disrupt connectivity by wind and water transport processes. Areas with low bare ground connectivity were manipulated by removing plants to increase bare ground connectivity and movement of soil by wind. Replicated non-manipulated areas serve as controls. Spatially distributed soil collectors are used within each study site to quantify the scale, direction, and rate of material transport by wind throughout each site through time. Data are collected in adjacent downwind areas to quantify connectivity among replicates and adjacent landscape and geomorphic units. Vegetation (cover, composition, spatial distribution) and soil properties are measured through time.

3.5.6 Step 6. Simulation modeling of responses

Simulation models can be used to investigate possible interactions between broad- and fine-scale drivers using a broader combination of environmental conditions than possible experimentally (e.g., Rastetter et al. 2003, Urban 2005). Model simulations can be conducted by modifying broad- and fine-scale drivers independently and in combination, and multiple regression analyses can be used to determine the relative importance of each driver. Models can also be used to push the system to novel conditions, such as directional changes in climate, and to determine the susceptibility of different patch types to regime shifts (Peters 2002).

Simulation modeling subsequently integrated with results from existing and planned field and greenhouse studies can be used to: (1) identify key processes and interactions driving ecosystem dynamics, (2) develop new testable hypotheses and guide experimental designs, and (3) predict future conditions under alternative management and climatic scenarios. Models of horizontal transport processes (wind, water, animal) are being linked with a vegetation–soil water dynamics model to simulate transfer of materials both within and among spatial units across a range of scales (Fig. 3.7). Modeling experiments will be designed to: (1) identify and assess plant-soil-water-nutrient interactions to determine their importance and the nature of limitations, co-limitations, contingencies, and feedbacks mechanisms, (2) determine the extent to which accounting for cross-scale interactions can improve predictions of ecosystem dynamics across a range of environmental conditions and management scenarios, (3) determine the conditions under which fluxes of water and windborne material between plants and interspaces at fine scales can cascade to impact ecosystem dynamics at broader scales, and (4) the conditions where broad scale drivers begin to overwhelm or constrain fine-scale processes.

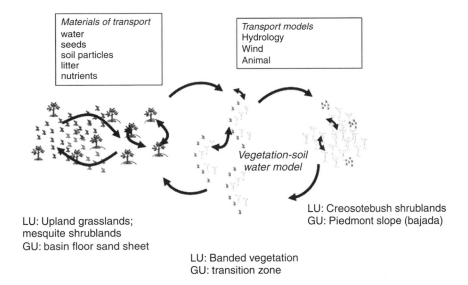

Materials of transport
water
seeds
soil particles
litter
nutrients

Transport models
Hydrology
Wind
Animal

Vegetation-soil
water model

LU: Creosotebush shrublands
GU: Piedmont slope (bajada)

LU: Upland grasslands;
mesquite shrublands
GU: basin floor sand sheet

LU: Banded vegetation
GU: transition zone

**Transfers within and among landscape units (LU) and
geomorphic units (GU)**

Fig. 3.7 Simulation models of vegetation and soil processes at the scale of individual plants and patches are being linked with transport models of wind, water, and animals to simulate the transfer of materials (water, seeds, soil particles, litter, and nutrients) both within and among landscape units (LU; e.g., upland grasslands) and geomorphic units (GU; e.g., basin floor sand sheet). Specific materials and transport vectors can differ for each landscape unit and geomorphic unit

3.6 Analytical Approaches to Identifying and Predicting Regime Shifts

There is an important role for conventional statistics to analyze data generated from hierarchically structured or spatially structured sampling. The key to such analysis lies in linking inferences gained from broad-scale, pattern-based studies with embedded fine-scale, mechanistic studies to identify transitions. For example, traditional statistics can be used to determine if nonlinearities detected in pattern–process relationships at fine scales (e.g., between grass cover and erosion rate or shrub establishment rate) translate to differences in the occurrence of alternative states at broad spatial and temporal scales. In addition, statistical analyses can be used to determine if the following criteria have been met to identify the presence and type of regime shift (from Scheffer and Carpenter 2003): (1) Is there a discrete step function or intervention in the time series? (2) Does the response variable have a bimodal or multi-modal distribution? (3) Is there a different functional relationship in different regimes? (4) Given different starting conditions, will the system go to different stable states? (5) Does the system shift to an alternative state when perturbed? (6) Does the system have a different trajectory when the forcing function increases compared to when it decreases? (7) Does the second derivative of the time series have peaks that indicate nonlinearities?

There are also a variety of novel statistical approaches to deal with pattern (survey) data gathered at multiple scales that show promise in predicting regime shifts (e.g., Lichstein et al. 2002, Keitt and Urban 2005). Three non-traditional approaches warrant discussion here.

First, quantile regression offers the ability to examine relationships across a range of quantiles of a dataset, especially at the limits of the 'data cloud' that often result when limiting factors (such as soils and climate) structure data (i.e., 5–10% and 90–95%; Cade and Noon 2003). Quantile regression is especially useful when the distribution of a response variable (e.g., grass cover) on environmental gradients in a reference state at a particular scale is obscured by events that cause transitions to alternative states (Bestelmeyer et al. in review). For example, deviations away from an upper limit to the data (defined by sites in a reference state) may be caused by transitions and disjunct clouds of points that represent alternative states.

Second, variability in temporal data may also indicate transitions. Rising variance in time series data of key proximate variables underlying transitions may provide a useful means of predicting transitions (Brock and Carpenter 2006). Increasing variability in grass cover (possibly detected at several scales) or erosion rates with variations in climate may similarly herald an increased risk of a transition to woodland-dominated state (see also Viglizzo et al. 2005 for other approaches).

Third, changes in the size distribution of vegetated or bare soil patches may be warning signs of an impending regime shift in arid systems. Cellular automata modeling suggests that transitions between patch types defined by simple rules can lead to a predictable departure from the power-law behavior of grasslands (Kéfi et al. 2007, Scanlon et al. 2007). Field observations of changing distribution of bare soil patches with increasing soil degradation that leads to transition to woody plant dominance support these theoretical analyses (J.E. Herrick unpublished data).

3.7 Conclusions

State changes from grasslands to woodlands provide the context for development of a multi-scale experimental approach combined with simulation modeling and statistical analyses to examine the key processes influencing these dynamics. This approach is sufficiently general to be applied to other systems where drivers and responses interact across scales, either with or without a state change (e.g., Scheffer et al. 2001, Allen 2007, Willig et al. 2007). In addition, this approach is relevant to broader questions regarding changes in global ecology, such as: how and under what conditions do dynamics and decisions made at fine scales influence dynamics at broader scales and how and under what conditions do broad-scale dynamics overwhelm fine-scale processes to influence landscape patterns? Addressing these questions is becoming increasingly important as ecologists expand the scope and spatial extent of interest beyond individual sites to regions, continents, and the globe (Peters et al. 2008). Existing and emerging networks of sites, such as the Long Term Ecological Research network (www.lternet.edu), the National Ecological Observatory Network (www.neoninc.org), AmeriFlux (public.ornl.gov/ameriflux), and the National Atmospheric Deposition Program (nadp.sws.uiuc.edu), are collecting observational and experimental data primarily at the site scale, yet are charged with addressing broad-scale questions. These networks necessitate a multi-scale approach such as the one described here.

Acknowledgments This study was supported by National Science Foundation awards to LTER programs at the Jornada Basin at New Mexico State University (DEB 0618210) and the Konza Prairie at Kansas State University (DEB 0218210). We thank Barbara Nolen for assistance in figure preparation. We thank Greg Okin, Tony Parsons, John Wainwright, Al Rango, Ed Fredrickson, Laurie Abbott, Osvaldo Sala, Steve Archer, Rhonda Skaggs, and Heather Throop for helpful discussions. We thank the reviewers and editors for helpful comments on the manuscript.

References

Allen, C.D. 2007. Interactions across spatial scales among forest dieback, fire, and erosion in northern New Mexico landscapes. Ecosystems 10:797–808.

Archer, S. 1994. Woody plant encroachment into south-western grasslands and savannas: rates, patterns and proximate causes. In Ecological implications of livestock herbivory in the West, ed. M, Vavra, W. Laycock, and R. Pieper, pp. 13–68. Denver, CO: Society of Rangeland Management.

Archer, S.R., Schimel, D.S., and E.H. Holland. 1995. Mechanisms of shrubland expansion: land use, climate or CO_2? Climate Change 29:91–99.

Bahre, C.J., and M.L. Shelton. 1993. Historic vegetation change, mesquite increase, and climate in southeastern Arizona. Journal of Biogeography 20:489–504.

Beaugrand, G. 2004. The North Sea regime shift: evidence, causes, mechanisms and consequences. Progress in Oceanography 60:245–262.

Bestelmeyer, B.T. 2006. Threshold concepts and their use in rangeland management and restoration: the good, the bad, and the insidious. Restoration Ecology 14:325–329.

Bestelmeyer, B.T., Trujillo, D.A., Tugel, A.J., and K.M. Havstad. 2006b. A multi-scale classification of vegetation dynamics in arid lands: what is the right scale for models, monitoring, and restoration? Journal of Arid Environments 65:296–318.

Bestelmeyer, B.T., Tugel, A.J., Peacock, G., Robinett, D., Shaver, P.L., Brown, J.R., Herrick, J.E., Sanchez, H., and K.M. Havstad. A cross-scale sampling strategy to develop ecological land classifications coupled to State-and-Transition models. Rangeland Ecology and Management (in review).

Bestelmeyer, B., Ward, J., Herrick, J.E., and A.J. Tugel, A.J. 2006a. Fragmentation effects on soil aggregate stability in a patchy arid grassland. Rangeland Ecology and Management **59**:406–415.

Briggs, J.M., Knapp, A.K., Blair, J.M., Heisler, J.L., Hoch, G.A., Lett, M.S., and J.K. McCarron. 2005. An ecosystem in transition: causes and consequences of the conversion of mesic grassland to shrubland. BioScience **55**:243–254.

Brock, W.A., and S.R. Carpenter. 2006. Variance as a leading indicator of regime shift in ecosystem services. Ecology and Society **11**(2):9. [online] URL: www.ecologyandsociety.org/vol11/iss2/art9/ .

Brown, J.R., and S. Archer. 1989. Woody plant invasion of grasslands: establishment of honey mesquite (*Prosopis glandulosa* var. *glandulosa*) on sites differing in herbaceous biomass and grazing history. Oecologia **80**:19–26.

Cade, B.S. and B.R. Noon. 2003. A gentle introduction to quantile regression for ecologists. Frontiers in Ecology and Environment **1**:412–420.

Carpenter, S.R. 2003. Regime shifts in lake ecosystems: pattern and variation. International Ecology Institute, 21385 Oldendorf/Luhe, Germany.

Carpenter, S.R., and M.G. Turner. 2000. Hares and tortoises: Interactions of fast and slow variables in ecosystems. Ecosystems **3**:495–497.

Chambers, J.C., Vander Wall, S.B., and E.W. Schupp. 1999. Seed and seedling ecology of piñon and juniper species in the pygmy woodlands of western North America. Botanical Review **65**:1–38.

Charney, J.G., Stone, P.H., and W.J. Quirck. 1975. Drought in the Sahara: a biogeophysical feedback mechanism. Science **187**:434–435.

Claussen, M. 1997. Modeling biophysical feedback in the African and Indian Monsoon region. Climate Dynamics **13**:247–257.

Davenport, D.W., Breshears, D.D., Wilcox, B.P. and C.D. Allen. 1998. Viewpoint: sustainability of piñon-juniper ecosystems – a unifying perspective of soil erosion thresholds. Journal of Range Management **51**:231–240.

Diffenbaugh, N.S., Pal, J.S., Trapp, R.J., and F. Giorgi, F. 2005. Fine-scale processes regulate the response of extreme events to global climate change. Proceedings National Academy of Sciences **102**:15774–8.

Epstein, H.E., Beringer, J., Gould, W.A., Lloyd, A.H., Thompson, C.D., Chapin, F.S. III, Michaelson, G.J., Ping, C.L., Rupp, T.S., and D.A. Walker. 2004. The nature of spatial transitions in the arctic. Journal of Biogeography **31**:1917–1933.

Foley, J.A., Coe, M.T., Scheffer, M., and G. Wang. 2003. Regime shifts in the Sahara and Sahel: interactions between ecological and climatic systems in Northern Africa. Ecosystems **6**:524–539.

Fredrickson, E.L., Estell, R.E., Laliberte, A., and D.M. Anderson. 2006. Mesquite recruitment in the Chihuahuan Desert: historic and prehistoric patterns with long-term impacts. Journal of Arid Environments **65**:285–295.

Fredrickson, E.L. Havstad, K.M., Estell, R., and P. Hyder. 1998. Perspectives on desertification: Southwestern United States. Journal of Arid Environments **39**:191–207.

Gibbens, R.P., McNeely, R.P., Havstad, K.M., Beck, R.F., and B. Nolen. 2005. Vegetation changes in the Jornada Basin from 1858 to 1998. Journal of Arid Environments **61**:651–668.

Grimm, E.C. 1983. Chronology and dynamics of vegetation change in the prairie-woodland region of southern Minnesota, U.S.A. New Phytologist **93**:311–350.

Groffman, P.M., Baron, J.S., Blett, T., Gold, A.J., Goodman, I., Gunderson, L.H., et al. 2006. Ecological thresholds: the key to successful management or an important concept with no practical application? Ecosystems **9**:1–13.

Gunderson, L., and C. Holling, Eds. 2002. Panarchy: understanding transformations in human and natural systems. Island Press, Washington DC.

Hall, S.A. 2005. Ice Age vegetation and flora of New Mexico. New Mexico Museum of Natural History and Sciences Bulletin **28**:171–183.

Hargrove, W.W., and J. Pickering. 1992. Pseudoreplication: a sine qua non for regional ecology. Landscape Ecology **6**:251–258.

Havstad, K.M., Huenneke, L.F., and W.H. Schlesinger, Eds. 2006. Structure and function of a Chihuahuan Desert ecosystem: the Jornada Basin Long Term Ecological Research site. Oxford University Press: Oxford.

Higgens, P.A.T., Mastrandrea, M.D., and S.H. Schneider. 2002. Dynamics of climate and ecosystem coupling: abrupt changes and multiple equilibria. Philosophical Transactions Royal Society of London B. **357**:647–655.

Houghton, R.A., Hackler, J.L., and J.T. Lawrence. 1999. The U.S. carbon budget: contributions from land-use change. Science **285**:574–578.

Howes, B.L., Dacey, J.W.H., and D.D. Goehringer. 1986. Factors Controlling the Growth Form of *Spartina alterniflora*: feedbacks Between Above-Ground Production, Sediment Oxidation, Nitrogen and Salinity. Journal of Ecology **74**:881–898.

Huenneke, L.F., Anderson, J.P., Remmenga, M., and W.H. Schlesinger. 2002. Desertification alters patterns of aboveground net primary production in Chihuahuan ecosystems. Global Change Biology **8**:247–264.

Humphrey, R.R. 1958. The desert grassland: a history of vegetational change and an analysis of causes. Botanical Review **24**:193–252.

Jaffe, D., McKendry, I., Anderson, T., and H. Price. 2003. Six "new" episodes of trans-Pacific transport of air pollutants. Atmospheric Environment **37**:391–404.

Kéfi, S., Rietkerk, M., Alados, C.L., Pueyo, Y., Papanastasis, V.P., ElAich, A., and P.C. de Ruiter. 2007. Spatial vegetation patterns and imminent desertification in Mediterranean arid ecosystems. Nature **449**:213–217.

Keitt, T.H. and D.L. Urban. 2005. Scale-specific inference using wavelets. Ecology **86**:2497–2504.

Knapp, A.K., Briggs, J.M., Collins, S.L., Archer, S.R., Bret-Harte, M.S., Ewers, B.E., Peters, D.P.C., Young, D.R., Shaver, G.R., Pendall, E., and M.B. Cleary. 2008. Shrub encroachment in North American grasslands: shifts in growth form dominance rapidly alters control of ecosystem carbon inputs. Global Change Biology **14**:615–623.

Laliberte, A.S., E.L. Fredrickson, and A. Rango. 2007. Combining decision trees with hierarchical object-oriented image analysis for mapping arid rangelands. Photogrammetric Engineering and Remote Sensing **73**:197–207.

Lett, M.S. and A.K. Knapp. 2003. Consequences of shrub expansion in mesic grassland: resource alterations and graminoid responses. Journal of Vegetation Science **14**:487–496.

Lett, M.S., Knapp, A.K., Briggs, J.M., and J.M. Blair. 2004. Influence of shrub encroachment on aboveground net primary productivity and carbon and nitrogen pools in a mesic grassland. Canadian Journal of Botany **82**:1363–1370.

Lichstein, J., T. Simons, S. Shriner, and K. Franzreb. 2002. Spatial autocorrelation and autoregressive models in ecology. Ecological Monographs **72**:445–463.

Ludwig, J.A., Bartley, R., Hawdon, A., and D. McJanner. 2007. Patchiness effects sediment loss across scales in a grazed catchment in north-east Australia. Ecosystems **10**:839–845.

Ludwig, J.A., Wiens, J.A., and D.J. Tongway. 2000. A scaling rule for landscape patches and how it applies to conserving soil resources in savannas. Ecosystems **3**:84–97.

Ludwig, J.A., Wilcox, B.P., Breshears, D.D., Tongway, D.J., and A.C. Imeson. 2005. Vegetation patches and runoff-erosion as interacting processes in semiarid landscapes. Ecology **86**:288–297.

Lyford, M.E., Jackson, S.T., Betancourt, J.L., and S.T. Gray. 2003. Influence of landscape structure and climate variability on a late Holocene plant migration. Ecological Monographs **73**:567–583.

Magnuson, J.J., Kratz, T.K., and B.J. Benson. 2006. Long-term dynamics of lakes in the landscape: long-term ecological research on North Temperate lakes. Oxford University Press, Oxford.

Marr, J.W. 1977. The development and movement of tree islands near the upper limit of tree growth in the southern Rocky Mountains. Ecology 58:1159–1164.

Martin, R.E., and G.P. Asner. 2005. Regional estimate of nitric oxide emissions following woody encroachment: linking imaging spectroscopy and field studies. Ecosystems 8:33–47.

McCarron, J.K. and A.K. Knapp. 2003. C_3 shrub expansion in a C_4 grassland: positive post-fire responses in resources and shoot growth. American Journal of Botany 90:1496–1501.

McCook, L.J. 1999. Macroalgae, nutrients, and phase shifts on coral reefs: scientific issues and management consequences for the Great Barrier Reef. Coral Reef 18:357–367.

McKergow, L.A., Prosser, I.P., Huges, A.O., and J. Brodie. 2005. Sources of sediment to the Great Barrier Reef World Heritage Area. Marine Pollution Bulletin 51:200–211.

Michener, W.K., Baerwald, T.J., Firth, P., Palmer, M.A., Rosenberger, J.L., Sandlin, E.A., and H. Zimmerman. 2001. Defining and unraveling complexity. BioScience 51:1018–23.

Monger, H.C., and B.T. Bestelmeyer. 2006. The soil-geomorphic template and biotic change in arid and semiarid ecosystems. Journal of Arid Environments 65:207–218.

Monger, H.C., Cole, D.R., Gish, J.W., and T.H. Giordano. 1998. Stable carbon and oxygen isotopes in Quaternary soil carbonates as indicators of ecogeomorphic changes in the northern Chihuahuan Desert, USA. Geoderma 82:137–172.

Monger, H.C., Mack, G.H., Nolen, B.A., and L.H. Gile. 2006. Regional setting of the Jornada Basin. Pages 15–43 In Havstad, K.M., Huenneke, L.F., and W.H. Schlesinger, eds. Structure and function of a Chihuahuan Desert ecosystem: the Jornada Basin Long Term Ecological Research site. Oxford University Press, Oxford.

Okin, G.S., Gillette, D.A., and J.E. Herrick. 2006. Multi-scale controls on and consequences of aeolian processes in landscape change in arid and semi-arid landscapes. Journal of Arid Environments 65:253–275.

Pacala, S.W., Hurtt, G.C., Baker, D., Peylin, P., Houghton, R.A., Birdsey, R.A., Heath, L., Sundquist, E.T., et al. 2001. Consistent land- and atmosphere-based U.S. carbon sink estimates. Science 292:2316–2320.

Peters, D.P.C. 2002. Plant species dominance at a grassland-shrubland ecotone: an individual-based gap dynamics model of herbaceous and woody species. Ecological Modelling 152: 5–32.

Peters, D.P.C., Bestelmeyer, B.T., Herrick, J.E., Monger, H.C., Fredrickson, E., and K.M. Havstad. 2006a. Disentangling complex landscapes: new insights to forecasting arid and semiarid system dynamics. BioScience 56:491–501.

Peters, D.P.C., Bestelmeyer, B.T., and M.G. Turner. 2007. Cross-scale interactions and changing pattern-process relationships: consequences for system dynamics. Ecosystems 10:790–796.

Peters, D.P.C., Gosz, J.R., Pockman, W.T., Small, E.E., Parmenter, R.R., Collins, S.L., and E. Muldavin. 2006b. Integrating patch and boundary dynamics to understand and predict biotic transitions at multiple scales. Landscape Ecology 21:19–33.

Peters, D.P.C., Groffman, P.M., Nadelhoffer, K.J., Grimm, N.B., Collins, S.L., Michener, W.K., and M.A. Huston. 2008. Living in an increasingly connected world: a framework for continental-scale environmental science. Front. Ecol. Environ 6:229–237.

Peters, D.P.C., I. Mariotto, K.M. Havstad, and L.W. Murray. 2006c. Spatial variation in remnant grasses after a grassland to shrubland state change: implications for restoration. Rangeland Ecology and Management 59:343–350.

Peters, D.P.C., Pielke, R.A. Sr, Bestelmeyer, B.T., Allen, C.D., Munson-McGee, S., and K.M. Havstad. 2004a. Cross scale interactions, nonlinearities, and forecasting catastrophic events. Proceedings of the National Academy of Sciences 101:15130–15135.

Peters, D.P.C., Urban, D.L., Gardner, R.H., Breshears, D.D., and J.E. Herrick. 2004b. Strategies for ecological extrapolation. Oikos 106:627–636.

Peters, D.P.C., J. Yao, and J.R. Gosz. 2006d. Woody plant invasion at a semiarid-arid transition zone: importance of ecosystem type to colonization and patch expansion. Journal of Vegetation Science 17:389–396.

Peters, D.P.C., J. Yao, L.F. Huenneke, K.M. Havstad, J.E. Herrick, A. Rango, and W.H. Schlesinger. 2006e. A framework and methods for simplifying complex landscapes to reduce uncertainty in predictions. Pages 131–146. In: J. Wu, B. Jones, H. Li, and O.L. Loucks, eds. Scaling and uncertainty analysis in ecology: methods and applications. Springer, Dordrecht, The Netherlands.

Petersen, J.E., Kemp, W.M., Bartleson, R., Boynton, W.R., Chen, C., Cornwell, J.C., Gardner, R.H., Hinkle, D.C., Houde, E.D., Malone, T.C., Mowitt, W.P., Murray, L., Sanford, L.P., Stevenson, J.C., Sundberg, K.L., and S.E. Suttles. 2003. Multiscale experiments in coastal ecology: improving realism and advancing theory. BioScience 53:1181–1197.

Pielke, R.A. Sr, Adegoke, J., Beltrán-Przekurat, A., Hiemstra, C.A., Lin, J., Nari, U.S., Niyogi, D., and T.E. Nobis. 2007. An overview of regional land-use and land-cover impacts on rainfall. Tellus DOI: 10.1111/j.1600-0889.2007.00251.x

Polley, H.W., Johnson, H.B., and C.R. Tischler. 2002. Woody plant invasion of grasslands: evidence that CO_2 enrichment indirectly promotes establishment of *Prosopis glandulosa*. Plant Ecology 164:85–94.

Rastetter, E.B., Aber, J.D., Peters, D.P.C. Peters, and D.S. Ojima. 2003. Using mechanistic models to scale ecological processes across space and time. BioScience 53:1–9.

Ricketts, T.R., Dinerstein, E., Olson, D.M., and Loucks, C.J., et al. 1999. Terrestrial Ecoregions of North America: A Conservation Assessment. Island Press, Washington DC.

Ricklefs, R.E. 1987. Community diversity: relative roles of local and regional processes. Science 235:167–171.

Rietkerk M., and J. van de Koppel. 1997. Alternate stable status and threshold effects in semi-arid grazing systems. Oikos 79:69–76.

Scanlon, T.M., Caylor, K.K., Levin, S.A., and I. Rodriguez-Iturbe. 2007. Positive feedbacks promote power-law clustering of Kalahari vegetation. Nature 449:209–211.

Scheffer, M., and S.R. Carpenter. 2003. Catastrophic regime shifts in ecosystems: linking theory to observation. Trends in Ecology and Evolution 18:648–656.

Scheffer, M., Carpenter, S.R., Foley, J.A., Folke, C., and B. Walker. 2001. Catastrophic shifts in ecosystems. Nature 413:591–596.

Schlesinger, W.H., Reynolds, J.F., Cunningham, G.L., Huenneke, L.F., Jarrell, W.H., Virginia, R.A., and W.G. Whitford. 1990. Biological feedbacks to global desertification. Science 247:1043–1048.

Scholes, R.J., and S.R. Archer. 1997. Tree-grass interactions in savannas. Annual Review of Ecology and Systematics 28:517–544.

Shaver, G.R., Bret-Harte, M.S., Jones, M.H., Johnstone, J., Gough, L., Laundre, J., and F.S. Chapin. 2001. Species composition interacts with fertilizer to control long-term change in tundra productivity. Ecology 82:3163–3181.

Stohlgren, T.J., Chong, G.W., Kalkhan, M.A., and L.D. Schell. 1997. Multiscale Sampling of Plant Diversity: Effects of Minimum Mapping Unit Size. Ecological Applications 7:1064–1074.

Tongway, D.J., Valentin, C., and J. Seghieri. 2001. Banded Vegetation Patterning in Arid and Semi-arid Environments: Ecological Processes and Consequences for Management. Ecological Studies No. 149, Springer Verlag, New York, 243 pp.

Urban, D.L. 2005. Modeling ecological processes across scales. Ecology 86:1996–2006.

Van Auken, O.W. 2000. Shrub invasions of North American semiarid grasslands. Annual Review of Ecology and Systematics 31:197–215.

Van Devender, T.R. 1990. Late Quaternary vegetation and climate of the Chihuahuan Desert, United States and Mexico. Pages 104–133 In: Betancourt, J.L., Van Devender, T.R., and P.S. Marton, eds. Packrat middens: the last 40,000 years of biotic change. University of Arizona Press: Tucson, AZ, USA.

Viglizzo, E.F., Pordomingo, A.J., Buschiazzo, D., and M.G. Castro. 2005. A methodological approach to assess cross-scale relations and interactions in agricultural ecosystems of Argentina. Ecosystems 8:546–558.

Wainwright, J. 2006. Climate and climatological variation in the Jornada Basin. Pages 44–80 In: Havstad, K.M., Huenneke, L.F., and W.H. Schlesinger, eds. Structure and function of a Chihuahuan Desert ecosystem: the Jornada Basin Long Term Ecological Research site. Oxford University Press: Oxford.

Weaver, J.E., and F.W. Albertson. 1940. Deterioration of Midwestern ranges. Ecology 21:216–236.

Wiens, J.A. 1989. Spatial Scaling in Ecology. Functional Ecology 3:385–397.

Willig, M.R., Bloch, C.P., Brokaw, N., Higgens, C., Thompson, J., and C.R. Zimmerman. 2007. Cross-scale responses of biodiversity to hurricane and anthropogenic disturbance in a tropical forest. Ecosystems 10:824–838.

Wright, R.A., and J.H. Honea. 1986. Aspects of desertification in southern New Mexico, USA: soil properties of a mesquite duneland and a former grassland. Journal of Arid Environments 11:139–145.

Young, D.R., Porter, J.H., Bachmann, C.M., Shao, G., Fusina, R.A., Bowles, J.H., Korwan, D., and T. Donato. 2007. Cross-scale patterns in shrub thicket dynamics in the Virginia barrier complex. Ecosystems 10:854–863.

Chapter 4
Integrating Multiple Spatial Controls and Temporal Sampling Schemes To Explore Short- and Long-Term Ecosystem Response to Fire in an Everglades Wetland

ShiLi Miao, Susan Carstenn, Cassondra Thomas, Chris Edelstein, Erik Sindhøj and Binhe Gu

4.1 Introduction

Ecologists, natural resource managers, and policy makers have faced the necessity of applying the findings of ecological research to solving real world problems resulting from both natural disasters and anthropogenic disturbances. In response, new disciplines in ecology have emerged, such as ecosystem, landscape, and restoration ecology. These disciplines and others reflect a critical paradigm shift in ecological research from classical small-scale mechanistic experiments to large-scale studies which attempt to integrate ecosystem complexity, landscape processes, and multiple response variables (Legendre 1993, Carpenter et al. 1995, Carpenter et al. 1998, Clark et al. 2001, Simberloff 2004, Melbourne and Chesson 2005). Ecological experimental design has progressed (Legendre et al. 2002, Petersen et al. 2003, Hewitt et al. 2007, Miao and Carstenn 2006, Miao et al. Chapter 1, Peters et al. Chapter 3) by recognizing and addressing ecosystem heterogeneity across spatial and temporal scales and incorporating response variables operating at multiple levels of biological organization (e.g., individuals, population, communities, and ecosystems) into design and analysis. While advances in large-scale and long-term ecological studies have been heralded, deliberately incorporating ecological complexity into experimental design is still in its infancy (Hewitt et al. 2007) in part because of the shortcomings of traditional analytical methods in dealing with complexity (Miao et al. Chapter 1, Grace et al. Chapter 2).

Experimentation at the ecosystem scale is a powerful approach to understanding the linkages among processes operating at multiple scales (Likens 1985, Carpenter et al. 1995, Clark et al. 2001, Mitsch and Day 2004). Ecosystem experiments are field studies incorporating multiple spatial and temporal scale including most, if not all

S. Miao (✉)
South Florida Water Management District, TA Management Division, 3301 Gun Club
Road, West Palm Beach, FL 33406, USA
e-mail: smiao@sfwmd.gov

S. Miao et al. (eds.), *Real World Ecology*, DOI 10.1007/978-0-387-77942-3_4,
© Springer Science+Business Media, LLC 2009

major ecological processes (Carpenter et al. 1995, Mitsch and Day 2004). However, increasing the spatial and temporal scales of an experiment in turn increases the extent of natural variation encompassed by the experiment. Furthermore, the extent of spatial variation is relative to the response variable. For example, in a nutrient-enriched wetland the variation in hydrology over space and time is much greater than the variation in vegetation. Therefore, it is important to link spatial variation and process response-scale when designing an ecosystem study.

Traditionally, scientists addressed increased spatial variability at larger scales by increasing the number of replicates; however, replication of large-scale ecosystem disturbances is not always feasible, logistically or financially, and might even be considered undesirable in many real-world circumstances (Carpenter 1990, 1996, Schindler 1998). Many contemporary designs and analytical approaches for nonreplicated experiments originated from experiments conducted as early as 1948 when Hasler and his colleagues applied a Treatment/Control or Reference approach (Hasler et al. 1951). In the decades to follow, these approaches were gradually refined and resulted in Intervention Analysis (IA; Box and Tiao 1965, 1975), Before–After (BA; Green 1979), Randomized Intervention Analysis (RIA; Carpenter et al. 1989), Impact–Control (IC; Pickett 1989) and Before–After Control–Impact (BACI; Stewart-Oaten 1996c).

Site selection is critical to accounting for spatial variability and detecting treatment impacts in nonreplicated experimental designs and will influence the robustness of ecological generalizations. Currently, there is no consensus on the requirements for selecting controls in a field study, neither what constitutes an ideal control nor how many control sites are best suited for detecting a response to a treatment. Underwood (1991) proposed that increasing the number of controls would increase the likelihood of *correctly* detecting a response to treatment. He further suggested that the location and similarity among control sites was not critical for detecting treatment impact, as long as a sufficient number of controls were selected and all controls fell within the expected range of natural variability (Underwood 1993, 1994). Thus, controls should be randomly located and increasing the number of control sites should reduce variability in the data and more accurately estimate parameter values, thereby increasing the likelihood of detecting a response. Yet, other scientists have suggested that control and treatment sites should be similarly influenced by factors affecting system changes and therefore display similar temporal trajectories, i.e., control and treatment sites should be highly correlated in the absence of a treatment (Osenberg and Schmitt 1996, Stewart-Oaten 1996c). Hewitt et al. (2001) carried this argument further and concluded that the greater the correlation between control and treatment sites in the absence of the treatment, the greater the chance of detecting treatment effects. Accordingly, controls should be selectively located to maximize the covariation between the control and treatment site, as described for Before-After-Control-Impact Paired Series (BACIPS) analysis (Stewart-Oaten 1996c, Urquhart et al. 1998).

Increasing the number of controls in nonreplicated ecosystem experiments has been argued to increase the inferential robustness of the BACI design

(Underwood 1993, 1994, Conquest 2000, Hewitt et al. 2001, Hewitt et al. 2007). Based on this assumption, Underwood (1993, 1994) presented an asymmetrical multiple-control BACI design (Beyond BACI), and Keough and Quinn (2000) proposed a multiple-control and multiple-impact BACI (MBACI) design. Other scientists, however, claimed that using multiple controls in the BACI model would hinder impact detection (Stewart-Oaten 1996c). Hewitt et al. (2001) went on to demonstrate that each successive control location added to the BACI model decreased the likelihood of detecting the impact. Ultimately, the logistical and financial limitations precluding experimental replication limit our ability to replicate regardless of whether we are replicating the control or the treatment; therefore, experimental results suggesting that replication may actually hinder impaction detection deserve further attention.

The potential for treatment effects extending beyond the borders of the treatment site (heretofore called regional effects or responses), such as downstream or downwind effects from a treatment, complicates appropriate treatment and control site selection. Most contemporary nonreplicated designs and associated analytical approaches focus primarily on temporal effects of a disturbance to the treatment site and do not address spatial questions, such as the distance from the disturbance that effects can be detected or the magnitude and duration of the regional responses. For example, the influence of flowing water and wind on downstream post-fire nutrient transport in riverine and wetland systems must be included when evaluating the effects of fire on aquatic systems. Ecologists and resource managers must account for unexpected or undesirable regional responses to management activities as part of comprehensive ecosystem management plans.

Addressing temporal variation is another challenge for experimental design and data analyses of ecosystem-scale studies. Temporal variation can result from daily, seasonal, and annual cyclical influences, but is also a function of scale-related processes, biological organizational levels, and feedback loops; each requires a unique sampling regime for adequate detection. The more scales at which processes occur and are measured, the more sampling schemes required. Temporal sampling schemes that are optimal for detecting response variables at longer temporal scales (months and years) are not necessarily adequate for detecting those at shorter temporal scales (hours or days). Similarly, optimal sampling schemes designed to detect pulsed responses are not logistically or economically practical over long periods of time, particularly when disturbances can generate both "pulsed" (short-duration) and "press or sustained" (long-duration) effects and responses (Bender et al. 1984, Underwood 1989). Thus, a single balanced sampling strategy, defined as an equal number of control and treatment samples, equal intervals between samples, and equal sampling duration, that is often required by traditional analytical approaches may not be suitable for an ecosystem-scale study. Instead, imbalanced temporal sampling schemes are needed with high frequency sampling events to detect pulsed responses and low frequency sampling events to detect seasonal and long-term trends. Not only do the sampling interval, frequency, and duration need to vary to detect responses at different time scales, but also they must vary by system response variables examined.

While incorporating spatial variation and imbalanced temporal sampling schemes into the design of large-scale ecosystem studies makes ecological sense, it significantly complicates data analysis. Many analytical procedures require equal variance, equal sampling intervals, and equal sampling duration, all of which are not practical when attempting to reach an integrated understanding of ecosystem response at multiple scales. Moreover, existing analytical approaches for assessing effects of environmental disturbances within a non-replicated design are largely designed to capture sustained treatment effects from long-term datasets. For example, time-series analysis may seem ideal for evaluating the onset, duration, and magnitude of an ecosystem response over time; however, it requires a long sequence of data collected at equal intervals, which would likely miss pulsed effects or include such a large number of samples as to be economically infeasible. Although the Beyond BACI approach using multiple controls was proposed to address both pulsed and sustained impacts of environmental disturbances (Underwood 1993, 1994), it was still based on equal sampling numbers and durations both before and after a disturbance. Thus, other scientists have suggested using linear regression to detect differences in temporal trends among sites by using the slope of the regression as the response variable (Gotelli and Ellison 2004, Murtaugh 2007). This regression approach requires replicate treatment, controls, and slope estimates for conducting either a t-test or an ANOVA and is therefore not appropriate for nonreplicated design. However, if no replicates are available the slopes among the sites can be compared using a test for homogeneity of slope or ANCOVA. Each of these analyses is based on different assumptions and is useful for detecting sustained effects, but none of these approaches are applicable for data that demonstrate pulsed response patterns. Therefore, there is an urgent need to refine an approach that will facilitate analyses of nonreplicated repeated measures data exhibiting both pulsed and sustained responses at multiple spatial and temporal scales.

Recently, Hewitt et al. (2007) suggested that researchers needed to embrace variation and incorporate it into experiments designed to detect and understand changes in variability (Thrush et al. 1998, Thrush et al. 1999, Hewitt et al. 2001). The present study is one of only a few empirical studies that have deliberately addressed the incorporation of multiple spatial and temporal scales into their design and analysis. We address this issue from a case-study perspective using our ongoing empirical study investigating the effects of a prescribed surface fire on the recovery of a nutrient-enriched and *Typha domingensis* (cattail)-dominated wetland ecosystem within the Florida Everglades. Our main objectives were to illustrate the following: (1) the integration of variability within and among multiple-spatial and temporal scales into design, experimental setup, and temporal sampling schemes; (2) the application of control sites at multiple spatial scales; and (3) novel, yet simple, analytical approaches to detect pulsed and sustained fire effects and system responses.

4.2 Rationale of a Large-Scale Fire Project

In northern portions of the Everglades, decades of increased phosphorus (P) loading, altered hydrology, and disrupted fire regimes have led to the degradation of the historical topography and the replacement of historical vegetation with dense monotypic stands of cattails. Recent management efforts, such as reduced nutrient loading to the Everglades through the creation of Stormwater Treatment Areas (STAs), Best Management Practices (BMPs) for agriculture, and partially restored hydrology (Chimney and Goforth 2001), were thought to eventually lead to natural recovery of the eutrophic areas. However, there was little understanding of processes and time scale involved in recovery of these areas. Therefore, the 2003 Long Term Plan for Achieving Water Quality Goals in the Everglades Protection Area and Tributary Basins (LTP; Burns and McDonnell 2003), an amendment to the Everglades Forever Act, required large-scale research into restoration options that may "accelerate" recovery of nutrient-enriched areas of the Everglades. The key emphasis of the LTP was accelerating the decline of dense stands of cattail and the return of historical Everglades vegetation.

Vegetation restoration is a major goal and approaches to accelerating recovery must contend with the approximately 20 cm thick layer of P-enriched soil, a legacy of past nutrient loading. Several approaches to large-scale cattail removal include herbicide application, mechanical removal, and fire. Given that fire is a natural forcing function in the Everglades, one could surmise it is a more ecologically sensitive approach than chemical or mechanical removal. The role of fire in creation and maintenance of the historic Everglades was recently reviewed (Miao and Carstenn 2005, and references therein); however, fire's ability to shift species composition and reduce cattail domination in the nutrient-enriched areas has yet to be rigorously investigated in the Everglades. Therefore, an empirical field study (the Fire project) driven by legislated environmental restoration goals was initiated to assess the use of repeated prescribed fires as a management tool to drive an ecosystem state change, thereby reducing the competitive advantage of cattail, and facilitating re-establishment of historical species, i.e., fire-induced accelerated recovery.

This ecosystem restoration study is investigating the potential role of fire in halting and reversing the transition from sawgrass to cattail in nutrient enriched areas of the Everglades. Three overall hypotheses guide the project. First, while single fires have been shown to facilitate cattail growth, at least temporarily (Ponzio et al. 2004), multiple prescribed fires may provide a recurrent disturbance and accelerate the decline of cattail communities. Second, natural recovery of the eutrophic wetlands, a response to improved water quality due to the STAs and BMPs, may follow similar long-term trajectories as fire-induced accelerated recovery, but each would progress at a different rate. Last, accelerated recovery trajectories may reflect response variable pulses (e.g., surface water P concentrations, cattail biomass and others) of diminishing magnitude following each subsequent fire.

4.3 Spatial Features of the System Studied

Foreknowledge of the natural history and spatial structure of the Everglades was essential to understanding the multiple organizational-, temporal- and spatial-scales of major driving forces and ecosystem processes affected by fire. This provided the necessary insight for the design of a study that would detect ecosystem response to prescribed fire and determine the magnitude and ecological significance of the response. The Fire project was conducted in the northern Everglades (Fig. 4.1) in Water Conservation Area 2A (WCA 2A), which is an impounded wetland with a distinct nutrient enrichment gradient extending from the north along the Hillsboro Canal, where water enters WCA 2A, southward about 10 km (Davis and Ogden 1994). The enrichment gradient was classified into three distinct ecological zones: a highly P-enriched area (H) dominated by monotypic cattail stands; a moderately P-enriched area (M) composed of a cattail and sawgrass mixed community; and a reference area (R) which was considered unenriched with low water and soil P concentrations and characterized by sawgrass ridges and open water sloughs dominated by *Nymphaea odorata* (water lily). Two transects (E and F) were established in 1994 for long-term monitoring of water quality, soil nutrient concentrations, and plant communities within the three zones (Fig. 4.1). Data from this ongoing monitoring study were used to assess temporal and spatial variability of multiple response variables and influenced the design of the Fire project (Miao and Carstenn 2005). A priori power analysis of the data collected at the transect monitoring stations indicated that, depending on the response variable, between 13 and 90 replicates were required for a classical block design using a two-way ANOVA to detect a 30% ($\alpha = 0.05$; $\beta = 0.2$) change in the response variable (unpublished data). Spatial replication required for a classical replicated experimental design exceeded that which could be efficiently and effectively implemented from a physical, financial, and logistical perspective. Furthermore, limited site access and the infrastructure construction and maintenance required to conduct sample collections via airboat and helicopter further limited our ability to replicate. These along with other constraints convinced us to pursue a nonreplicated experimental design.

While there was evidence of a north-south nutrient gradient in water, soil, and vegetation resulting from a long history of nutrient enriched runoff, there also existed a short-term (days to weeks) east-west variation in water quality and water depth. Surface water inputs to WCA 2A from three main water control structures located along the Hillsboro Canal (Fig. 4.1) influenced water quantity and quality in WCA 2A. Under certain conditions, surface

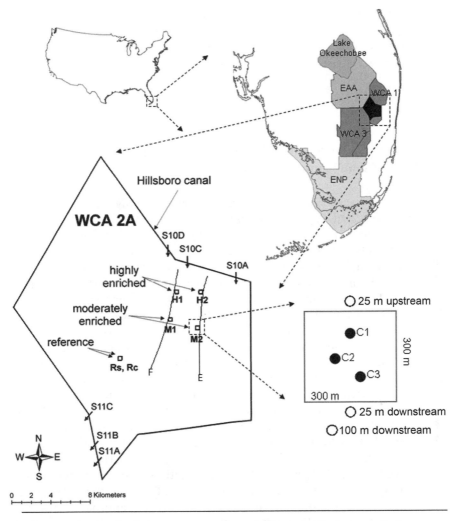

Fig. 4.1 The study area included the experimental plots in the highly enriched (H), moderately enriched (M), and reference (R) regions along a nutrient gradient in Water Conservation Area 2A (WCA 2A). Each of the six plots have a 300 × 300 m boundary with the same layout; one upstream, three within plot (C1, C2 and C3), and two downstream sampling stations at 25 and 100 m

water loading into WCA 2A differed greatly from east to west even between the two transects. For example, during extreme weather conditions, control structures were used to move large quantities of water into or out of water conservation areas within the Everglades to avert urban flooding. Events such as these periodically resulted in localized preferential flows which lead to hydraulic and nutrient loading differences among areas that normally had similar surface water characteristics. For a study proposed to span a large spatial area and evaluate short- and long-term ecosystem responses to fire, the east–west variation had to be accounted for at a finer temporal scale in the overall design and analysis.

4.4 Overall Experimental Design Incorporating Ecosystem Spatial and Temporal Features

We integrated several common nonreplicated designs along a gradient of nutrient enrichment and vegetation dominance into the Fire project (Fig. 4.1). Briefly, two paired plots adjacent to sampling stations on the long-term monitoring transects were established. One pair of plots (H1 and H2) was established in the highly P-enriched areas, a second pair (M1 and M2) in the moderately P-enriched areas. Two additional plots (Rc, cattail-dominated, and Rs, sawgrass-dominated) were located in the reference area (Fig. 4.1). The eastern plots in the enriched zones, H2 and M2, were subjected to fire and the western plots, H1 and M1 (approximately 2.4 km west of the burned plots), served as controls or covariates. The two unburned plots, hereafter called distant controls, are similar to the "controls" used in typical BACI or BACIPS designs. Neither Rc nor Rs were burned; instead, they were used primarily as benchmarks for comparing natural and accelerated recovery trajectories. As a result, this overall design strategy consists of Control/Impact (H1 vs H2; M1 vs M2), Reference/Impact (Rc vs H2; Rc or Rs vs M2), BACI (H1 vs H2; M1 vs M2), BACIPS (H1 vs H2; M1 vs M2 incorporating fire project and long-term monitoring data), and Beyond BACI (H2 vs H1, M1, Rc and Rs; M2 vs H1, M1, Rc and Rs). This multi-faceted design provides an opportunity to compare strengths and limitations of existing analytical designs and contributes to the development of novel analytical approaches for these designs using real-world datasets. In the present chapter, we focus on the results from the paired plots (H1 and H2) established in the highly nutrient-enriched area.

All six plots were 300 m × 300 m and four of the six contained six sampling stations: one 25 m upstream of the plot, three within, and two downstream (25 m and 100 m; Fig. 4.1). The reference plots did not have upstream and downstream sampling stations because they were not subjected to fire. The three within-plot sampling stations captured small-scale spatial variation; the upstream station served as an adjacent control for water quality parameters. The two downstream stations were used to assess the regional effects of fire impacts beyond the burned areas, a critical issue for evaluating ecosystem-scale responses to fire. Incorporating the second spatial control directly adjacent to the treatment plot (25 m upstream) was critical for assessing pulsed effects of fire on water quality parameters. Moreover, depending on the response variable analyzed, the six sampling stations of the distant control were used independently or in combination to test hypotheses related to the number of controls used in the analysis of a BACI design (Underwood 1993, 1994, Hewitt et al. 2001, Hewitt et al. 2007).

Planning and implementing data collection, analysis, and interpretation of processes operating at different spatial and temporal scales were challenging and exacerbated by the interplay of feedbacks between processes. The initial challenge was determining which response variables captured the system's response to fire across abiotic, individual, population, and community organizational levels. The next challenge was to hypothesize which response variables would experience the immediate effects of fire, and which response variables would display delayed ecosystem responses following the fire. These hypothesized ecosystem response scenarios were imperative to designing appropriate temporal sampling schemes. Immediate impacts were defined as those alterations in response variables that occurred during or immediately after the fire (Fig. 4.2A). Delayed ecosystem responses were defined as alterations in response variables that occurred any time other than immediately after the fire (Fig. 4.2B). Immediate and delayed responses could be either directly caused by the fire, or indirectly caused by changes in processes or parameters or a function of internal ecosystem feedbacks and therefore would vary by response variable. In fact, some response variables demonstrated both immediate responses to the fire and a delayed response (Fig. 4.2C). There were some processes and variables for which no response to the treatment was detected (Fig. 4.2D). Regardless of whether immediate or delayed, a response that continued over a long period of time (months, years) was referred to as a sustained effect or response and a short duration response (days, weeks) was defined as a pulsed effect or response (Fig. 4.2E).

System response variables operating at different organizational levels often differ greatly in onset and duration of response; therefore a classical temporally balanced sampling approach did not meet the objectives of our study. Sampling schedules became complex as we attempted to capture both short-term pulsed and long-term sustained responses of the critical variables. Figure 4.2E and Table 4.1 define the extensive and intensive sampling schemes for each of the multiple response variables under investigation. Extensive sampling was intended to detect seasonal and long-term trends and was conducted monthly, bimonthly or quarterly. The intensive sampling scheme, however, was intended to detect pulsed effects and included hourly, daily, and weekly sampling. Similar to BACI designs reported elsewhere, all data collected before the fire were referred to as "Before" data, and all data collected after the fire were referred to as "After" data. In contrast to the BACI temporal sampling scheme, the unique goal of the imbalanced sampling scheme used in this study was to distinguish a "Response" period within the After period, defined as the total time elapsed from the onset of an effect or response until a response variable's return either to pre-fire conditions (Table 4.1) or to a new equilibrium. As a result, the After fire period was

Fig. 4.2 Hypothetical scenarios of disturbance impacts and ecosystem responses to fire. The *solid and broken lines* represent two hypothetical measured responses of a particular parameter. The *solid line* represents an immediate and delayed pulse response followed by recovery in (**A**) and (**B**), respectively. The *broken line* represents an immediate and delayed sustained response in (A) and (B), respectively, (**C**) represents possible combinations of parameter responses, (**D**) represents no measured response of a parameter, and (**E**) shows visually that if pulsed responses are anticipated sampling intensity should increase to measure the pulse. The dots in (E) represent sampling events

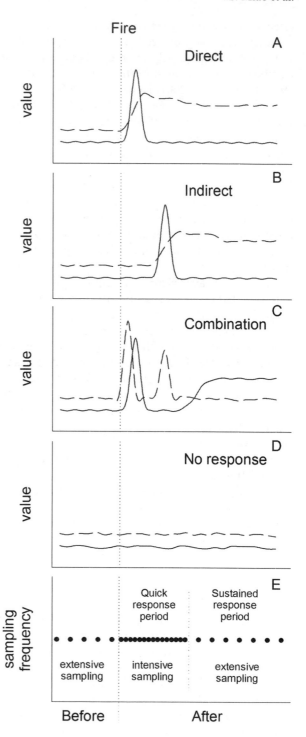

Table 4.1 Extensive and intensive sampling schemes for multiple ecosystem parameters and processes in a cattail-dominated Florida Everglades wetland along with the onset and duration of ecosystem response to prescribe fire. Extensive sampling intervals were intended to monitor seasonal and long-term trends. Intensive sampling schemes were designed to detect pulsed events from the prescribed fire. For the intensive sampling frequency, "d" refers to days, "m" to months, "pre" to samples collected within 2 weeks prior to the fire, and "post" to any time period after the fire. "NR" refers to no response detected and "TBD" means to "to be determined" following additional research

Ecosystem parameters	Sampling frequencies		Responses to fire	
	Extensive	Intensive	Onset	Duration
Surface water				
Phosphorus (TP, TDP, SRP)	Monthly	Hourly for 2 d pre and 9 d post	<1 d	7 d
TKN, TDKN, TC, DIC	Monthly	Hourly for 2 d pre and 9 d post	NR	NR
pH	Monthly	30min for 2 d pre and 9 d post	<30min	<3 w
Temp (logger data)	Hourly	Hourly for 2 d pre and 9 d post	<1 d	>1 y
Dissolved Oxygen (DO)	Monthly	30min for 2 d pre and 9 d post	4 d	>1 y
Pore water				
Phosphorus (TDP, SRP)	Monthly	Days 2, 9 & 20	<2 d	6 m
TDKN	Monthly	Days 2, 9 & 20	NR	NR
pH	Monthly	Days 2, 9 & 20	NR	NR
Temp	Monthly	Days 2, 9 & 20	<2 d	6 m
Periphyton				
Biomass	Annual	Pre, 1 and 3 m post	<1 m	>3 m
Nutrients (TCa, TC, TOC, TN)	Annual	Pre, 1 and 3 m post	<1 m	>3 m
Soil floc				
Nutrients (TP, Oslen P, TC, TN, TS)	Quarterly	Pre, weeks 2, 3 and 5 post	NR	NR
IP fractionation	Annually	Pre, weeks 2 and 5 post	NR	NR
TFe	Quarterly	Pre, weeks 2, 3 and 5 post	<3 w	<5 w
TCa		Pre and week 2 post	<2 w	>2 w

Table 4.1 (continued)

Ecosystem parameters		Sampling frequencies		Responses to fire	
		Extensive	Intensive	Onset	Duration
Soil profile (0-5cm)					
Nutrients (TP, Oslen P, TC, TN, TS)		Quarterly	Pre, weeks 2, 3 and 5 post	NR	NR
IP fractionation		Annually	Pre, weeks 2 and 5 post	NR	NR
TFe		Quarterly	Pre, weeks 2, 3 and 5 post	<3 w	<5 w
TCa			Pre and 2 weeks post	<2 w	>2 w
Temp (logger data)		Hourly	Hourly	<1 d	NR
Soil profile (0-30)					
Nutrients (TP, TC, TN)		Annually	—	NR	NR
Redox		Quarterly	2 d pre and 1, 3, 7, 10 & 11 d post	NR	NR
Vegetation					
Community biomass & nutrients:					
Live	Leaf biomass	Quarterly	Pre, 1 d, 1 and 5 m	<1 d	>1 y
	Leaf nutrients (TP, TN, TC)	Quarterly	Pre, 1 d, 1 and 5 m	<1 m	<5 m
	Root biomass	Quarterly	Pre, 1 and 5 m	<1 m	<1 y
	Rhizome biomass	Quarterly	Pre, 1 and 5 m	<5 m	<1 y
	Root and rhizome nutrients TP	Quarterly	Pre, 1 and 5 m	<1 m	<5 m
	TN	Quarterly	Pre, 1 and 5 m	<1 m	>5 m
	TC	Quarterly	Pre, 1 and 5 m	NR	NR
	Shoot base biomass	Quarterly	Pre, 1 and 5 m	<1 m	<1 y
	Shoot base nutrients (TP, TN, TC)	Quarterly	Pre, 1 and 5 m	NR	NR
Dead	Leaf mass	Quarterly	Pre, 1 d, 1 and 5 m	<1 m	>1 y
	Leaf nutrients TP	Quarterly	Pre, 1 d, 1 and 5 m	<1 m	<5 m
	TN	Quarterly	Pre, 1 d, 1 and 5 m	<5 m	>5 m
	TC	Quarterly	Pre, 1 d, 1 and 5 m	NR	NR
	Belowground mass	Quarterly	Pre, 1 and 5 m	<5 m	>1 y
	Belowground (TP, TN, TC)	Quarterly	Pre, 1 and 5 m	NR	NR

Table 4.1 (continued)

Ecosystem parameters	Sampling frequencies			Responses to fire	
	Extensive	Intensive		Onset	Duration
Community production					
Leaf Area Index	Weekly	Quarterly		<1 d	>1 y
Litter production	Monthly	Monthly		<1 d	>1 y
Community structure					
Soil seed bank	Quarterly	Pre and 2 days post		<2 d	TBD
Transect vegetation	Annual			TBD	TBD
Aerial photography	Annual	Pre, 3 d, 3, 4 and 5 m post		TBD	TBD
Population demograph					
Cattail density	Quarterly	Weeks 1, 2, 3 and 4		<1 d	6 m
Mean plant size	Quarterly	Pre, 1 d, 1 and 5 m		<1 d	>1 y
Individual plant					
Leaf height growth	Quarterly	Weeks 1, 2, 3 and 4		<1 d	>5 m
Root growth	Bi-monthly			TBD	TBD
Plant decomposition	Bi-monthly			TBD	TBD
Environmental parameters					
Wind speed	2 minutes	2 minutes		<1 d	1 y
Water depth				NR	NR
Ash					
Nutrients (TP, PO$_4$, TN, TC, NH$_4$, NO$_3$)		Immediately post-fire		<1 d	2 w
pH		Immediately post-fire		<1 d	2 w

divided into Response and Recovered periods of different lengths depending on the response pattern of the variable. In general, extensive sampling regimes were implemented during the Before and After periods to monitor pre-fire conditions, sustained responses to fire, and eventual system recovery, while intensive sampling schemes were nested within the extensive sampling scheme directly after the fire (Fig. 4.2E).

Based on our foreknowledge of the Everglades, as well as general ecological and biological principles, multiple temporal scales were established for the array of ecosystem variables in Table 4.1. For example, surface water was collected on an hourly basis for several days immediately before and after the fire, while soil was collected once pre-fire and approximately bimonthly post-fire. The organizational level of each response variable influenced the intensive sampling frequency. Leaf growth and plant density (individual and population levels) were sampled weekly post-fire, while biomass (community level) was sampled at 1 day, 1 month, and 5 months post-fire. Intensive sampling was conducted primarily during the first month after the fire.

Stochastic temporal events often confound data analysis. The probability of confounding events increases with spatial and temporal scales; therefore, spatial and temporal scales are an important consideration when selecting control locations. This was our motivation for monitoring response variables exhibiting rapid and slow turnover with adjacent (upstream) and distant control sites, respectively. Indeed, a confounding event was detected with our intensive post-fire sampling of surface water quality variables (Fig. 4.3). Two distinct temporal trajectories were found at the distant and upstream controls (Fig. 4.3). Surface water total phosphorus (SWTP) at the upstream control was stable during post-fire intensive sampling, while the distant control SWTP exhibited a pulsed increase. This suggests that the distant plot experienced a perturbation at the time of the prescribed fire treatment. Although the timing (including onset and duration) of the event at the distant control was similar to the prescribe fire, it occurred two days after the fire and increased slower than at the burned plot, not reaching its peak until five days after the fire. On the other hand, the SWTP of the burned plot increased instantaneously following the fire, reaching its peak within one day after the fire and then decreasing sharply (Fig. 4.3). The SWTP pulse at the distant control was most likely the

Fig. 4.3 (continued) Temporal trends in eight selected ecosystem parameters characteristic of different temporal scales or turnover times. The first panel includes data collected before the prescribed fire conducted on July 25, 2006. The second panel includes data collected from July 25 to August, 2006. The third panel includes data collected from August 26 2006 through September 2007. Note the difference in the X-axis scale among panels. Monthly intervals are used in panels 1 and 3 whereas panel 2 used 2 day intervals. Floc refers to unconsolidated peat. Sample sizes of all parameters were given in Table 4.1

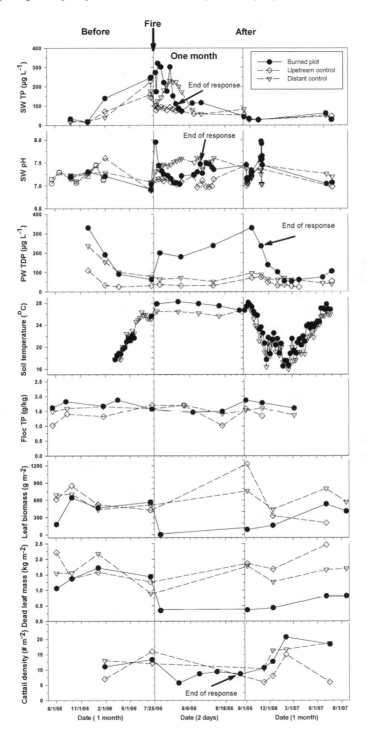

Fig. 4.3 (continued)

result of a large water release into WCA 2A from the Hillsboro Canal. This hydrological event increased spatial and temporal variation between the distant control and the burned plot to such a degree that the distant control was inappropriate for distinguishing fire effects for many rapidly responding variables.

Differences in SWTP, between the two controls and the burned plot, resulting from differential hydrologic loading can be elucidated with our year-long data set. The coefficients of correlation of SWTP were 0.67 (p = 0.01) between the burned plot and the distant control, but 0.95 ($p < 0.001$) between the burned and the upstream control, suggesting that the SWTP at the upstream control tracked the burned plot better than the distant control. The same was also true for surface water pH (SWpH); correlation coefficients between the distant and burned plots were 0.60 ($p=$ 0.01) and 0.73 ($p<$ 0.001) for the upstream and burned plots. However, there was minimal difference in correlation of pore water total dissolved phosphorus (PWTDP) between the distant control and burned plots (0.87; p < 0.01) and the upstream control and burned plots (0.89; p < 0.01). In addition, we conducted single-control BACI-ANOVA and BACIPS t-test (Table 4.2) for SWTP, SWpH, and PWTDP using the distant control and upstream control separately. The results showed that the likelihood of detecting SWTP and SWpH impacts increased using the upstream plot as the control, whereas both control locations detected significant fire impacts for PWTDP (Table 4.2). In this particular case, proximity of the upstream control to the burned plot reduced the confounding factor of water quantity differences at our distant control site during the time of the impact allowing us to detect fire effects on surface water variables. When response variables are temporally or spatially variable, strong correlations between a specific control and treatment site may be the best criterion for choosing which control site to use for detecting an effect.

Furthermore, we took advantage of our integrated design and real-world dataset to address the effects of control number on a BACI model designed to detect fire impacts. Several models, single control/single impact BACI-ANOVA, Beyond BACI-ANOVA using multiple control sites, and BACIPS t-test (Stewart-Oaten 1996c), were selected to test for fire's effect on SWTP, SWpH, and PWTDP (Table 4.2). The models examined the distant control, the upstream control, and the two downstream sites using their Recovered period data only. For the single- and multiple-control BACI-ANOVA models, two terms, fire effects and temporal variation, were collectively used to detect fire effects. There were four control scenarios, each with an Upstream and Distant control comparison. The first scenario, with a single control, compared either the Upstream or the Distant control to the burned plot using either BACI-ANOVA (Table 4.2; case 1 and 2) or BACIPS models (Table 4.2; cases 8 and 9). In both cases, fire impacts on surface water variables were detected using the Upstream control only. An impact on PWTDP was detected with a BACI-ANOVA using either control. The second scenario examined the ability of two controls to detect an impact (Table 4.2; cases 3 and 4), the third scenario examined the effect of three controls (Table 4.2; cases 5 and 6), and the fourth

Table 4.2 P-values using BACI, Beyond-BACI, and BACIPS t-tests to test for sensitivity of the number and location of control sites using surface and pore water parameters. The four control sites included a Distant control (2.4 km west), an Upstream control (25 m north), and two Downstream controls (25 m and 100 m south) of the burned plot (see Fig. 4.1 for spatial locations). Downstream sites were analyzed as controls only after parameters had returned to pre-fire conditions based on moving regressions. The remaining comparisons were based on the Before–Response period as defined in Table 4.3. Boldfaced p-values were significant at $\alpha = 0.05$. Sample sizes for the two Downstream and Upstream sites were $n = 1$, while the Burned and Distant plots were $n = 3$. Surface water total phosphorus (SWTP) and pore water total dissolved phosphorus (PWTDP) were log transformed

Model	Plots		SWTP		Surface water pH		PWTDP	
	Control	Impact	Fire effect	Temporal variation	Fire effect	Temporal variation	Fire effect	Temporal variation
Single Control/ Impact BACI- ANOVA								
1. Distant	Burned plot		0.618	0.112	0.266	**0.001**	**0.013**	0.889
2. Upstream	Burned plot		**0.026**	0.121	**0.036**	**0.002**	**0.041**	0.996
Multiple Control/Impact BACI-ANOVA								
3. Distant, Upstream	Burned plot		0.605	0.519	0.919	**0.022**	0.190	0.113
4. Upstream, 25 m Downstream	Burned plot		0.660	0.756	0.846	0.051	0.347	0.104
5. Distant, 25 m Downstream 100 m Downstream	Burned plot		0.373	0.652	0.892	0.826	0.593	0.307
6. Upstream, 25 m Downstream 100 m Downstream	Burned plot		0.511	0.684	0.884	0.155	0.204	0.662
7. Distant, Upstream, 25 m Downstream 100 m Downstream	Burned plot		0.464	0.500	0.780	0.921	0.588	0.564
BACIPS								
8. Distant	Burned plot		0.781	–	0.202	–	0.176	–
9. Upstream	Burned plot		**0.045**	–	**0.030**	–	0.544	–

scenario (Table 4.2; case 7) considered four control sites together (the two downstream sites were used as control sites only after response parameters had returned to pretreatment values). Regardless of whether the Distant or Upstream control was used, none of the combinations of more than one control were able to detect fire impacts. This was especially interesting for the PWTDP since individually both the Distant and Upstream controls were able to detect impacts with the BACI-ANOVA model. Our results suggest that increasing the number of controls increases background variability, which inhibits the ability to detect a treatment effect. Hewitt et al. (2001) reported similar results and attributed their finding to dissimilarities between controls and impact sites over time. Thus, to assess environmental impacts using a BACI design, control sites must be in proximity to each other and the impact site (Osenberg and Schmitt 1996) and display similar temporal trajectories in the absence of a treatment (Stewart-Oaten 1996c).

The validity of this conclusion may, of course, depend on the parameter measured and the variability of the study area. A more holistic interpretation could be that well-matched, single control/impact comparisons were the most sensitive methods for detecting an effect, while multiple control/impact comparisons were better suited for examining the magnitude of the response in comparison to overall ecosystem variability. For example, as assessed using the upstream control, fire did impact SWTP and P moved downstream; however, this event was not outside the range of normal ecosystem variability, because the release of P-enriched water from the Hillsboro Canal into WCA 2A caused a similar magnitude P peak at the distant control. Our results demonstrated that although multi-spatial scale controls are necessary for placing the magnitude and duration of an impact within context of natural ecosystem variation, adding more controls may reduce impact detection.

4.5 Applying Moving Regressions to Determine Onset and Duration of Fire Impacts and Magnitude of Ecosystem Responses

As mentioned above, one of the underlying assumptions of a nonreplicated design is that treatment and control plots are similarly influenced by factors affecting system change (Stewart-Oaten 1996b). Thus, a quantitative relationship should exist between each response variable measured at the control and treatment sites before a disturbance. Furthermore, if the relationship changes after the disturbance, then it is not unreasonable to attribute the change to the disturbance. A regression or correlation depicting differences in this relationship before and after a disturbance is a useful analysis only if the change in the relationship is sustained. When a trend changes with time, as is the case with pulsed events, multiple estimates based on data from sequential time intervals are necessary to detect the change. Trends in time series data are sometimes analyzed using a moving average, a method for smoothing a time series by averaging a fixed number of consecutive data points (Box and Tiao 1975). The average "moves" over time in that each data point of the series is sequentially included in the averaging, while the oldest data point in the span of the average is removed. A similar approach, moving regression (not average), was applied to generate a series of linear regressions, in which each data point of the series was sequentially included in the regression, while the oldest data point in the span of the regression was removed. The resulting series thus depicts how the relationship changes over time.

To generate the first in a series of regressions, the first four data points in the time series were used. For the second regression in the series, the first data point was removed and the fifth data point was added; maintaining $n = 4$ for each regression. This continued until a series of regressions had been modeled using all the data available. Slopes and intercepts among the series of regressions were visually compared and grouped by similarity of linear features into Before, Response, and Recovered periods. Once the periods were established, the slopes and intercepts of the three periods were (Table 4.3) then compared using ANCOVA to determine whether significant differences existed between periods.

For our study, the series of control vs. impact regressions were useful particularly for those variables exhibiting strong correlative relationships between the treatment and control sites before the fire. Thus, we applied this regression approach to the SWTP relationship between the treatment and control sites during the Before period; this relationship is depicted as a solid black line ($n = 4$) in Fig. 4.4A. For example, the slope of the relationship between SWTP (Fig. 4.4A) at the control and treatment site increased less than 24 h after the fire (Response) and then returned to the pre-fire slope within 9 days (Recovered). Changes in the slope of the relationship between SWpH (Fig. 4.4B) at the control and treatment sites were not as clear. A significant relationship existed between SWpH at the control and treatment sites before the fire ($p = 0.03$; $r^2 = 0.58$). Although the slope increased less than 30 min after the fire (Response) and remained greater through most of the first week post-fire, at which time SWpH slope gradually became negative (Delayed response) before returning to a slope similar to pre-fire (Recovered), none of relationships were significant ($p > 0.05$). The lack of significance may simply be a reflection of system variability or inadequate sample size, or perhaps this indicates the system is still responding. Pore water TDP (Fig. 4.4C) responded similarly to SWTP including a brief Response period followed by a return to conditions similar to the Before period. Again these relationships were not significantly different ($p > 0.05$). The regression relationship between soil temperature (Fig. 4.4D) at the control and treatment sites differed before and after the fire and had not recovered even after one year.

For those variables not exhibiting strong correlations between treatment and control plots, e.g., live leaf biomass, dead leaf biomass, and cattail density (Fig. 4.4E, F, G, H) before the fire another regression approach was used. The

Table 4.3 Slopes and Y-intercepts of surface water, pore water, and soil response variables. Surface water total phosphorus (TP) and pH; pore water total dissolved phosphorus (TDP); soil temperature, and unconsolidated peat (floc TP) measured at the burned plot were regressed against the control data. Live and dead leaf biomass and cattail density deltas (difference between the response variable measured at the burned and control plots) were regressed against time. Two or three response periods, Before, Response, and Recovered, were identified depending on the parameter. The Before time period started on July 1, 2005 and ran through July 24, 2006. The Response period coincided with the onset of the fire or immediately after depending on the response variable and ended when the response variable could no longer be distinguished from pre-fire conditions. The recovery period began when the response period ended. Control data used for each regression was dependent upon the response variable ([1]Burned plot regressed against upstream control, [2]burned plot regressed against distant control and [3]delta calculated by subtracting the burned plot from the distant control and regressed against time)

Time period	Slope	Slope p value	Y intercept	Y inter p value	Power	r^2	N
Surface water TP[1]							
Before	1.64	0.001	1.68	0.91	0.96	0.98	5
Response	7.38	0.01	−407.75	0.06	0.77	0.7	8
Recovered	1.44	0.02	−14.95	0.6	0.65	0.61	8
Surface water pH[1]							
Before	0.48	0.03	3.7	0.02	0.61	0.58	8
Response 1	2.18	0.63	−8.16	0.8	0.07	0.06	6
Response 2	−0.22	0.81	8.78	0.21	0.04	0.01	7
Recovered	0.35	0.2	4.78	0.02	0.24	0.1	17
Pore water TDP[1]							
Before	2.72	0.09	33.05	0.6	0.33	0.82	4
Response	3.2	0.08	99.66	0.15	0.36	0.85	4
Recovered	3.28	0.004	−32.97	0.32	0.87	0.84	7
Soil temperature[2]							
Before	0.83	<0.0001	3.29	0.02	1	0.91	12
Response	1	<0.0001	1.23	0.07	1	0.96	50
Floc TP[2]							
Before	0.48	0.7	906	0.65	0.04	0.09	4
No response	0.03	0.97	1592.06	0.25	0.03	0.001	5
Leaf biomass[3]							
Before	1.27	0.28	−300.62	0.21	0.15	0.52	4
Response	0.96	0.07	−559.95	0.01	0.43	0.73	5
Dead leaf mass[3]							
Before	2.74	0.13	−529.08	0.13	0.26	0.75	4
Response	0.14	0.89	−936.34	0.03	0.03	0.01	5
Cattail density[3]							
Before	0.02	0.44	−6.65	0.23	0.1	0.21	5
Response	0.29	0.2	−10.3	0.05	0	0.91	3
Recovered	0.02	0.21	−7.39	0.05	0.22	0.29	7

dependent variable, the differences (delta; Osenberg et al. 1994, Stewart-Oaten 1996a, 1996c, 2003) between the parameter measured at the treatment and control sites, was regressed against time. Soil floc TP did not show any response to the fire. Although there is no doubt that fire affected both live and dead leaf biomass, no significant differences ($p > 0.5$) in the delta-time relationship were detected between the Before and After periods. The data suggest that these relationships may still be responding. The delta-time relationship for stem density, on the other hand, demonstrated both a response, although not statistically significant, and a recovery to pre-fire deltas within one year. The Response period differed in duration among response variables. Plant density appears to have recovered within 7 months, whereas live and dead leaf biomass and soil temperature were still in the Response period 1 year post-fire.

After distinguishing between the Before, Response, and Recovered periods using moving regression analyses, we conducted a standard regression using prediction intervals to assess fire impacts Fig 4.4. The prediction interval distinguishes between observed and expected values in the dependent variable, which can lead to the discovery of outliers caused by measurement errors, lack of homogeneity of samples, or samples that exhibit a treatment effect (Sokal and Rohlf 1981). We calculated 95% prediction intervals for the regression between the burned plot and the control after combining the Before and Recovered period data (Fig. 4.5A, B, C, D). All data collected during the Before and Recovered (open symbols) periods for these four parameters fell within the 95% prediction interval, while many data points collected during the Response period (Fig. 4.5A, B, C, D; We suggest closed symbols) were located beyond the upper bound of the prediction intervals. All data points falling within the bounds of the 95% prediction intervals represented expected values given the

───▶

Fig. 4.4 (continued) Regression analyses of eight ecosystem parameters reflecting turnover times ranging from days to months or years. Moving regressions were used to determine transition from Response period (a difference between burned and control) to Recovered period (no longer a difference between control and burned plot). The independent variable is the response variable measured at the control and the dependent variable is the response variable measured at the burned plot for surface water TP, pH and porewater total dissolved P (TDP), soil temperature, and unconsolidated peat (floc) TP. Control data used for each regression was dependent upon the response variable. For surface water variables, the burned plot was regressed against the upstream control; for soil variables, the burned plot was regressed against the distant control. Live and dead leaf biomass and cattail density deltas (difference between the response variable measured at the burned and control plots) were regressed against time. Time on the X-axis represents the number of days in each of the individual moving average regression periods. *Thick black lines* represent the pre-fire conditions while *long dashed lines* represent the direct effect time period; *short dashed lines* represent the indirect response time and a *dotted line* represents recovery to pre-fire conditions. Before the fire $n = 5$ for surface water TP, pH and density; $n = 4$ for pore water TDP, floc TP and live and dead leaf biomass; and $n=12$ for soil temperature. After the fire $n = 4$ for surface water TP and plant density; $n = 6$ for pH; $n = 8$ for pore water TDP; $n = 13$ for soil temperature; and $n = 3$ for floc TP and live and dead leaf biomass

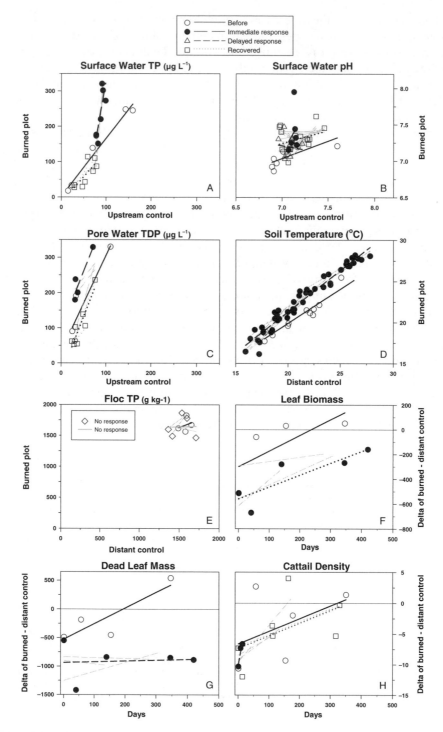

Fig. 4.4 (continued)

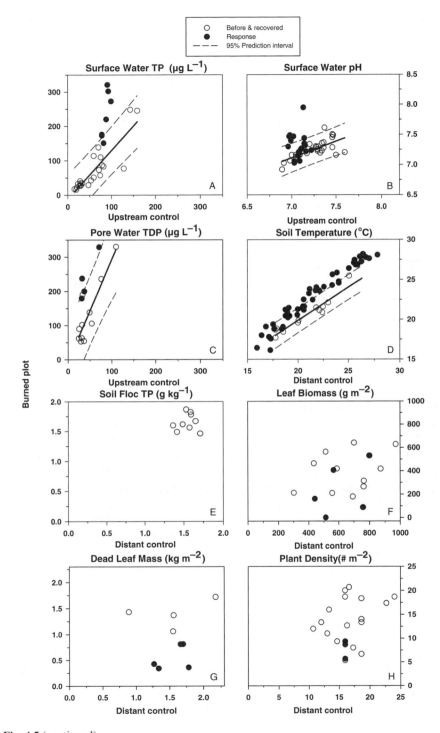

Fig. 4.5 (continued)

Table 4.4 P-values of a three-factor BACI-ANOVA, Site (control vs. impact), Period (Before vs. After), and Time nested within Period, testing for fire effect and temporal variation for eight ecosystem variables. The model was run twice for each variable using two sampling periods determined by moving regressions in Table 4.3, Before–Response and Before–After. The "Response" period defined as when the parameter was different from the Before period based on moving regressions, while the "After" period included the entire time period after fire (1 year for this study) regardless of variable response. The Response period differed in duration among parameters. The surface water total phosphorus (SWTP) response period was 7 days; surface water pH was 1 month; pore water total dissolved phosphorus (PWTDP) was 2 months; plant density was 7 months; whereas live and dead leaf biomass and soil temperature were still in Response period 1 year post-fire. Soil floc TP did not respond to fire. SWTP, pH, and PWTDP analyses used the Upstream as the control site, whereas soil and vegetation analyses used the Distant control site. Boldfaced p-values were significant at $\alpha = 0.05$. Sample sizes for the Upstream control was $n = 1$, while the Burned plot and Distant control were $n = 3$. Surface water TP and pore water TDP were log-transformed, while temperature was Box-Cox Y transformed

Variables	Before–Response		Before–After	
	Fire effect	Temporal variation	Fire effect	Temporal variation
SWTP	**0.026**	0.121	0.797	**0.000**
pH	**0.036**	**0.002**	0.816	**0.000**
PWTDP	**0.041**	0.996	0.631	0.997
Soil temperature	**0.000**	0.976	–	–
Floc TP	–	–	0.913	0.249
Leaf live biomass	0.421	0.494	–	–
Leaf dead biomass	**0.018**	0.842	–	–
Plant density	0.420	0.286	0.523	0.574

variation of the data, whereas those lying beyond the upper or lower bounds of the predict interval represent responses that may be attributable to fire.

In addition to regression analysis, we conducted a BACI-ANOVA and compared analyses using the After period and the Response period only. BACI-ANOVA detected significant fire effects on all variables except live leaf biomass and plant density using the Before-Response period but did not detect any significant effects when the entire After period was considered (Table 4.4). Significant temporal variation was found in the BACI-ANOVA using the Before–After period for SWTP and SWpH, due largely to the pulsed responses these variables exhibited following the fire. Increased variation during the After period masked the pulsed responses and resulted in no significant fire impacts

Fig. 4.5 (continued) Linear regressions with 95% prediction intervals for eight ecosystem parameters representing different temporal scales or turnover times measured before the fire and after the variable had recovered (*open circles*) at the control and burned plot. The independent variable was the response variable measured at the control plot and the dependent variable is the response variable measured at the burned plot. *Solid circles* represent the response variable measured at the control plot after the fire regressed against the response variable measured at the burned plot on the same date. A regression line was included only if it was significant at alpha <.05.

detected for these response variables (Table 4.4), a typical case of Type I error. Obviously, the magnitude and duration of a pulsed response would influence the ability of an ANOVA model to detect an impact if the model was based solely on the Before–After temporal periods. Thus, we encourage investigators to evaluate graphs of the response variables and utilize regressions to identify the response phase within the After period prior to running the BACI-ANOVA models. All data collected after the impact or treatment should not be routinely combined into a single period if short-term, pulsed impacts and responses are to be detected.

4.6 Approaches to Determining the Onset and Duration of Downstream Impacts of Fire

The potential for regional effects of a localized impact raises a number of critical issues for experimental design and analytical approaches. Many disturbances, such as air pollution, non-point source agricultural runoff, point-source industrial pollution, oil spills, and wildfires, potentially affect regions downwind or downstream of the areas directly affected by the disturbance. While regional effects that extend beyond the immediate vicinity of the disturbance should diminish with distance, they are nonetheless relevant to evaluating ecosystem response. A nonreplicated research design permits the reallocation of resources from replicating the disturbance to addressing the question of potential regional effects. For example, what is the magnitude, duration, and distance from a disturbance at which effects can be detected, and how many sampling stations are necessary to capture the full regional impact of the disturbance? Essentially, an experimental design intended to detect regional effects of an impact is the reverse of the Beyond BACI design (Underwood 1993). Whereas the Beyond BACI design includes one treatment and multiple controls, the "regional effects" design includes at least one control and multiple impacted sites. The impacted sites are not replicate treatments sites; they all experience the effects of the treatment to varying degrees as a function of distance from the treatment. Using SWTP as an example, we applied the same approaches previously used to assess fire effects, descriptive statistics, regression, BACI-ANOVA, and BACIPS, to explore downstream post-fire effects during the response period.

Surface water TP delta values between the upstream control plot and the two downstream sampling stations (25 and 100 m) showed a pulsed response that quickly returned to pre-fire levels and had a similar pattern as the burned plot (Fig. 4.6A). The SWTP deltas of both the burned plot and the downstream plots experienced two peaks only a few days apart. The first peak at the downstream sites occurred one day after the peak within the burned plot. A few days later, shortly after a rain, a second SWTP delta peak occurred in the burned plot. Ash caught on the litter and leaves that were not consumed by the fire was likely washed into the surface water by rain. As with the first peak, this effect at the downstream sites was also delayed one day and did not occur at the upstream

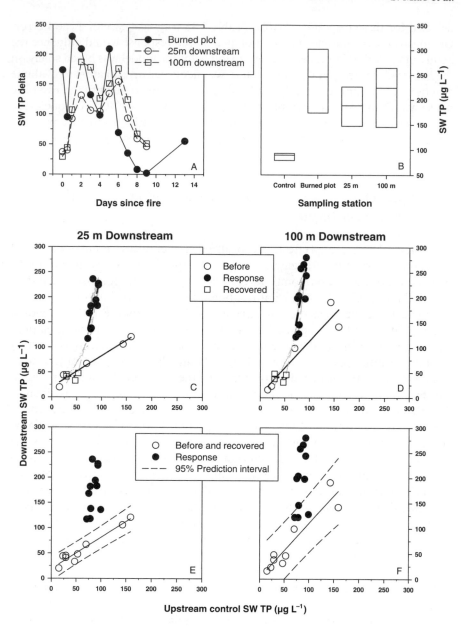

Fig. 4.6 Multiple analyses applied to detect downstream effects of prescribed fire on surface water TP. Graph A displays differences (deltas) between the burned plot and upstream control, between 25 m downstream and upstream control, and between 100 m downstream and upstream control over the first 15 days following the prescribed fire. Graph B displays a box plot of the upstream control, the burned plot, and 25 m and 100 m downstream sampling stations during the Response period (7 days post-fire) defined by moving regression. Graph C displays the response and recovered periods as detected using a moving regression. The *open circles* and *solid line* represent Before (Sept. 05–July 19, 2006), *closed circles* and *long dashed*

control. While the SWTP delta peaks of the downstream pulses were delayed and slightly less than the burned plot pulse, the duration of the downstream delta pulse was several days longer (Fig. 4.6A and Table 4.1).

Moreover, the downstream data were regressed against the upstream (adjacent) control to determine the Response and Recovered period (Fig. 4.6C and D). The Before period regressions were characterized by positive slopes between the upstream control and the 25 and 100 m downstream stations. The coefficients of variations were 0.62 ($p = 0.002$) and 1.07 ($p = 0.02$) respectively. These Before slopes were significantly different from the Response period slopes, but were not different from the slopes of the Recovered period. Once the Response and Recovered periods were determined for each downstream station, the mean SWTP during the Response period were calculated for at the burned plot, 25 m downstream, 100 m downstream, and upstream ($240 \pm 68, 188 \pm 43, 213 \pm 61$, and $88 \pm 8 \, \mu g \, TP \, L^{-1}$, respectively) (Fig. 4.6B). In addition, simple regressions of each response period and their 95% prediction intervals were calculated (Fig. 4.6E and F). For both downstream stations, all sampling points during the Response period fell outside of the 95% prediction interval of the Before and Recovered combined regression.

Finally, using the Before and Response periods established by regression analysis, several BACI-ANOVA models and BACIPS t-tests were used to explore downstream post-fire nutrient pulse effects at two downstream sites. Downstream sites were analyzed as separate impacted sites, and also averaged together as one impacted site (Table 4.5). Several interesting results became obvious after completing these different combinations of statistical analyses. First, regardless of whether the BACI-ANOVA or BACIPS t-tests were used, a significant SWTP pulse was detected when compared to the upstream control whereas the pulse was considered insignificant when compared to the distant control or the burned plot. Second, using the upstream control, the single control/impact model BACI-ANOVA detected downstream post-fire SWTP effects regardless of whether the two downstream plots were analyzed as separate impacted sites, or averaged together.

Using the burned plot for comparison allowed us to determine the similarity of the burned plot's SWTP concentration with downstream plots. The p-values for the BACI-ANOVA and BACIPS models comparing the burned and

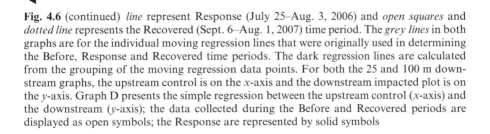

Fig. 4.6 (continued) *line* represent Response (July 25–Aug. 3, 2006) and *open squares* and *dotted line* represents the Recovered (Sept. 6–Aug. 1, 2007) time period. The *grey lines* in both graphs are for the individual moving regression lines that were originally used in determining the Before, Response and Recovered time periods. The dark regression lines are calculated from the grouping of the moving regression data points. For both the 25 and 100 m downstream graphs, the upstream control is on the *x*-axis and the downstream impacted plot is on the *y*-axis. Graph D presents the simple regression between the upstream control (*x*-axis) and the downstream (*y*-axis); the data collected during the Before and Recovered periods are displayed as open symbols; the Response are represented by solid symbols

Table 4.5 P-values of BACI-ANOVA and BACIPS t-test analyses of the downstream effects of fire on surface water total phosphorus concentrations. Downstream plots were analyzed separately (25 m downstream or 100 m downstream) and in combination (25 m downstream and 100 m downstream). The analyses used the Before–Response period defined in Table 4.3. Bolded p-values are significant at $\alpha = 0.05$. Sample sizes for the two downstream and upstream control were $n = 1$, whereas for Burned and Distant control were $n = 3$

Model	Control	Impact	Fire effects	Temporal variation	Spatial variation
Single control/ impact BACI ANOVA	Upstream control	25 m downstream	**0.032**	**0.041**	–
		100 m downstream	**0.008**	0.122	–
		Total downstream	**0.008**	0.226	–
	Distant control	Total downstream	0.890	**0.007**	–
	Burned plot	25 m downstream	0.551	**0.041**	–
		100 m downstream	0.815	**0.035**	–
		Total downstream	0.665	**0.002**	–
Single control/ multi-impact BACI ANOVA	Upstream control	25 m and 100 m downstream	0.061	**0.021**	0.103
		Burned, 25 m and 100 m downstream	0.051	**0.008**	**0.017**
		Burned and Total downstream	0.214	0.087	0.084
	Distant control	25 m and 100 m downstream	0.641	**0.014**	0.223
		Burned, 25 m and 100 m downstream	0.715	**0.003**	0.088
		Burned and Total downstream	0.751	0.057	0.191
BACIPS	Upstream plot	25 m downstream	**<0.001**	–	–
		100 m downstream	**<0.001**	–	–
		25 m and 100 m downstream	**<0.001**	–	–
		Total downstream	**<0.001**	–	–
	Distant plot	Total downstream	0.104	–	–
	Burned plot	25 m downstream	0.863	–	–
		100 m downstream	0.979	–	–
		Total downstream	0.918	–	–
MBACI ANOVA	Distant and Upstream	Burned and Total downstream	0.326	**0.017**	0.131

downstream plots were large and insignificant (p values ranged from 0.551 to 0.979), indicating the increase in SWTP in the downstream plots could not be distinguished from the increase in the burned plot (Table 4.5). Overall, fire caused an immediate increase in SWTP within the burned plot and P was transported at least 100 m downstream of the burned plot. Again, defining the response period using a series of regressions was critical for statistical validation of detecting impacts in a nonreplicated design. Increasing the number of impact sites seemed to increase temporal variability in the data, which led to reduced impact detection sensitivity, as was the case with multiple control sites.

The results reported in sections 4.5 and 4.6 suggest that the regression approach was more effective than ANOVA for detecting the effects of fire. The BACI-ANOVA determined only whether there was an impact, while regression provided information on the onset, duration, or recovery of the response variable. Although ANOVA may represent a contemporary method, we believe that using regression has additional advantages in large scale studies. Regression avoids the necessity of having to subjectively categorize continuous variables (i.e., After, Response, or Recovered) to construct the ANOVA model and possibly toss away hard-earned information. One major advantage of analyses of continuous variables is that measures of potentially important sources of variation can be easily incorporated and variability can then be separated into that associated with the main variable of interest, other measured variables, and noise, allowing clearer identification of effects (Hewitt et al. 2007). In our study, the regression analyses addressed spatial variation by regressing response variables at the impact sites against the same variables at the control sites and fire effects by evaluating changes in slope between Before and After and/or among the Before, Response and Recovered periods. The regression models were not confounded by variation; instead, the variation benefited the analyses. By comparing slopes rather than just mean responses, we were able to detect emerging patterns and properties in the landscape. In addition to slope comparisons, the simple regressions also provided prediction intervals for detecting impact effects. Furthermore, the moving regression analysis defined intervals including the onset and duration of the response, neither of which could be accomplished by BACI-ANOVA, thereby avoiding a Type I error, particularly for the disturbances creating pulsed effects.

4.7 Ecosystem Synthesis: Fire Effects on Wetland Phosphorus Pools

While detailed analyses are necessary to assess specific effects of prescribed fire, synthesis of key system responses is the ultimate goal of ecosystem-scale studies. Fire's effect on a cattail community in the highly enriched area is shown in a simplified depiction in Fig. 4.7A. Based on this, we developed a conceptual model of TP pools (g TP m^{-2}) for the major wetland components studied at our site (Fig. 4.7B) to give an overview of ecosystem response to fire for one year. The data used in this analysis were based primarily on Fig. 4.3 and Table 4.1.

The fire consumed most of the aboveground dead leaves, but only a small portion of the live leaf biomass. Combustion converted previously organic-bound P in the biomass to inorganic P in the ash. Soluble components of the ash

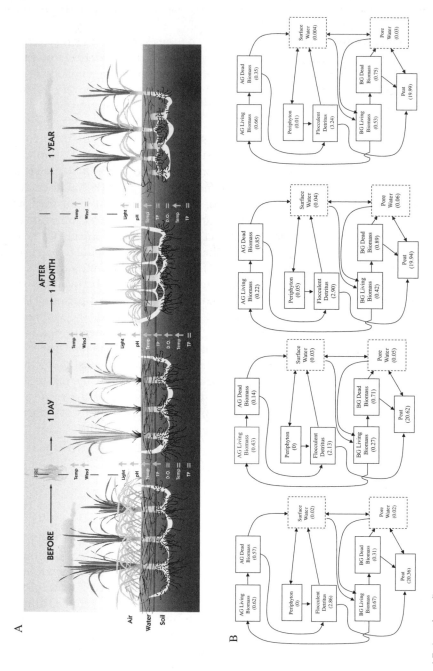

Fig. 4.7 (continued)

dissolved upon entering the water and SWTP increased immediately; however, the SWTP pulse only lasted about 10 days (Fig. 4.3). The SWTP pool was greatest one month after the fire, but this was likely the consequence of a large release of water into WCA 2A since SWTP increased simultaneously at the distant control. During the SWTP pulse following the fire, plant regrowth and periphyton production at the burned plot were minimal indicating these processes contributed little to the quickly diminishing SWTP after the fire. Pore water TDP (PWTDP) increased after the fire; however, it remained elevated long after SWTP returned to pre-fire concentrations, suggesting that processes other than P fluxing below-ground were responsible for the sustained increase in PWTDP. Calcium-bound P concentrations increased in the floc and top few cm of soil immediately after the fire; however, this accounted for a relatively small portion of the TP pulse after the fire and the change in soil TP was not significant (unpublished data). One day after the fire, a SWTP pulse was detected at both downstream sampling locations, while upstream SWTP did not change. The magnitude of the downstream pulse was slightly less than in the burned-plot pulse, yet the duration was one day longer. Considering the downstream pulse together with the other processes discussed above suggests that the bulk of P released by the fire was transported downstream by water and as ash in the air.

Live leaves were generally not consumed in the fire, but were scorched by the heat and died (Fig. 4.7A). These newly dead leaves had greater nutrient concentrations than naturally senesced cattail leaves, explaining the TP increase in the aboveground dead biomass pool one month post-fire when the dead leaf litter mass was less than before the fire. Fire opened the canopy, which allowed more light to reach the soil surface and increased soil and water temperatures even one year after the fire. Periphyton production was noted two weeks after the fire in response to increased light and was still greater than pre-fire or control after one year. Cattail leaves began growing soon after the fire and

Fig. 4.7 (continued) (A) Chronological pictorial illustrating the fire impacts and temporal responses of three main ecosystem components (water quality, soil biogeochemistry, and vegetation dynamics) to a prescribed fire in a cattail-dominated wetland. Arrows indicate increases ($\uparrow$), no change ($\rightarrow$), or decreases ($\downarrow$) in the corresponding parameter. Green cattail leaves indicate live leaves; light blue leaves at 1 day after the fire represent the live leaves that were scorched and killed by the fire but not consumed, and the brown leaves represents the general dead leaf litter. (B) Temporal changes in a conceptual model of relevant wetland ecosystem total P (g m^{-2}) pools and transfer pathways in the air, surface water, and soil after a prescribed fire. Each of the four models corresponded to each of the time periods in A. AG and BG stands for above and below ground respectively. The red AG pool 1 day after the fire represents the scorched leaves not consumed by the fire. Pore water TDP pools were estimated to a depth of 30 cm based samples taken from the 15–25 cm depth range and measured soil bulk density profiles. The Peat TP pools were to a depth of 30 cm. Only the top 5 cm were measured on the 1 day and 1 month after the fire, and therefore we assumed the 5–30 cm layer to be similar to the before fire sample.

within one month the TP of the aboveground biomass pool was 50% of pre-fire pool, even though live leaf biomass was less than 20% Fig 4.7B. This seemed to be due to significantly greater TP concentrations in new leaves compared to older leaves collected before the fire (unpublished data). One-year after the fire, the aboveground biomass TP pool had recovered and was similar to before the fire; however, aboveground dead leaf TP was still 40% less than pre-fire.

Pulsed and sustained responses of the belowground biomass TP pools were more difficult to determine, so some of the results and conclusions stated here were theoretically derived. Fire caused direct mortality of approximately 60% of cattail ramets (Sindhøj and Miao, submitted) which would have resulted in a decrease in the TP pool of belowground live biomass and a subsequent increase in the TP pool of belowground dead biomass. The relatively large TP increase in the belowground live biomass one month after the fire was attributed mostly to the belowground components of new ramets, which accounted for over 50% of individual cattail plants at this time (Sindhøj and Miao, submitted) and should have greater nutrient concentrations compared to older plants, as described above for the leaves. By the end of the first year, TP of the belowground live biomass had almost recovered to pre-fire levels, but there was still more than twice as much TP in the belowground dead biomass than before the fire. This input of belowground dead biomass that could result in increased long-term P storage belowground may be offset by increased peat surface temperatures that could enhance decomposition and accelerate nutrient cycling, so it is unclear whether fire will lead to increased long-term storage of nutrients below-ground.

Overall, fire resulted in a SWTP pulse of short duration at the burned site. Downstream transport of the fire-released P exceeded 100 m and the total extent of downstream transport remains unknown. However, taking multiple factors into consideration (e.g., the short TP pulse, no increase in soil TP, and high background TP level), it appears that P mobilized from prescribed fires at the scale investigated in this study would not exceed variations in P transport resulting from other system-wide events, processes or perturbations. Furthermore, while cattail populations had almost recovered after one year, community structure still exhibited impacts which would affect ecosystem processes.

4.8 Summary

Large-scale ecosystem restoration presents unprecedented challenges, for which current methods of ecological research and analysis are not well-prepared, indicating a need for a paradigm shift in experimental design and analysis that address spatial- and temporal-scales and variances. At the same time, ecosystem restoration projects provide an ideal framework for ecologists to question, investigate, and develop important ecological theory and philosophy. Driven by a legislative mandate to restore the historical vegetation in nutrient enriched areas of the Everglades, we designed and implemented the ecosystem-scale Fire project to investigate the effects of fire on wetland ecosystem structure and function and to address issues related to nonreplicated experimental design and analysis. Preliminary data analysis of the Fire project's initial years clearly shows the challenges and benefits of integrating multiple spatial scales and unbalanced temporal designs for detecting responses to fire amid large-scale natural variability.

Replication of ecosystem-scale environmental disturbances or treatments is often not possible or desirable under real world circumstances. While there should be no moral or ethical reasons to limit control site replication, financial, and logistical restraints to replication in general are often a reality. The Fire project demonstrated that a few carefully chosen controls at multiple spatial scales (such as adjacent and distant controls) provided analytical flexibility. Considering spatial structure of the system studied and multiple ecosystem response variables operating at different temporal scales, we suggest that at least one carefully chosen control site should closely co-vary with the impact site to maximize the potential for impact detection for variables with rapid response times, and one control site should be ecologically similar but separated at a greater spatial scale providing a context of natural ecosystem variability for variables with relatively longer response times. Thus, we echo Carpenter's (1990) recommendation that "resources that could be invested in duplicates [minimal replication] might be better spent on more detailed mechanistic analyses conducted within the context of a large-scale experiment."

Purposely selecting controls at the spatial scales of major system driving forces seemed an effective approach to simultaneously detect a fire impact at a small scale while placing the magnitude of the impact in the context of natural variation at a larger scale. However, it may be argued that this approach is perilously close to predetermining one's results, i.e., choosing a control that will show the intended effect. Continuing with this argument, it is recommended that using different controls yields different results and therefore multiple random controls are required to ensure analytical rigor and robustness. These are rational and elegant arguments from the traditional ecological paradigm and directly relate to the inferential power of nonreplicated large-scale ecological studies. Inference, however, is not always necessarily the goal of ecosystem restoration or disturbance studies. One fundamental difference between null hypothesis testing and large-scale ecosystem restoration or disturbance studies is the former merely seeks to determine whether an impact has occurred, whereas the later seeks to elucidate the direction, duration, relevance, and magnitude of the impact (Canham and Pace, Chapter 8). Foreknowledge of the systems studied, the potential scales of impacts, and the forcing functions behind natural variation should shape experimental design adequately to detect and quantify impacts. In this study, fire did cause a nutrient pulse in surface waters; however, it was indeed no greater than the overall natural long-term temporal variability of the system. In addition, the multiple spatial control design had advantages to assess post-fire downstream nutrients effects, a critical issue in aquatic and wetland systems.

The multiple, unbalanced design of temporal sampling schemes for response variables operating at different biological organization scale is important when assessing both short- and long-term effects on a whole ecosystem. When pulsed responses are anticipated, sampling regimes should be adjusted accordingly to capture the changes even at the expense of not meeting standard statistical assumptions. This strategy, however, requires the development of a wide range of novel analytical tools that encompass the assessment of ecosystem components operating at different spatial and temporal scales.

The two regression approaches, moving and simple regressions, outlined here appear to effectively determine ecosystem response characteristics such as direction (increase, decrease, or no change), onset, duration, and magnitude. For both regressions, data collected at the control site were defined as the independent variable and the data from the impact site as the dependent variable. The moving regression identifies the onset and duration of impacts, i.e. Before, Response or Recovered periods. After which a simple regression can elucidate the slopes, confidence and prediction intervals, and scatter of individual data points sampled at specific times during these periods. Overall, these two regression-analysis models incorporate spatial (Control and Impact) and temporal (Before and Response) variation in a correlative context and eliminate many analytical challenges of dealing with variation in nonreplicated design. However, accurate use of these regression models requires that both Control and Impact sites are sampled at the same time.

These regression models also advanced our insights into the ecological context of the After period. In the BACI literature, the system response period is represented by the entire After period and is solely decided by the timing of the treatment. Analyzing ecosystem response according to these assumptions limits the likelihood of impact detection when pulsed effects occur. Clearly, widening the After period definition to include both Response and Recovered periods enhances the applicability of both BACI and BACIPS comparisons and consequently reduces the likelihood of a Type I error. While common graphical analysis of data may obviously indicate a response, the regression approaches offer statistical support for the characterization of Before, Response, and Recovered periods of disturbance impacts and provide an overall framework for further analysis of system responses. We will continue to investigate the application of regression approaches from other analytic tools to address new issues that arise along with our ongoing Fire project.

Acknowledgments We thank the South Florida Water Management District for supporting the Fire project and Hawaii Pacific University for providing release time for S. Carstenn. We greatly appreciate many former (D. Pisut, H. Chen, J. Creasser, D. Salembier, M. Tapia, and D.Condo) and current members (Robert Johnson, Christina Stylianos, D. Monette) of the Fire project team who conducted field samplings. We thank C. Stow, D. Hui, J. Hewitt, J. Grace, S. Hill, M. Nungesser and D. Drum for their valuable comments on early drafts of the chapter. S.L. Miao contributed to overall idea, structure, and preparation of the manuscript; S.Carstenn contributed to analysis of moving regression and manuscript preparation; C. Thomas conducted BACI-ANOVA analyses related to control similarity and control numbers for detecting fire impacts; C. Edelstein was responsible for field setup and sampling, data management and figure and table creation; E. Sindhøj developed the model for nutrient pool analysis used in the synthesis section and contributed to manuscript preparation; and Binhe Gu contributed to the development of the nutrient pool analysis.

References

Bender, E. A., T. J. Case, and M. E. Gilpin. 1984. Perturbation experiments in community ecology: theory and practice. Ecology **65**:1–13.

Box, G. E. P. and G. C. Tiao. 1965. A change in level of a nonstationary time series. Biometrika **52**:181–192.

Box, G. E. P. and G. C. Tiao. 1975. Intervention analysis with applications to economic and environmental problems. Journal of the American Statistical Association **70**:70–79.

Burns and McDonnell. 2003. Everglades Protection Area Tributary Basins Long-Term Plan for Achieving Water Quality Goals. South Florida Water Management District, West Palm Beach, FL.

Carpenter, S., N. F. Caraco, D. L. Correll, R. W. Howarth, A. N. Sharpley, and V. H. Smith. 1998. Nonpoint pollution of surface waters with phosphorus and nitrogen. Ecological Applications **8**:559–568.

Carpenter, S. R. 1990. Large-scale perturbations: Opportunities for innovation. Ecology **71**:2038–2043.

Carpenter, S. R. 1996. Microcosm experiments have limited relevance for community and ecosystem ecology. Ecology **77**:677.

Carpenter, S. R., S. W. Chisholm, C. J. Krebs, D. W. Schindler, and R. F. Wright. 1995. Ecosystem experiments. Science **269**:324–327.

Carpenter, S. R., T. M. Frost, D. Heisey, and T. K. Kratz. 1989. Randomized intervention analysis and the interpretation of whole-ecosystem experiments. Ecology **70**:1142–1152.

Chimney, M. J. and G. Goforth. 2001. Environmental impacts to the Everglades ecosystem: a historical perspective and restoration strategies. Water Science and Technology **44**:93–100.

Clark, J. S., S. R. Carpenter, M. Barber, S. Collins, A. Dobson, J. A. Foley, D. M. Lodge, M. Pascual, R. Pielke Jr., W. Pizer, C. Pringle, W. V. Reid, K. A. Rose, O. Sala, W. H. Schlesinger, D. H. Wall, and D. Wear. 2001. Ecological forecasts: an emerging imperative. Science **293**:657 660.

Conquest, L. L. 2000. Analysis and interpretation of ecological field data using BACI designs: discussion. Journal of Agricultural, Biological and Environmental Statistics **5**:293–296.

Davis, S. M. and J. C. Ogden. 1994. Toward ecosystem restoration. Pages 769–796 *in* S. M. Davis and J. C. Ogden, editors. Everglades: The Ecosystem and Its Restoration. St. Lucie Press, Delray Beach, Florida, USA.

Gotelli, N. J. and A. M. Ellison. 2004. A Primer of Ecological Statistics. Sinauer Associates, Sunderland, MA.

Green, R. H. 1979. Sampling Design and Statistical Methods for Environmental Biologists. John Wiley & Sons, University of Western Ontario.

Hasler, A. D., O. M. Brynildson, W. T. Helm. 1951. Improving conditions for fish in brown-water bog lakes by alkalization. Journal of Wildlife Management **15**:347–352.

Hewitt, J. E., S. E. Thrush, and V. J. Cummings. 2001. Assessing environmental impacts: effects of spatial and temporal variability at likely impact scales. Ecological Applications **11**:1502–1516.

Hewitt, J. E., S. F. Thrush, P. K. Dayton, and E. Bonsdorff. 2007. The effect of spatial and temporal heterogeneity on the design and analysis of empirical studies of scale-dependent systems. The American Naturalist **169**:398–408.

Keough, J. M. and G. P. Quinn. 2000. Legislative vs. practical protection of an intertidal shoreline in Southeastern Australia. Ecological Applications **10**:871–881.

Legendre, P. 1993. Spatial autocorrelation: trouble or new paradigm? Ecology **74**: 1659–1673.

Legendre, P., M. R. T. Dale, M.-J. Fortin, J. Gurevitch, M. Hohn, and D. Meyers. 2002. The consequences of spatial structure for the design and analysis of ecological field surveys. Ecography **25**:601–615.

108   S. Miao et al.

Likens, G. E. 1985. An experimental approach for the study of ecosystems: The Fifth Tansley Lecture. Journal of Ecology 73:381–396.

Melbourne, B. A. and P. Chesson. 2005. Scaling up population dynamics: integrating theory and data. Oecologia 145:179–187.

Miao, S. and S. Carstenn. 2006. A new direction for large-scale experimental design and analysis. Frontiers in Ecology 4:227.

Miao, S. L. and S. Carstenn. 2005. Assessing Long-Term Ecological Effects of Fire and Natural Recovery in a Phosphorus Enriched Everglades wetlands: Cattail Expansion Phosphorus Biogeochemistry and Native Vegetation Recovery. West Palm Beach, Florida.

Mitsch, W. J. and J. W. Day Jr. 2004. Thinking big with whole-ecosystem studies and ecosystem restoration- a legacy of H.T. Odum. Ecological Modeling 178:133–155.

Murtaugh, P. A. 2007. Simplicity and complexity in ecological data analysis. Ecology 88:56–62.

Osenberg, C. W. and R. J. Schmitt, editors. 1996. Detecting Ecological Impacts Caused by Human Activities. Academic Press, Inc., San Diego, CA.

Osenberg, C. W., R. J. Schmitt, S. J. Holbrook, K. E. Abu-Saba, and A. R. Flegal. 1994. Detection of environmental impacts: natural variability, effect size, and power analysis. Ecological Applications 4:16–30.

Petersen, J. E., W. M. Kemp, R. Bartleson, W. R. Boynton, C.-C. Chen, J. C. Cornwell, R. H. Gardner, D. C. Hinkle, E. D. Houde, T. C. Malone, W. P. Mowitt, L. Murray, L. P. Sanford, J. C. Stevenson, K. L. Sunderburg, and S. E. Suttles. 2003. Multiscale experiments in coastal ecology: improving realism and advancing theory. BioScience 53:1181–1197.

Pickett, S. T. A. 1989. Space-for-time substitution as an alternative to long-term studies. Pages 110–135 in G. E. Likens, editor. Long-term studies in ecology. Springer-Verlag, New York.

Ponzio, K. J., S. J. Miller, and M. A. Lee. 2004. Long-term effects of prescribed fire on Cladium jamaicense crantz and Typha domingensis pers. densities. Wetlands Ecology and Management 12:123–133.

Schindler, D. W. 1998. Replication versus realism: The need for ecosystem-scale experiments. Ecosystems 1:323–334.

Simberloff, D. 2004. A rising tide of species and literature: a review of some recent books on biological invasions. BioScience 54:247–254.

Sindhøj, E. and S. Miao. Submitted. Recovery dynamics and underlying mechanisms of Typha domingensis communities following fire in a highly- and moderately-nutrient enriched Everglades wetland Ecosystem Restoration.

Sokal, R., R. and F. J. Rohlf. 1981. Biometry. 2nd edition. W.H. Freeman and Company, New York.

Stewart-Oaten, A. 1996a. Goals in environmental monitoring. Pages 17–27 in R. J. Schmitt and C. W. Osenberg, editors. Detecting Ecological Impacts: Concepts and Applications in Coastal Habitats. Academic Press, Inc.

Stewart-Oaten, A. 1996b. Problems in the analysis of environmental monitoring data. Pages 109–131 in R. J. Schmitt and C. W. Osenberg, editors. Detecting Ecological Impacts: Concepts and Applications in Coastal Habitats. Academic Press, San Diego.

Stewart-Oaten, A. 1996c. Problems in the analysis of environmental monitoring data. Pages 109–131 in R. J. Schmitt and C. W. Osenberg, editors. Detecting Ecological Impacts: Concepts and Applications in Coastal Habitats. Academic Press, Inc.

Stewart-Oaten, A. 2003. Using Before-After-Control Impact in Environmental Assessment: Purpose, Theoretical Basis, and Practical Problems. Coastal Research Center, Marine Science Institute, University of California, Santa Barbara, California.

Thrush, S. F., J. E. Hewitt, V. J. Cummings, P. K. Dayton, M. Cryer, S. J. Turner, G. Funnell, R. Budd, C. Milburn, and M. R. Wilkinson. 1998. Disturbance of the marine benthic habitat by commercial fishing: impacts at the scale of the fishery. Ecological Applications 8:866–879.

Thrush, S. F., S. M. Lawrie, J. E. Hewitt, and V. J. Cummings. 1999. The problem of scale: uncertainties and implications for soft-bottom marine communities and the assessment of human impacts. Pages 185–210 *in* J. S. Gray, editor. Biogeochemical Cycling and Sediment Ecology. Kluwer Academic, Netherlands.

Underwood, A. J. 1989. The analysis of stress in natural populations. Biological Journal of the Linnean Society **37**:51–78.

Underwood, A. J. 1991. Beyond BACI: experimental designs for detecting human environmental impacts of temporal variations in natural populations. Australian Journal of Freshwater Research **42**:569–587.

Underwood, A. J. 1993. The Mechanics of Spatially replicated sampling programs to detect environmental impacts in a variable world. Australian Journal of Ecology **18**:99–116.

Underwood, A. J. 1994. On beyond BACI: sampling designs that might reliably detect environmental disturbances. Ecological Applications **4**:3–15.

Urquhart, N. S., S. G. Paulsen, and D. P. Larsen. 1998. Monitoring for policy-relevant regional trends over time. Ecological Applications **8**:246–257.

Chapter 5
Bayesian Hierarchical/Multilevel Models for Inference and Prediction Using Cross-System Lake Data

Craig A. Stow, E. Conrad Lamon, Song S. Qian, Patricia A. Soranno and Kenneth H. Reckhow

Abbreviations AIC – Akaike's Information Criterion, ANOVA – Analysis of Variance, BCART – Bayesian classification and regression tree, BIC – Bayesian Information Criterion, CART – Classification and Regression Tree, DIC – Deviance Information Criterion, LIL – Log Integrated likelihood, MCMC – Markov Chain Monte Carlo, MLE – Maximum Likelihood Estimator, SBC – Schwarz's Bayesian criterion, TP – Total phosphorus

5.1 Introduction

Cross-system studies are commonly used for large-scale ecological inference (Cole et al. 1991). Many processes change slowly within a particular ecosystem, thus long time periods can be required to measure how changes in one process may influence changes in another. By using data from many systems researchers essentially substitute space for time, assuming commonality among the systems being compared. Comparing characteristics among systems helps researchers identify patterns that provide clues for understanding ecosystem function, generate testable hypotheses, and isolate cause–effect relationships.

In limnology, cross-system studies have been widely applied, in part because lakes are relatively discrete ecosystems with tangible boundaries, making their properties straightforward to evaluate and compare. Additionally, many important lake attributes, such as trophic status, can be well-approximated by a small number of quantitative measures. In the 1960s and 1970s cross-system lake comparisons were influential in resolving the limiting nutrient debate. The work of Richard Vollenweider, in particular, popularized this approach, and led to nutrient loading concepts that are still widely applied in aquatic ecosystem management (Vollenweider 1968, 1969, 1975, 1976).

C.A. Stow (✉)
NOAA Great Lakes Environmental Research Laboratory, 2205 Commonwealth Blvd., Ann Arbor, MI 48105, USA
e-mail: craig.stow@noaa.gov

S. Miao et al. (eds.), *Real World Ecology*, DOI 10.1007/978-0-387-77942-3_5,
© Springer Science+Business Media, LLC 2009

Early efforts to quantify relationships among lakes may have been largely ad hoc; electronic calculators were not widely available until the mid-1970s and access to fast computers was limited. Thus, even relatively simple models like linear regression and ANOVA were tedious to compute for more than a small number of observations. With the increasing availability of fast, inexpensive computing power researchers began to use cross-system lake data to develop mathematical relationships useful for quantifying limnological processes and to build models for decision-making (Reckhow and Chapra 1983).

The use of cross-system data for lake-model development has also been fostered by a general lack of extensive long-term data from individual lakes. However, individual lake behavior can be idiosyncratic – highly dependent on features such as the landscape setting in which the lake is located (Stow et al. 1998). Thus, models developed from cross-system data may not accurately capture the behavior of a particular lake. Because management decisions are typically made on a lake-by-lake basis, it is arguably preferable to construct lake-specific models to support management actions such as Total Maximum Daily Loads (National Research Council 2001). Nevertheless, with some exceptions (Stow et al. 1997, Lathrop et al. 1998), relatively few lakes have been sufficiently well-monitored to provide adequate data for individual model development. This is particularly true in lake-rich states in the US and other regions of the world where limited resources prohibit intensive data collection on more than a few lakes.

Additionally, data representing a range of conditions are necessary to assess the functional form of a model and estimate model parameters. Individual lakes in approximately steady-state conditions are unlikely to generate data of sufficient variance for accurate model estimation even over long time periods. In such cases extrapolation beyond the observed conditions may be required to predict the likely changes that will occur under alternative management actions, resulting in considerable predictive uncertainty. Thus, even if many years of data are available for a particular lake, augmenting the lake-specific data with data from other lakes can increase predictive accuracy for the lake under consideration.

Building models for prediction from cross-system data is based on the supposition that the relationship between the response and predictor variables is the same for all the lakes used to estimate the model. While this presumption is never exactly correct, most modelers implicitly hope that it is close enough to the truth to make useful inferences and decisions. A popular strategy to increase the likelihood of similar behavior is to group lakes based on common attributes. Features used to categorize lakes have included lake type (natural vs. reservoir), geographic setting, and geomorphology (Canfield and Bachman 1981, Reckhow 1988, Malve and Qian 2006) and models based on such groupings are then used for individual lake forecasts (Hession et al. 1995). The goal of this strategy is to "borrow" information from other similar lakes to increase the accuracy of prediction for a particular lake, so that the risk of making a bad decision is acceptably low.

But are models developed from a cross-section of lakes sufficiently accurate to make good management decisions for an individual lake? Moreover, how uncertain can a model be and still be useful for effective management decision-making? The answer to these questions is context specific, depending on the stakes associated with the pending decisions. If the consequences of a poor decision are not very severe, high model uncertainty may not matter much. Alternatively, in high-stakes decision situations, models with large uncertainties may not differentiate the likely outcomes of one management alternative from another, providing little basis for selecting a management option.

Given a choice of models, a decision-maker is likely to use the model with the lowest uncertainty. A model that provided correct forecasts 98% of the time would be a clear choice over one that was accurate only 75% of the time. With 98% accuracy, management actions could be chosen based only on the societal value of the consequences of those actions. A model with higher uncertainty may still be informative, but applying this model requires decision-makers to hedge their decisions by considering a range of possible outcomes, based on knowing how uncertain the model is. Therefore, while uncertainty may arise from a lack of knowledge, quantified uncertainty provides information that is useful for both picking the best model and applying that model for decision support.

Often, however, decision-makers are given models or model results with no supporting information regarding uncertainty – providing an illusion of precision (Pappenberger and Beven 2006). How is this information useful for high-stakes decisions?

Uncertainty is typically expressed as a probability statement. The term probability has several interpretations; the most widely appreciated is the long-term relative frequency of a particular event. Coin-flipping is a common example of this probability notion. The probability of flipping a coin and obtaining "heads" is approximately 50% – meaning that if a coin is flipped enough times approximately 50% of those flips will result in the coin landing heads-side up. Another interpretation of probability is that it represents a degree-of-belief (Winkler 2003). If an individual is presented with a coin and asked the probability that the coin will land heads-side up when flipped, it is likely that the response will be: "approximately 50%". In this instance there is no long-term relative frequency involved, in fact the coin has not been flipped even once yet. The response "50%" represents the respondent's degree-of-belief that the coin will land heads-side up. This belief is based on prior experience resulting in knowledge of what usually happens when a coin is flipped. If the respondent had reason to believe that they were being tricked with a coin that had been engineered to produce heads they might indicate 90 or 95%. Thus, the degree-of-belief notion of probability is a quantification of the confidence an individual has in the occurrence of a particular result.

These two notions of probability are not mutually exclusive; in fact they can be reconciled using Bayes theorem:

$$\pi(\theta|y) = \frac{\pi(\theta)f(y|\theta)}{\int_\theta \pi(\theta)f(y|\theta)\mathrm{d}\theta} \tag{5.1}$$

The interpretation of Bayes theorem is that prior beliefs (those held before the experiment), represented by $\pi(\theta)$, are combined with new information from the experiment, contained in the likelihood function, which is represented by $f(y \mid \theta)$, to obtain posterior beliefs, represented by $\pi(\theta \mid y)$. For most practical applications, the denominator on the right side of the equation can be regarded as a scaling constant and essentially ignored.

If, for example, before flipping a coin we believed that the probability of heads was 50%, and in our experiment of 100 flips we obtained 50 heads and 50 tails our belief would be unchanged; the posterior belief would be the same as the prior belief. If, however, the coin-flipping experiment resulted in 95 heads it is likely that our posterior belief in the probability of heads would move toward 95%, depending on whether or not we thought the 95 head outcome was typical or accidental.

While Bayesian approaches have been known in ecology and the environmental sciences for some time (Reckhow 1990), they are being increasingly used (Ellison 2004), in part because they have only become practically feasible since the widespread availability of fast, inexpensive computers. Many models are analytically intractable in a Bayesian context, particularly nonlinear models or models with more than just a few parameters to be estimated. Using a Bayesian approach in such cases was virtually impossible until recently. But the advent of cheap computing has fostered the development of algorithms that provide precise numerical approximations for most problems, making the routine application of Bayes theorem a practical option.

Bayesian approaches are not new; their use predates what are often referred to as "frequentist" or "classical" methods, which were developed in the early-mid 1900s (Salsburg 2001). Classical approaches are what most ecologists have been trained in since the 1960s, and usually involve setting up a null hypothesis, then performing a "significance test" to evaluate the validity of the null hypothesis. Many textbooks present classical methods as a coherently conceived approach to scientific inference, and their application has become deeply engrained in ecology. Thus, many ecologists perceive them to be inviolate rules that govern the way science is properly conducted, and finding "significant" results has become an end in itself rather than a means to an end. In fact, classical significance testing arose from a somewhat rocky fusion of two contrasting schools of thought, contributing to considerable confusion among applied scientists when interpreting their results, as p-values and significance levels are often interpreted to be synonyms (Hubbard and Bayarri 2003, Hubbard and Armstrong 2006).

Working in a Bayesian framework offers distinct advantages over a classical null hypothesis testing framework. A Bayesian approach provides a posterior probability distribution for the model parameters, $\pi(\theta \mid y)$, which indicates the relative probabilities of all possible values of θ, conditional on the observed data, y. This posterior distribution can be used to support a wide range of decisions applying many possible decision criteria. Predictions for y are made by evaluating the model over the entire posterior parameter distribution, resulting in a predictive probability distribution for future y values.

Classical null hypothesis testing approaches are much more constrained. The decision to be made is to either accept or reject an *a priori* null hypothesis, and the decision criterion is chosen to minimize the chance of accepting a false null hypothesis. Acceptance or rejection of the null hypothesis is based on a *p*-value, which is the probability of obtaining results at least as extreme as the observed data, conditional on the null hypothesis being true. Many researchers find *p*-values (symbolically expressed as $\pi(\geq y \mid \theta)$ where θ is typically zero) confusing, often misinterpreting them to convey the same information as a posterior distribution.

Bayesian approaches are sometimes disparaged as subjective because of the need to provide prior information regarding parameter values (Dennis 1996), but in fact modelers often fix parameters at specific values based on precedent, published literature values, or expert-judgment, and this practice is rarely criticized. Bayes theorem offers modelers considerably more flexibility. If a parameter value is well-known then a fixed value can be used, but if the modeler has very little prior information regarding a parameter value, or wants the posterior distribution to be determined by the data, then a non-informative or vague prior distribution can be chosen. Non-informative priors typically have a large variance, thus they convey minimal information about the value of a model parameter. Alternatively, semi-informative priors, with intermediate variances, can be used for parameters that are not known precisely, but are likely to be within a limited range (Stow and Scavia 2008). In any case, prior specification in Bayes theorem is explicit, allowing other researchers the opportunity to evaluate the sensitivity of model-based inferences to the choice of the prior.

5.2 Multilevel/Hierarchical Models

Although lake management decisions are usually made on a lake-by-lake basis, the scientific information for such decisions is often based on cross-system data. On one hand, a model developed using data from multiple lakes will almost certainly be less accurate for a specific lake, because the model represents the average of the lakes. On the other hand, a model based only on lake-specific data will have a large uncertainty because of the small lake-specific sample size. Multilevel models provide a rigorous framework to systematically combine information from several sources and appropriately weight the group-specific or lake-specific information depending on the degree of similarity to other groups or lakes in the data set.

In a multilevel model, parameters representing individual lakes, or groups of lakes, are given a probability structure. For example, within a lake, observations of a response variable, such as chlorophyll a concentration, exhibit a lake-specific mean, and a certain level of within-lake variability. If the within-lake variability is expected to be similar for lakes belonging to the same group, then the between-lake variability can be expressed in terms of the variability among the lake-specific means. This variability among lake-specific means can be modeled with a normal (or other) distribution and the parameters of this distribution can be estimated from the cross-system data. In some cases, observations may be nested – there may be distinct basins within lakes, lakes within similar groups, groups within a region, etc., and the probability model capturing each of these levels of variability assumes a hierarchical configuration.

To explain the difference between the multilevel approach and more traditional approaches, we use a simple example of estimating lake-specific means of a response variable. Mathematically, we would model this between and within-lake variability using a model similar to an ANOVA model

$$y_{ij} \sim N(\mu_i, \sigma^2), \tag{5.2}$$

where y_{ij} is the jth observation from the ith lake, μ_i is the mean of the ith lake and σ^2 is the variance of the observations. When individual lakes are modeled separately, each μ_i is estimated to be a lake-specific mean. When cross-system data are used, all lakes implicitly share the same mean. The multilevel approach is a compromise between these two extremes. It treats each individual lake as a separate entity, but the lake-specific means are structured to share the same prior distribution:

$$\mu_i \sim N(\mu, \tau^2), \tag{5.3}$$

where μ and τ^2 are the mean and variance, respectively, of the μ_i. By using a shared prior distribution of all lake-specific means, the multilevel model results in at least two improvements over more conventional approaches: first, the shared prior distribution provides a connection between individual lakes. Because of the common prior, the lake specific means are now estimated as a weighted average of individual lake mean and the overall mean.

$$\bar{\mu}_i = \frac{(n_i/\sigma^2)\bar{y}_i + (1/\tau^2)\bar{y}_{\text{all}}}{(n_i/\sigma^2) + (1/\tau^2)} \tag{5.4}$$

The weights are determined by individual lake sample size (n_i), individual lake variance (σ^2), and the variance among lake-specific means (τ^2). When the sample size of a lake is large, the multilevel estimated lake specific mean will be close to the sample mean. When the sample size is small, the multilevel estimated lake specific mean will be close to the overall mean. If there are no data from a lake, the multilevel estimate of the lake specific mean is the overall mean. In addition to the sample size, the multilevel estimate also considers the levels of between and within lake variances. If the between-lake variance (τ^2) is high, the weight on overall mean will be small, individual lake mean will be weighted heavier, and vice versa. This partial pooling of the cross-system data allows system specific estimates without requiring a large sample size for each system. Second, by pooling data from multiple lakes, the uncertainty associated with lake-specific prediction is reduced. If we refer to the cross-system approach, where data from multiple lakes are pooled, as complete pooling of all lakes, and refer to the separate analysis of individual lake approach as no pooling, the multilevel modeling approach is called partial pooling of the data. By partial pooling, we balance the information from individual lakes and the overall average from all lakes. The individual lake data are more variable (more uncertain) while the overall average is less variable (more certain).

Compared to the separate lake means, the multilevel-based estimates of individual lake means are pulled towards the overall mean. Lakes with larger sample sizes are pulled less, while lakes with smaller sample sizes will be pulled more to the overall mean. This pulling (also known as Bayesian shrinkage) represents a form of information discounting. If a lake with few observations, or a large variance, has a mean chlorophyll *a* concentration that is much higher than the overall mean of all the lakes, we are likely to doubt the validity of the anomalously high estimate. The multilevel estimate provides a sensible way of discounting the information that is less trustworthy. This is consistent with the way that Bayes theorem pools information from different sources. The information represented in the overall mean can be seen as the prior for an individual lake, while data from the lakes are treated as observations. Under the hierarchical framework, prior distributions are typically needed for the model parameters that are of secondary interest, thus their influence on the parameters of primary interest is indirect and often minimal. In our case, we need to supply prior distributions for the mean of group means of model coefficients and the within and between group standard deviations of model coefficients. When there are many groups, the mean of group means is of less interest and we usually do not have much information on the within and between group standard deviations of model coefficients. As a result, non-informative priors are used.

The multilevel approach can be readily extended to more complex models where the variable of interest is a linear or nonlinear function of one or more predictor variables, with unknown model parameters. In our example, lake chlorophyll *a* concentration is the variable of interest and we model the log of chlorophyll *a* as a simple linear function of the log of total phosphorus concentration. The multilevel model partially pools the group data by introducing common prior distributions for model coefficients. With this approach, the model structure is:

$$\log(\text{Chla})_{ik} \sim \text{N}(\mu_{ik}, \tau^2) \tag{5.5a}$$

and

$$\mu_{ik} = \alpha_k + \beta_k \log(\mathrm{TP}_{ik}), \tag{5.5b}$$

$$a_k \sim N(\alpha, \sigma_\alpha^2), \tag{5.5c}$$

$$\beta_k \sim \mathrm{N}(\beta, \sigma_\beta^2), \tag{5.5d}$$

where $\log(\mathrm{Chla})_{ik}$ is the natural log of chlorophyll a concentration from lake i in group k, μ_{ik} is the mean of the ith lake in group k, τ^2 is the variance of $\log(\mathrm{Chla})$ at a given value of $\log(\mathrm{TP})$ (assumed to be the same for all lakes), α_k and β_k are the intercept and slope parameters, respectively, for group k, $\boldsymbol{\alpha}$ is the mean of the α_ks, $\boldsymbol{\beta}$ is the mean of the β_ks, and σ_α^2 and σ_β^2 are the respective variances of the α_ks and β_ks.

Computing a multilevel model can be done in either a Bayesian or non-Bayesian context. The non-Bayesian implementation is usually referred to as a random, fixed, or mixed-effect model. Typical random, fixed, and mixed-effect models use a maximum likelihood estimator (MLE) producing a result similar to a Bayesian model using vague prior distributions for σ^2, $\boldsymbol{\alpha}$, $\boldsymbol{\beta}$, σ_α^2, and σ_β^2. However, estimating the variances can be difficult using MLE when the number of lakes or number of groups is small (single digit). The Bayesian method is more flexible, especially when proper prior information is available. A Bayesian implementation of multilevel modeling is typically calculated using Markov chain Monte Carlo (MCMC) simulation (Qian et al. 2003) and can readily be programmed in WinBUGS (Lunn et al. 2000) which is a free downloadable software. Multilevel/hierarchical models are similar in concept to random coefficient and empirical Bayes models which have been previously demonstrated with cross-system lake data (Reckhow 1993, Reckhow 1996). They can be very useful for organizing ecological data and have been used to synthesize information in a cross-system data set of Finnish lakes (Malve and Qian 2006).

5.3 Finding Groups in Data

As noted earlier, lake water quality models may be classified by the level of pooling used in the estimation of their parameters, and have historically either been developed in a lake-specific context (no pooling) or from a cross-system study (complete pooling). The simple linear structure of a regression model does not always apply to an entire cross-system dataset in the complete pooling situation. There are many reasons that this may be true for lake water quality data, including factors such as regional and local differences in climate, geology, morphometry, land use, land cover, and food web structure. An open question with regard to partially pooled lake models is "How do we determine which of the above mentioned factors to use to create lake groups for pooling?"

In some situations, a natural hierarchical structure is present in the data and may be exploited for the purpose of assigning lakes to groups for pooling. Lakes occur in watersheds, watersheds in ecoregions, etc. These are categorical or factor variables, in which observations take multiple discrete values or levels with in each factor. In such instances, it may be of interest to determine if one or more levels of the factor may be combined to reduce model complexity. Further, there may be thresholds, or change points, along the axes of continuous variables that may be used to define groups (above/below threshold) for pooling.

Chipman et al. (2002) provide an algorithm to obtain models that may better describe simple structure in cross-system data, by sub-setting the data and then fitting separate sub-models for each subset. We used Bayesian Treed models (Chipman et al. 2002) to determine if model predictive performance could be improved by fitting our simple linear regression model to subsets of cross-system lake data. Bayesian CART and Bayesian Treed models are both enhancements to the more familiar Classification and Regression Tree (CART) procedure (Breiman et al. 1984). Bayesian Treed Models select subsets of observations on partitions of the matrix of predictor variables for which linear models are estimated. Further, they identify subsets of the predictor variables for which linear model performance is improved as measured by predictive log integrated likelihood (LIL). Chipman et al. (2002) provide full details regarding Bayesian Treed models.

Use of Bayesian CART (Chipman et al. 1998) and Bayesian Treed models (Chipman et al. 2002) allows predictors to work together in a nonlinear, non-additive fashion by virtue of their tree-based structure, because the disjoint partitions on the predictor space provide for global non-linearity and non-additivity, even though model structure may be locally linear (and additive). We begin our discussion of methods with a general discussion of tree based methods, followed by a more specific description of Bayesian CART and Bayesian Treed models.

5.3.1 Tree-Based Models

Tree-based models are useful for classification and regression problems in which the analyst cannot (or does not want to) specify *a priori,* the form of important interactions between explanatory (or independent) variables (Clark and Pregibon 1992). Tree models are easy to interpret, invariant to monotone transformations of the predictors, and able to capture interaction effects among the independent variables (non-additivity). Tree-based models made their first appearance in the statistical literature due to Sonquist and Morgan (1964). Much of their recent development is due to the work of Breiman et al. (1984). Because they are a computationally intensive procedure, their use has grown concurrently with the advent of the personal computer. Clark and Pregibon (1992) provide a description of classification and regression trees, along with examples. Classification and regression trees have been used in numerous ecological studies (e.g., Magnuson et al. 1998, Lamon and Stow 1999, Qian and Anderson 1999). Bayesian Treed Models have been used with cross sectional data to link nutrients to chlorophyll *a* concentrations in lakes of the continental U.S. (Lamon and Stow 2004, Freeman et al. 2008), and in Finnish lakes (Lamon et al. 2008).

To fit a regression tree, the algorithm begins with the root (or parent) node. The root node contains all the observations and their associated variability. The data are split by binary recursive partitioning into increasingly homogeneous subsets until within-node variability is below some user-specified value. A terminal node is one that cannot be split further according to the user specified rules. Terminal nodes are also called leaves, consistent with the tree analogy. For every split made, the algorithm uses all unique values along the axes of each predictor variable as candidate values for splitting the dataset.

The process starts by defining the deviance of the root node (all of the data) as:

$$D(\mu) = \sum (y_i - \mu)^2 \tag{5.6}$$

where y_i are the observations within the node and μ is the node mean. Then each candidate predictor variable is examined to find a point that splits the response variable into two new nodes, a left and right, where:

$$D(\mu_L) = \sum (y_i - \mu_L)^2 \tag{5.7a}$$

and

$$D(\mu_R) = \sum (y_i - \mu_R)^2 \tag{5.7b}$$

are the deviances of the left (μ_L) and right (μ_R) nodes. The split that maximizes the deviance reduction, defined as:

$$\Delta D(\mu) = D(\mu) - \{D(\mu_L) + D(\mu_R)\} \tag{5.8}$$

is chosen, and the process begins again at the left and right nodes. The result is analogous to a dichotomous key where successive choices are made regarding the value of the response variable, based on predictor characteristics.

A comparison to ANOVA provides some qualitative insight into the CART procedure. ANOVA attempts to answer the question "Is the mean response different among the various levels of the chosen factors?" (Chambers et al. 1992). CART searches for the levels of the factors (as defined by splits) that have different means, and does not restrict the search to additive, globally linear models. When there is just one predictor variable, tree models are step functions, and with two predictors the partitions of a tree model may be plotted on a bivariate plot of the pair of predictor variables.

Some limitations of the method include the fact that it results in discontinuities at the partition boundaries rather than providing smooth transitions between partitions, as well as the inability to provide good approximations to linear and additive functions. Uncertainty regarding selection of the predictor variable upon which to partition the data, as well as the value at which to make the split once the variable is chosen, is dealt with in CART models only one node at a time. This decision is based on maximizing ΔD. Because ΔD is calculated considering only the current node and the resulting daughter nodes, there may be a split of the current node that yields a less than maximum ΔD in the two daughter nodes, but provides a much better (i.e., larger) ΔD in subsequent splits.

5.3.1.1 Bayesian CART and Bayesian Treed Models

These limitations are addressed using the Bayesian approach to the classification and regression tree algorithm (Chipman et al. 1998, Chipman et al. 2002). Instead of the binary recursive partitioning approach of conventional CART models, Bayesian CART (BCART) uses MCMC methods to explore the tree structure. The MCMC approach is computationally more demanding than the recursive partitioning used by conventional CART, but produces trees of lower overall deviance.

The difference between BCART and Bayesian Treed models is the specification of the end node model. BCART (and conventional CART) uses the mean of y value in each leaf as the end node model, while a Bayesian Treed model uses a simple linear regression. By using a richer structure on the terminal nodes, we transfer model structure from the tree to the terminal nodes. We therefore expect smaller, more interpretable trees to result.

It is difficult to find the "best" tree using any tree-based method. The conventional approach to this problem is to use a "greedy" algorithm to "grow" a tree, then "prune" it back to avoid overfitting, as in conventional CART. These greedy algorithms usually grow a tree by sequentially choosing splitting rules for nodes based on maximizing some fitting criterion (Chipman et al. 1998). This approach produces a sequence of trees, all of which are refinements of the previous tree in the sequence severely limiting the exploration of all possible trees. In contrast, the Bayesian approach to CART consists of a prior specification and stochastic search (Chipman et al. 1998), exploring a much richer set of candidate trees. The Bayesian Treed models presented here were fit using software available by download free of charge: (http://faculty. chicagogsb.edu/robert.mcculloch/research/code/CART/index.html).

5.4 Comparing Models

Several criteria have been proposed for use in model comparison and selection. Many proposed criteria have a component that quantifies goodness of model fit, along with a component that penalizes model complexity. Among these criteria are the Akaike Information Criterion (AIC, Akaike 1973), the Bayesian Information Criteria (BIC, Schwarz 1978), and the Deviance Information Criterion (DIC, Spiegelhalter et al. 2002). The goodness of fit component for all these make use of the deviance, $D(y,\theta) = -2\log[p(y|\theta)]$, where y is response data associated with a parameterized model, θ are the parameters of that model, and $p(y|\theta)$ is the likelihood. Note that the likelihood is associated with a specific set of data, y. For this reason, criteria that are based on deviance (and therefore, on the likelihood) should be based on the same data y for all models considered.

The Akaike Information Criterion (AIC) is calculated with the formula

$$AIC = D(y,\theta) + 2p, \qquad (5.9)$$

where p represents the number of parameters in the fitted model. Note that the penalty, $2p$, increases with increasing complexity (number of model parameters), and a model with a lower AIC is preferred over a model with a larger AIC.

The Bayesian Information Criterion, also known as Schwarz's Bayesian criterion (SBC), is calculated according to the formula

$$BIC = D(y,\theta) + p \times \log(n), \qquad (5.10)$$

where p again represents the number of parameters and n the number of observations in the fitted model. As with AIC, a model with a lower BIC is preferred over a model with a larger BIC. Functions for calculation of AIC and BIC are readily available in the R statistical graphics package.

In the context of a Bayesian hierarchical model, the number of independent parameters included in the model is difficult to determine, which makes the use of AIC or BIC problematic. DIC has been proposed for model comparison in this context. Spiegelhalter et al. (2002) show that the effective number of parameters in a complex hierarchical model, p_D, can be computed as

$$p_D = \overline{D(y,\theta)} - D(y,\bar{\theta}). \qquad (5.11)$$

The first term on the right is the mean deviance of the model using all of our estimates (samples) of θ, and the second term on the right is the deviance of the model using only the mean value of our estimate of θ (mean of our samples). The deviance information criterion is then

$$DIC = \overline{D(y,\theta)} + p_D. \qquad (5.12)$$

As with the other criteria, a lower value of DIC is preferred over a higher value. Spiegelhalter et al. (2002) offer guidelines for using DIC to compare competing models similar to those suggested by Burnham and Anderson (1998) for interpretation of differences in AIC between models, such that models within 1–2 of the "best" deserve further consideration and 3–7 have considerably less support. Calculation of DIC is relatively simple in MCMC, and a function for calculating DIC is implemented in WinBUGS.

5.5 Our Analyses

The goal of our analysis is to illustrate the utility of Bayesian and multilevel approaches for constructing models using cross-system data. This particular model is developed to help managers assess target phosphorus concentrations in lakes where data may be sparse. We used data from 382 Michigan lakes and reservoirs (≥ 0.20 km^2), with public access, sampled by the MI-Department of Environmental Quality (Fig. 5.1). Water data were collected from each lake on a single date during the summer stratified season (July–September) for chlorophyll *a*, Secchi depth, total phosphorus, total nitrogen, alkalinity, chloride, dissolved silica, and true water color. We separated the lakes into two types: natural lakes (lakes with or without a water control structure on it, but with little change in water level), and reservoirs (lakes that were created by damming of a river with a range in water residence time of several days to years). We also characterized each lake by surface area, shoreline development factor (ratio of shoreline perimeter to the perimeter of a circle of area equal to the lake), average depth, and catchment area defined as the cumulative catchment (including all connected upstream lakes and streams). Finally, we included ecoregion

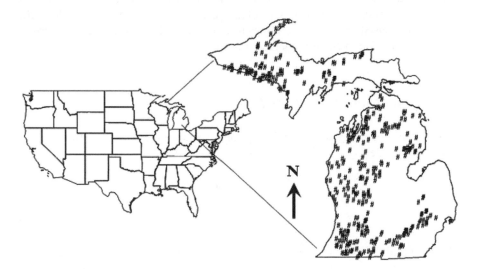

Fig. 5.1 Study area with locations of the 382 Michigan lakes

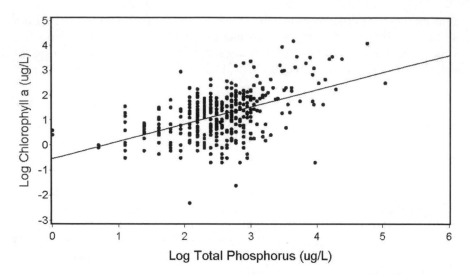

Fig. 5.2 Log Chlorophyll *a* as a function of log Total Phosphorus for the 382 Michigan lakes

classifications including Omernik Level III ecoregion (Rohm et al. 2002), Great Lakes Basin, and landscape position (Martin and Soranno 2006).

Overall, the data exhibit an approximately linear relationship between log of chlorophyll *a* concentration and log of total phosphorus concentration, though there is considerable scatter about a simple regression line fit to the data (Fig. 5.2). This residual scatter will lead to increased uncertainty in the model parameter estimates and predictions, thus, it may be advantageous to subset the data into logical groups that reduce the scatter. We present the results from three models: a completely pooled model with no groups; a model with four groups based on Omerniks ecoregion; and a model with two groups based on the results from a Bayesian Treed analysis.

5.5.1 Completely Pooled Model

Our first model, using complete data pooling, is analogous to the simple linear regression model:

$$\log(\text{chla}) = \alpha + \beta \log(\text{TP}) + \varepsilon \tag{5.13}$$

$$\varepsilon \sim N(0, \sigma^2) \tag{5.13a}$$

where log(chla) is the natural log of the chlorophyll *a* concentration, log(TP) is the natural log of the total phosphorus concentration, α and β are the intercept and slope parameters, respectively, each to be estimated from the data, and ε is an additive model error term that is normally distributed with mean = 0 and variance = σ^2 where σ^2 is also estimated from the data.

To estimate this model using Bayes theorem requires specification of three prior distributions, one for each of the model parameters, α, β, and σ^2. For this example, we programmed WinBUGS using non-informative priors so that the results would not be influenced by information outside of the data. For both α and β, we used a normal prior distribution with mean = 0 and variance = 10,000, and for $1/\sigma^2$ we used a gamma prior with scale and shape parameters both = 0.001. Variances in WinBUGS are specified by their inverses (hence the use of $1/\sigma^2$ which is referred to as the precision) and gamma distributions are a common non-informative prior for the precision. This choice has some caveats as we will illustrate in our second example model.

Because we used non-informative priors, the resulting posterior densities for α and β (Fig. 5.3) convey information closely analogous to what would be obtained using a classical approach for a simple linear regression model. The important difference, however, is that rather than confidence intervals, which cannot be used to infer the probability of any particular result, we can make intuitive probability statements from these posterior distributions. Summary statistics for α, for example, indicate that the mean and median of α's posterior are both –0.55 and the standard deviation of the posterior is 0.16 (Table 5.1). The 2.5 and 97.5 percentiles indicate that there is a 95% probability, given this model and these data, that α has a value between -0.86 and -0.23. Similar inference can be made for β, and σ, as well as for predictions of chlorophyll a in both the log and natural metrics.

Fig. 5.3 Parameter estimates and Chlorophyll predictive distribution for the fully pooled model. The circle represents the estimated mean, the *thick black line* is the 50% credible interval and the *thin black line* is the 95% credible interval

Table 5.1 Summary statistics for completely pooled model

Quantity	mean	median	sd	2.5%	97.5%
α	−0.55	−0.55	0.16	−0.86	−0.23
β	0.69	0.69	0.06	0.57	0.81
σ	0.82	0.81	0.03	0.76	0.88
Log predicted chlorophyll a	1.17	1.17	0.82	−0.42	2.79
Predicted chlorophyll a	4.50	3.23	4.4	0.66	16.3
DIC	930.2				

Figure 5.3 also depicts a predictive distribution for chlorophyll a at a log TP value of 2.5 (the mean value of the data – chosen for convenience). At this log TP value chlorophyll a has 95% probability of being between 0.66 and 16.3 µg/L (Table 5.1). The distribution exhibits a right skew and median and mean values of 3.23 and 4.50 µg/L, respectively, illustrating the separation of the mean and median that occurs when models estimated under a log-transformation are retransformed back to the natural metric (Stow et al. 2006).

5.5.2 Ecoregion Model

Our second model has four groups based on Omernik's ecoregion (Rohm et al. 2002). Ecoregion might be considered a reasonable *a priori* basis for assigning groups, assuming that the ecoregion designation implies features consistent with distinctions in lake behavior. In this case, the model is

$$\log(\text{chla}) = \alpha_j + \beta_j \log(\text{TP}) + \varepsilon \tag{5.14}$$

$$\varepsilon \sim N(0, \sigma^2) \tag{5.14a}$$

where $j = 1$–4 with 175, 27, 179, and only 1 lake in groups 1–4, respectively (note: 1–4 correspond to ecoregions 50, 51, 56, and 55, respectively). A hierarchical structure results by imposing normal prior distributions on the α_j and β_j with respective means μ_α and μ_β and respective variances $\sigma^2\alpha$ and $\sigma^2\beta$. Both μ_α and μ_β were assigned non-informative priors, normal distributions with mean $= 0$ and variance $= 10,000$, while the priors for $1/\sigma^2_\alpha$ and $1/\sigma^2_\beta$ were gamma with both scale and shape parameters $= 0.001$. Note that though each ecoregion has a distinct intercept and slope, a common variance, σ^2, was assigned. We also used a gamma prior with scale and shape parameters $= 0.001$ for $1/\sigma^2$. Additionally, to remove the correlation between the intercepts and slopes we centered the predictor variables, log TP, on their respective means, a step that aids convergence of the algorithm.

Though the means differ, the resultant posterior distributions for α_{1-4} exhibit considerable overlap (Fig. 5.4) as do the posterior distributions for β_{1-4} (Fig. 5.4), and the respective summary statistics for these parameters confirm these similarities among the group-specific parameters (Table 5.2). Predictive distributions for chlorophyll a (using a log TP value of 2.5) in each ecoregion also overlap (Fig. 5.5) with similar summary statistics (Table 5.2).

Additionally, a comparison of the summary statistics for the posterior density of σ from the fully pooled model (Table 5.1) and the ecoregion multilevel model (Table 5.2) reveals that they are almost identical. Together, this body of evidence suggests that the ecoregion multilevel

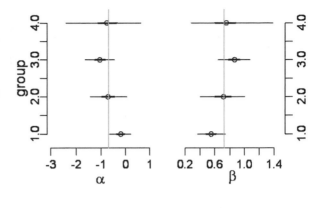

Fig. 5.4 Parameter estimates for the ecoregion multilevel model. The *circle* represents the estimated mean, the *thick black line* is the 50% credible interval and the *thin black line* is the 95% credible interval

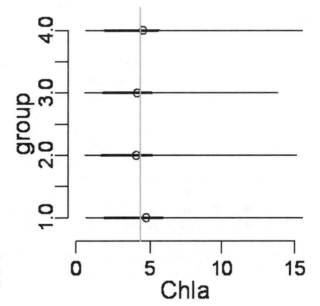

Fig. 5.5 Chlorophyll *a* predictive distribution for the ecoregion multilevel model. The *circle* represents the estimated mean, the *thick black line* is the 50% credible interval and the *thin black line* is the 95% credible interval

model might not represent much improvement over the fully pooled model (Tables 5.1 and 5.2). However, respective DICs of 930.2 and 926.0 for the fully pooled and ecoregion-based models indicate that a slight advantage is conferred by the more complex structure of this particular multilevel model. While this should not be interpreted to mean that the ecoregion-based model is a better forecasting tool, it does imply that the ecoregion model does somewhat better at predicting the existing data set than the fully pooled model does.

Although the ecoregion-based multilevel model does not represent a major improvement over the fully pooled model, it does illustrate some potentially important capabilities of multilevel models. In this model,

Table 5.2 Summary statistics for ecoregion model – gamma prior

Quantity	Mean	Median	Sd	2.5%	97.5%
α_1	−0.20	−0.20	0.23	−0.64	0.22
α_2	−0.71	−0.72	0.38	−1.43	0.06
α_3	−1.04	−1.04	0.30	−1.64	−0.45
α_4	−0.77	−0.72	0.72	−2.41	0.61
β_1	0.55	0.55	0.09	0.37	0.74
β_2	0.72	0.72	0.16	0.41	1.00
β_3	0.86	0.86	0.11	0.65	1.07
β_4	0.76	0.74	0.26	0.29	1.38
σ	0.81	0.81	0.03	0.75	0.86
Predicted chlorophyll a 1	4.25	3.13	4.06	0.74	14.08
Predicted chlorophyll a 2	4.02	3.04	3.57	0.55	13.32
Predicted chlorophyll a 3	4.39	3.24	4.19	0.64	14.12
Predicted chlorophyll a 4	4.33	2.94	4.52	0.60	16.69
DIC	926.0				

ecoregion group 4 (ecoregion 55) contains data from only one lake; therefore the group 4 parameters are based on extremely sparse group-specific data. This is possible because partial pooling in multilevel estimation permits information sharing among the groups. In this case, considerable information is borrowed from the other groups to estimate the group 4 parameters. Thus, multilevel models represent a powerful approach to estimate group specific, or in this example even lake specific models with limited group or lake specific information. However, there are some important caveats associated with this capability, some of which are subtle. Conceptually, it seems apparent that estimating a model with one or only a few observations could be risky, particularly if the observations represent unusual circumstances for that group or lake. And because this is a relatively new approach, it has not been extensively applied and tested to evaluate whether it provides better predictions in this particular context. But in addition, Gelman (2006) warns that some non-informative priors for the variance can actually be quite informative when groups contain sparse data or when there are only a few groups in the multilevel model. In such instances, Gelman (2006) recommends using a uniform prior for σ instead of a gamma prior for $1/\sigma^2$ to ensure that the prior is truly non-informative. When this change is made the differences, in this instance, are slight, but worth noting.

    With the new non-informative prior, the estimated means and medians for α_{1-4} and β_{1-4} differ slightly (Table 5.3) from those resulting from the gamma prior (Table 5.2) while σ is essentially unchanged. However, the standard deviation of α_4 increases from 0.72 (Table 5.2) to 1.14 (Table 5.3). The overall change in the parameter distributions is visually more apparent by comparing Figs. 5.4 and 5.6. As a result of the increased uncertainty in α_4, the width of the 95% credible interval for a prediction

Fig. 5.6 Parameter estimates for the ecoregion multilevel model, using a uniform prior distribution for sigma. The *circle* represents the estimated mean, the *thick black line* is the 50% credible interval and the *thin black line* is the 95% credible interval

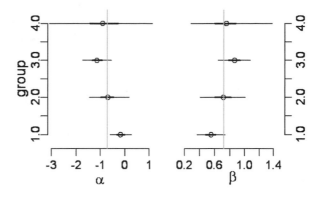

Fig. 5.7 Chlorophyll *a* predictive distribution for the ecoregion multilevel model, using a uniform prior distribution for sigma. The *circle* represents the estimated mean, the *thick black line* is the 50% credible interval and the *thin black line* is the 95% credible interval

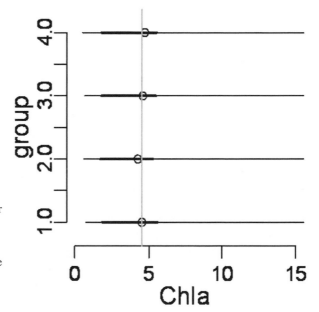

of chlorophyll *a* also increases for group 4, while the prediction for groups 1–3 change little (Table 5.3 and Fig. 5.7). Thus this example illustrates that the gamma prior, in groups with few observations, tends to be somewhat informative and may cause the parameter and predictive uncertainty to be under-estimated in those groups. Additionally, the DIC goes up from 926.0 using the gamma prior (Table 5.2) to 926.5 with the uniform prior (Table 5.3), indicating that the improved performance of the ecoregion-based model relative to the completely pooled model (DIC = 930.2, Table 5.1), may have been slightly overstated.

Table 5.3 Summary statistics for ecoregion model – uniform prior

Quantity	Mean	Median	sd	2.5%	97.5%
α_1	−0.17	−0.16	0.21	−0.56	0.23
α_2	−0.70	−0.7	0.41	−1.51	0.15
α_3	−1.12	−1.14	0.30	−1.70	−0.50
α_4	−0.85	−0.80	1.14	−3.1	1.1
β_1	0.55	0.55	0.09	0.37	0.74
β_2	0.72	0.72	0.16	0.41	1.01
β_3	0.86	0.86	0.11	0.65	1.07
β_4	0.76	0.74	0.26	0.29	1.38
σ	0.81	0.81	0.03	0.75	0.87
Predicted chlorophyll a 1	4.44	3.26	3.91	0.71	15.29
Predicted chlorophyll a 2	4.13	2.30	3.96	0.72	15.15
Predicted chlorophyll a 3	4.27	3.20	3.73	0.60	14.09
Predicted chlorophyll a 4	4.70	3.17	5.01	0.61	19.05
DIC	926.5				

5.5.3 Bayesian Treed Model

The Bayesian Treed model search resulted in the following tree structure:

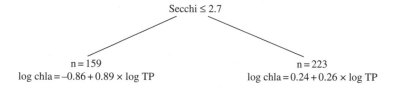

This tree structure indicates that the 382 lakes in the data set can be partitioned into two groups, 159 with a secchi depth of ≤ 2.7 meters and 223 with a secchi depth of > 2.7 meters, and that these two groups differ in their log chlorophyll a:log TP relationship (Fig. 5.8). This outcome is consistent with results presented by Webster et al. (2008) which showed that regression models predicting chlorophyll *a* from TP are improved with water color (a strong determinant of secchi depth) included as a predictor variable.

Using this result we developed the model:

$$\log(\text{chla}) = \alpha_j + \beta_j \log(\text{TP}) + \varepsilon \tag{5.15}$$

$$\varepsilon \sim \text{N}(0, \sigma^2) \tag{5.15a}$$

where j = 1–2, with 159 and 223 lakes in groups 1 (turbid) and 2 (clear), respectively. We built a multilevel model, using these two groups, in a manner analogous to the structure used for the ecoregion-based model. For this example we, again, centered the predictor variables, log TP, on their respective means

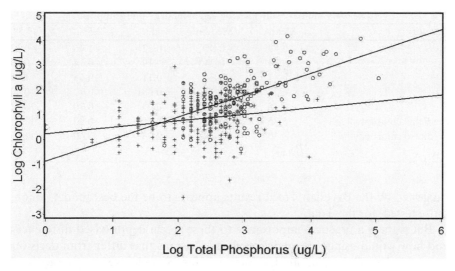

Fig. 5.8 Log Chlorophyll *a* as a function of log Total Phosphorus for the turbid lakes (*n* = 159, *circles, steeper line*) and clear lakes (*n* = 223, *plus signs, less steep line*)

(note this may cause the parameter estimates to differ slightly from those depicted in the displayed tree) and used a uniform prior for σ instead of the gamma prior on $1/\sigma^2$.

In this case, a comparison of the parameter posterior distributions indicates that α_1 and α_2 have dissimilar means (Table 5.4) and minimal overlap (Fig. 5.9). Similarly, the posterior distributions for β_1 and β_2 overlap minimally (Fig. 5.9) which is confirmed by comparing their respective means and standard deviations (Table 5.4). The mean value of σ in this model = 0.75 (Table 5.4), a reduction from 0.82 and 0.81 (Tables 5.1 and 5.3) for the fully pooled and ecoregion-based models, suggesting a better fit to the data. This is supported by a DIC of 874.6 (Table 5.4) as compared to DIC values of 930.2 and 926.5 for the first two models. Based on this evidence, the secchi depth-based model

Fig. 5.9 Parameter estimates for the multilevel model based on secchi depth, identified with the Bayesian Treed Model search. The *circle* represents the estimated mean, the *thick black line* is the 50% credible interval and the *thin black line* is the 95% credible interval

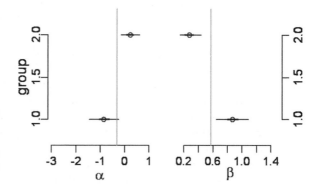

Table 5.4 Summary statistics for secchi model – uniform prior

Quantity	Mean	Median	sd	2.5%	97.5%
α_1	−0.86	−0.87	0.32	−1.46	−0.23
α_2	0.24	0.243	0.19	−0.14	0.60
β_1	0.87	0.88	0.11	0.65	1.08
β_2	0.27	0.27	0.081	0.12	0.44
Σ	0.75	0.75	0.03	0.70	0.81
Predicted chlorophyll a 1	5.36	4.13	4.55	0.93	17.51
Predicted chlorophyll a 2	3.30	2.46	2.93	0.60	11.09
DIC	874.6				

suggested by the Bayesian Treed results appears to be the best model, among those tested, for these data.

But is there a practical importance to these secchi depth-based differences, and how would using this model result in decisions that differ from decisions based on the fully pooled model? Suppose, for example, we are charged with evaluating whether or not a log TP concentration of 2.5 (12.2 μg/L in the natural metric) is sufficient to meet a state chlorophyll a criterion of 10 μg/L. Guidance provided by the U.S. Environmental Protection Agency suggests that numerical criteria should be regarded as an acceptable 90th percentile, recognizing that ambient concentrations vary spatially and temporally (U.S. Environmental Protection Agency 2000). Thus, to answer this question, we need to evaluate if these models differ in predicting whether the probability of exceeding 10 μg/L chlorophyll a is greater than 10%.

Both the mean and median predicted chlorophyll a values for secchi groups 1 and 2 are well below 10 μg/L (Table 5.4), as are the mean and median values for the fully pooled model (Table 5.1). Considering only these values of central tendency, which is often done in practice, would result in the conclusion that both models support the notion that 12.2 is an acceptable target TP concentration to meet the chlorophyll a criterion. Further, chlorophyll a predictive distributions for groups 1 and 2 from the secchi depth-based model exhibit considerable uncertainty, signifying a wide range of possible chlorophyll a values. They also display considerable overlap (Fig. 5.10) suggesting minimal differentiation of predicted outcomes. However, the probability of exceeding 10 μg/L for group 1 is ~12% while for group 2 this probability is ~3%. The fully pooled model, on the other hand, predicts an 8% probability that chlorophyll a will exceed 10 μg/L at a 12.2 μg/L TP level. Thus, according to the fully pooled model 12.2 μg/L is an acceptable target TP concentration for all lakes, while the secchi-based model indicates that only lakes in group 2 will meet that criterion at this TP concentration. Based on the much lower DIC value for the secchi-based model we would likely place more credence in this model's predictions.

Suppose, however, we want to evaluate a lake for which secchi depth measurements are not readily available, so that group membership is unclear. Using

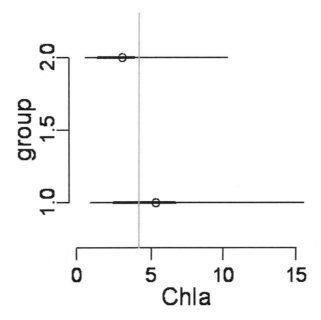

Fig. 5.10 Chlorophyll *a* predictive distribution for multilevel model based on secchi depth, identified with the Bayesian Treed Model search. The *circle* represents the estimated mean, the *thick black line* is the 50% credible interval and the *thin black line* is the 95% credible interval

the fully pooled model this would not be a problem, because knowing group membership is unnecessary. But the preferred model in this case is the secchi-based model and there are two possible parameter sets available – which of these is more appropriate for a lake with unknown group membership?

In principle, the answer to this question is that we would use the hyper-parameters, which are defined by the prior distributions on α_j and β_j. In practice, Gelman and Hill (2007) warn that, with only two groups, the hyper-parameters will not be well-estimated and are likely to be overly variable. Thus, the advantage of a two-group multilevel model is limited in this respect. For-tunately, at least for this example, secchi depth is probably the easiest limnolo-gical measurement to obtain.

5.6 Summary

Our example illustrates some advantages as well as some limitations of using Bayesian multilevel models for inference and prediction with cross-system lake data. Multilevel modeling is a rigorous basis for partial pooling of information among similar, but non-identical groups, and Bayesian approaches provide a framework for uncertainty analysis, an important ingredient for environmental decision support. The explicit inclusion of quantified uncertainty, expressed probabilistically, provided by Bayesian predictive distributions, underscores the reality that model predictions are probably better regarded as testable hypotheses rather than forecasts of the future.

The importance of models and their limitations were explicitly recognized in the development of the Adaptive Management concept (Holling 1978, Lee 1993). Adaptive Management views environmental stewardship as an ongoing process; models offer guidance for decision-making, and management actions serve as an ecosystem-scale experiment to learn more about system behavior. Measuring the response of the ecosystem to this management experiment supplies data to confirm and update the model and refine future projections. The adaptive process is not well-facilitated by the rather static view of classical null hypothesis-testing in which decisions are limited to accept or reject. However, these ideas are consistent with Bayesian inference, and Bayes theorem provides a rigorous framework for model updating and refinement to implement the Adaptive Management paradigm.

References

Akaike, H., 1973. Information theory and an extension of the maximum likelihood principle, In Proc. 2nd Int. Syp. Information Theory (eds. B.N. Petrov and F. Csáki), pp. 267–281. Budapest: Akadémiai Kiadó.

Breiman, L., J. H. Friedman, R. A. Olshen, and C. J. Stone. 1984. Classification and Regression Trees. Chapman & Hall. New York, NY.

Burnham, K. P., and D. R. Anderson. 1998. Model Selection and Inference, New York: Springer.

Canfield, D. E., and R. W. Bachmann. 1981. Prediction of total phosphorus concentrations, chlorophyll *a*, and secchi depths in natural and artificial lakes. Canadian Journal of Fisheries and Aquatic Sciences **38**: 414–423.

Chambers, J. M., A. E. Freeny, and R. M. Heiberger 1992. Analysis of Variance: designed experiments, in /Statistical Models in S/, J. M. Chambers and T. J. Hastie (eds.), Wadsworth and Brooks/Cole Advanced Books and Software, Pacific Grove, California.

Chipman, H. A., E. I. George, and R. E. McCulloch. 1998. Bayesian CART model search. Journal of the American Statistical Association **93**: 935–948.

Chipman, H. A., E. I. George, and R. E. McCulloch. 2002. Bayesian treed models. Machine Learning **48**: 299–320.

Clark, L. A., and D. Pregibon, 1992. Tree Based Models, in *Statistical Models in S*, J. M. Chambers and T. J. Hastie (eds.), Wadsworth and Brooks/Cole Advanced Books and Software, Pacific Grove, California.

Cole, J., G. Lovett, and S. Findlay (eds.). 1991. Comparative analyses of ecosystems: patterns, mechanisms and theories. Springer-Verlag. 375 pp.

Dennis, B. 1996. Discussion: Should ecologists become Bayesians? Ecological Applications **6**: 1095–1103.

Ellison, A. M. 2004. Bayesian inference in ecology. Ecology Letters **7**: 509–520.

Freeman, A. M., E. C. Lamon, and C. A. Stow. 2008. Regional nutrient and chlorophyll *a* relationships in lakes and reservoirs: A Bayesian TREED model approach. Ecological Modelling. in press.

Gelman, A. 2006. Prior distributions for variance parameters in hierarchical models. Bayesian Analysis 1: 515–533.

Gelman, A., and J. Hill. 2007. Data Analysis Using Regression and Multilevel/Hierarchical Models, Cambridge University Press, NY.

Hession, W. C., D. E. Storm, S. L. Burks, M. D. Smolen, R. Lakshminarayanan, and C. T. Haan. 1995. Using Eutromod with a GIS for establishing total maximum daily loads to Wister Lake, Oklahoma. Pages 215–222 in K. Steele, editor. Animal Waste and the Land-Water Interface. Lewis, Boca Raton.

Holling, C. S. ed. 1978. Adaptive Environmental Assessment and Management. John Wiley & Sons. NY.

Hubbard, R., and M. J. Bayarri. 2003. Confusion over measures of evidence (p's) versus errors (α's) in classical statistical testing. American Statistician 57: 171–178.

Hubbard, R., and J. S. Armstrong. 2006. Why we don't really know what "statistical significance" means: A major educational failure. Journal of Marketing Education 28: 114–120.

Lamon E. C., O. Malve, and O-P. Pietiläinen. 2008. Lake classification to enhance prediction of eutrophication endpoints in Finnish lakes, Environmental Modeling and Software. 23: 947–948.

Lamon, E. C., and C. A. Stow. 1999. Sources of variability in microcontaminant data for Lake Michigan salmonids: Statistical models and implications for trend detection. Canadian Journal of Fisheries and Aquatic Sciences 56, Supplement 1: 71–85.

Lamon, E. C., and C. A. Stow. 2004. Bayesian methods for regional-scale lake eutrophication models. Water Research 38: 2764–2774.

Lathrop, R. C., S. R. Carpenter, C. A. Stow, P. A. Soranno, and J. C. Panuska. 1998. Phosphorus loading reductions needed to control blue-green algal blooms in Lake Mendota. Canadian Journal of Fisheries and Aquatic Sciences 55: 1169–1178.

Lee, K. N. 1993. Compass and Gyroscope. Island Press. Washington, DC.

Lunn, D. J., A. Thomas, N. Best, and D. Spiegelhalter. 2000. WinBUGS - A Bayesian modelling framework: Concepts, structure, and extensibility. Statistics and Computing 10: 325–337.

Magnuson J. J., W. M. Tonn, A. Banerjee, J. Toivonen, O. Sanchez, M. Rask. 1998. Isolation vs. extinction in the assembly of fishes in small northern lakes. Ecology 79: 2941–2956

Malve, O., and S. S. Qian. 2006. Estimating nutrients and chlorophyll a relationships in Finnish lakes. Environmental Science & Technology 40: 7848–7853.

Martin, S. L., and P. A. Soranno. 2006. Defining lake landscape position: Relationships to hydrologic connectivity and landscape features. Limnology and Oceanography 51: 801–814.

National Research Council. 2001. Assessing the TMDL Approach to Water Quality Management. National Academy Press, Washington, D.C.

Pappenberger, F., and K. J. Beven. 2006. Ignorance is bliss: Or seven reasons not to use uncertainty analysis. Water Resources Research 42. Article number W05302.

Qian, S. S., C. A. Stow, and M. E. Borsuk. 2003. On Monte Carlo methods for Bayesian inference. Ecological Modelling 159: 269–277.

Qian, S. S., and C. W. Anderson. 1999. Exploring factors controlling the variability of pesticide concentrations in the Willamette River Basin using tree-based models. Environmental Science & Technology 33: 3332–3340.

Reckhow, K. H. 1988. Empirical-models for trophic state in Southeastern United-States lakes and reservoirs. Water Resources Bulletin 24: 723–734.

Reckhow, K. H. 1990. Bayesian-inference in non-replicated ecological studies. Ecology 71: 2053–2059.

Reckhow, K. H. 1993. Random coefficient model for chlorophyll nutrient relationships in lakes. Ecological Modelling 70: 35–50.

Reckhow, K. H. 1996. Improved estimation of ecological effects using an empirical Bayes method. Water Resources Bulletin 32: 929–935.

Reckhow, K. H., and S. C. Chapra. 1983. Engineering approaches for lake management, Volume 1: Data Analysis and Empirical Modeling. Butterworth Publishers. Boston.

Rohm C. M., J. M. Omernik, A. J. Woods, and J. L. Stoddard. 2002. Regional characteristics of nutrient concentrations in streams and their application to nutrient criteria development. Journal of the American Water Resources Association **38**: 213–239.

Salsburg, D. 2001. The Lady Tasting Tea. Henry Holt and Company. NY.

Schwarz, G., 1978. Estimating the dimension of a model, Annals of Statistics **6**:461–464.

Sonquist, J. N., and J. N. Morgan. 1964. The Detection of Interaction Effects. Monograph 35, Survey Research Center, Institute for Social Research, University of Michigan, Ann Arbor, MI.

Spiegelhalter, D. J., Best, N. G., Carlin, B. P., and van der Linde, A., 2002. Bayesian measures of model complexity and fit. Journal of the Royal Statistical Society, Series B **64**: 583–639.

Stow, C. A., S. R. Carpenter, and R. C. Lathrop. 1997. A Bayesian observation error model to predict cyanobacterial biovolume from spring total phosphorus in Lake Mendota, Wisconsin. Canadian Journal of Fisheries and Aquatic Sciences **54**: 464–473.

Stow, C. A., S. R. Carpenter, K. E. Webster, and T. M. Frost. 1998. Long-term environmental monitoring: Some perspectives from lakes. Ecological Applications **8**: 269–276.

Stow, C. A., K. H. Reckhow, and S. S. Qian. 2006. A Bayesian approach to retransformation bias in transformed regression. Ecology **87**: 1472–1477.

Stow, C. A., and D. Scavia. 2008. Modeling Hypoxia in the Chesapeake Bay: Ensemble estimation using a Bayesian hierarchical model. Journal of Marine Systems. In press.

U.S. Environmental Protection Agency. 2000. Nutrient Criteria Technical Guidance Manual, Lakes and Reservoirs. Office of water. EPA 822-B00-001.

Vollenweider, R. A. 1968. The Scientific Basis of Lake and Stream Eutrophication with Particular Reference to Phosphorus and Nitrogen as Eutrophication Factors. Technical Report DAS/DSI/68.27. OECD, Paris.

Vollenweider, R. A. 1969. Possibilities and limits of elementary models concerning the budget of substances in lakes. Archiv für Hydrobiologie **66**: 1–36.

Vollenweider, R. A. 1975. Input-output models with special reference to the phosphorus loading concept in limnology. Schweizerische Zeitschrift fur Hydrologie **37**: 53–84.

Vollenweider, R. A. 1976. Advances in defining critical loading levels for phosphorus in lakes eutrophication. Mem. Ist. Ital. Idrobiol. **33**:53–83.

Webster, K. E., P. A. Soranno, K. S. Cheruvelil, M. T. Bremigan, J. A. Downing, P. Vaux, T. Asplund, L. C. Bacon, and J. Connor. 2008. An empirical evaluation of the nutrient color paradigm for lakes. Limnology and Oceanography. **53**:1137–1148.

Winkler, R. L., 2003. An Introduction to Bayesian Inference and Decision, Second Edition, Probabilistic Publishing, Gainesville, FL, USA.

Chapter 6
Avian Spatial Responses to Forest Spatial Heterogeneity at the Landscape Level: Conceptual and Statistical Challenges

Marie-Josée Fortin and Stephanie J. Melles

6.1 Introduction

Forest ecosystems are increasingly fragmented directly by human activities, such as agriculture, forestry, transportation networks (Forman et al. 2003) and urbanization (Fahrig 2003, Melles et al. 2003). These practices interact with additional indirect anthropogenic impacts on biodiversity such as climate change (Thomas et al. 2004), pollution, and species homogenization (Olden 2006). These global disturbances can cause local extirpation and global species extinction (Gaston 2003). While land-use change fragments, or completely destroys species habitat, climate change modifies, shifts, reduces, and destroys the amount of habitat available to species (Pyke 2004). The major types of land-use change that affect species' survival include those that alter land cover, reducing suitable habitat (e.g., from forest to crop land or urban areas), and those that fragment species' habitat (e.g., road networks), increasing mortality rates and constraining animal movements (Forman et al. 2003). Climate change can affect both the size of a species' geographical range through range expansion and contraction as well as the limits of a species' geographic distribution by shifting their boundaries (Parmesan 1996, Jetz and Rahbek 2002, Gaston 2003, Fortin et al. 2005). Habitat alteration resulting from manifold global changes occurs at several spatial scales and has the potential to negatively impact wildlife diversity and species persistence.

To help conserve species, we need to understand how species respond to landscape spatial heterogeneity at multiple spatial scales. This implies that it is necessary to study species spatial responses to habitat loss and fragmentation locally (Bélisle et al. 2001) and regionally (Kerr and Cihlar 2004). The local distribution of a species is the end result of composite processes, including population demographics, behavioral traits, and physiological tolerances. These processes create inherent spatial patterns that result from biotic factors,

M.-J. Fortin (✉)
Department of Ecology and Evolutionary Biology, University of Toronto, 25 Harbord St., M5S 3G5, Toronto, Ontario, Canada
e-mail: mariejosee.fortin@utoronto.ca

S. Miao et al. (eds.), *Real World Ecology*, DOI 10.1007/978-0-387-77942-3_6,
© Springer Science+Business Media, LLC 2009

referred to as spatial autocorrelation, and abiotic factors, referred to as spatial dependence (Fortin and Dale 2005, Wagner and Fortin 2005). Spatial auto-correlation and spatial dependence are always confounded and the only way to tease them apart is to undertake intensive field work (Bélisle 2005, Ewers and Didham 2006). When the underlying abiotic and biotic processes are unknown, spatial structure is typically referred to as spatial autocorrelation regardless of its source and this is the convention that we will use. At the regional scale, the geographic distribution of a species is often most strongly related to physical characteristics such as climate, habitat availability, and landuse (i.e., spatial dependence is most apparent at large scales).

The goal of this chapter is to explore the relationship between the spatial distribution of ovenbirds (*Seiurus aurocapilla*) and a gradient of forest cover on a regional scale. To do so, we first summarize the conceptual and statistical challenges that emerge with spatially autocorrelated data. We then use a series of spatial statistics to determine the inherent spatial structure of the data. Equipped with these findings, two regression methods are highlighted, geographically weighted regression (GWR) and regression kriging (RK), to compare how they account for spatial autocorrelation and the insights they provide. Spatial autocorrelation is a feature that can be captured in a model to provide a better understanding of the relationships within an ecosystem.

6.2 Conceptual and Statistical Issues

6.2.1 Several Processes: Several Spatial Scales Versus Sub-regions

Without a process there would be no pattern. But what is a spatial pattern exactly? The term "spatial pattern" has several connotations ranging from the presence of structure to spatial heterogeneity (i.e., a non random spatial arrangement of quantitative or qualitative data in a repeating or characteristic way). The notion of spatial heterogeneity is better contrasted, however, with spatial homogeneity, which is in turn quite rare and more of a conceptual null model than something that exists in reality (Fischer and Lindenmayer 2007). Spatial heterogeneity, like stationarity, is a concept that depends on scale (Fortin and Dale 2005). At large extents and coarse resolutions, a pattern may appear to be homogeneous, whereas at small extents and finer spatial resolutions, heterogeneity emerges. Ironically, in the absence of pattern there is no spatial, or temporal scale, so examination of the end result of a process, i.e., analysis of a spatial pattern, helps to determine the spatial scale of that process (Dungan et al. 2002).

When several processes act together, what is the resulting spatial pattern? On one hand, the processes can interact to create non-additive spatial patterns characteristic of non-stationary processes. Stationarity implies that the under-lying process(es) are constant across the entire region of interest and therefore that the species–environment relationship does not change with location or scale (Fortin and Dale 2005). Most modeling approaches that account for residual spatial autocorrelation assume that the underlying processes are stationary, such that parameter values are constant over the study area (Fortin and Dale 2005). While it is impossible to test directly whether or not an underlying process is stationary, it is possible to examine and map model residuals across the region of interest. If there is a heterogeneous variance structure to the residuals, or they are otherwise non-normal, then this structure often indicates that the assumption of stationarity is violated. In this case, it may be necessary to subdivide the data into spatially homogeneous subregions in terms of mean or variance so that each subregion can be analyzed separately (see Section 6.4.2).

On the other hand, if we are lucky, the resulting spatial pattern is additive (i.e., the sum of patterns generated by each process), such that there are large scale trends due to environmental gradients, patches resulting from species dispersal, and some random noise. Additive spatial patterns can be analyzed by removing or detrending the effects of each key spatial pattern in dependent and independent variables (if appropriate) and then quantifying the relationships using the residuals (Fortin and Gurevitch 2001). For example, a large scale trend due to an environmental gradient can be removed. If some spatial pattern remains in the residuals, i.e., if patches are present, then the data can be detrended again with another type of structure (quadratic, cubic, etc.) so that only random noise remains. While in theory detrending is an elegant solution, in practice it is difficult to be certain that only information about the trend and no other patterns were removed. This is the reason that, in the absence of prior knowledge or hypothesis about the process(es) and their resulting pattern, several authors recommend avoiding detrending the data (Osborne and Suarez-Seoane 2002, Fortin and Dale 2005). In general, spatial autocorrelation in the dependent variable should not be viewed as a problem that needs to be corrected prior to analysis. Effectively, spatial autocorrelation typically includes the variation that we are most interested in understanding and modeling. However, autocorrelation in the residuals is often overlooked, resulting in misleading inference using classical parametric estimators and statistical tests.

Another way to analyze spatially autocorrelated data is to treat space as a predictor (Dray et al. 2006). To do so, spatial autocorrelation is decomposed using a Fourier spectral decomposition approach (Legendre and Legendre 1998) or a wavelet approach (Keitt and Urban 2005). Both of these approaches require complete coverage of evenly spaced data. Uneven or sparsely sampled data that is not on a regular grid (i.e., incomplete coverage) can be analyzed using principal coordinates of neighbor matrices (Borcard and Legendre 2002, Borcard et al. 2004, Dray et al. 2006), spatial eigenvector mapping (Griffith and Peres-Neto 2006) or nested kriging (Wackernagel 2003, Bellier et al. 2007). In either case (i.e., with or without complete spatial coverage data), when the region under study is large, the number of sampled locations is usually immense, such that decomposition methods will generate as many spatial predictors as there are samples. It is therefore necessary to determine the relevant spatial scales based either on a priori hypotheses of the key processes or statistical methods (Keitt and Urban 2005, Dray et al. 2006, Bellier et al. 2007).

6.2.2 Which Methods?

It is important to select appropriate statistical method(s) when the data are spatially autocorrelated (Dale and Fortin 2002, Dungan et al. 2002, Fortin and Dale 2005). Spatial statistics, however, cannot discriminate between induced spatial dependence (i.e., marginal autocorrelation due to environmental gradients) and inherent spatial autocorrelation (i.e., conditional autocorrelation due to species dispersal). Only prior knowledge and hypotheses tests of the underlying processes responsible for spatial patterns can help to differentiate them.

Spatial exploratory data analysis is a crucial first step that helps to determine the scale(s) of the pattern(s), which in turn aid in the selection of appropriate statistical methods. Most, if not all, ecological data show some degree of spatial autocorrelation due to the inherent spatial structure of biotic and induced abiotic processes (Wagner and Fortin 2005). In such circumstances, the independent variable (e.g., forest cover gradient) in a parametric linear regression, like ordinary least squares regression, may not capture all the spatial structure of the dependent variable (e.g., species abundance) such that the residuals still have some degree of spatial autocorrelation, i.e., they are not independent. This lack of residual independence violates the assumption of parametric statistics that errors should be independent (Dale and Fortin 2002). Residual autocorrelation is a problem for many classical estimators that are conditioned on the assumption of independence because if the assumption is not met (or nearly met) then the properties of the estimators, on which inference is based, may not hold and parameter significance can be inflated. However, it is the presence of spatial pattern(s) that can lead to greater understanding and insights about the data and the underlying processes that generate them (Fortin and Dale 2005).

We need, therefore, to use models that can explicitly account for spatial autocorrelation present in the data, either by modeling the spatial structure or by removing it. Several spatially explicit methods (e.g., conditional autoregressive models, simultaneous autoregressive models) that account for autocorrelated residuals are available (Keitt et al. 2002, Lichstein et al. 2002, Guisan and Thuiller 2005, Austin 2007, Dormann et al. 2007). There is also a family of methods (e.g., CART, random forest, boosted regression) that can deal with non-stationarity such as the regression tree approaches (Hastie et al. 2001, De'Ath 2002, Elith et al. 2006) and Bayesian spatial methods (Clark and Gelfand 2006, Diggle and Søren 2006, Latimer et al. 2006).

Here, we compare and contrast two spatial methods which may be of particular interest to ecologists: regression kriging, RK (Odeh, McBratney and Chittleborough 1995, Hengl et al. 2004, Hengl et al. 2007), and geographically weighted regression, GWR (Fotheringham et al. 2002). We show how the results of these spatially explicit modeling methods compare with an aspatial equivalent (ordinary linear regression by least squares). We have selected to highlight these two spatially explicit methods for a few reasons: (1) they do not require data with continuous spatial coverage and therefore they are spatially flexible, and (2) each of these methods has its own particular advantages, which will become clear in our examination of the case study below.

6.3 Case Study: Distribution of *S. aurocapilla* in Relation to Forest Cover

For conservation reasons, it is becoming increasingly important to understand how species respond to changes in the environment over a large region. Our overall goal here is to understand how the spatial distribution of the ovenbird is affected by a forest cover gradient in southern Ontario, Canada. To do this, we need first to explore the statistical and spatial characteristics of forest cover and bird abundance data separately (Section 6.4). Once these characteristics have been determined, appropriate statistics can be used to quantify the relationship between species distribution, abundance patterns, and forest cover (Section 6.5).

6.3.1 Study Area

Southern Ontario has undergone a process of deforestation over the last century and a half, which has resulted in a gradient of forest cover. Our case study area covers southern Ontario ($\sim$42° to 46°N and 73° to 83°W, Fig. 6.1). This area consists of three main ecoregions characterized by differences in climatic variables and vegetation: Lake Erie-Lake Ontario, Lake Simcoe-Rideau, and Georgian Bay. The geographical range limits for several species occur in southern Ontario, and climate warming is expected to have an impact on many of these species distributions (Bowman et al. 2005, Melles 2007). Monitoring the distributions of species at regional scales will therefore play an important role in future conservation efforts. Extreme southern Ontario in the Lake Erie-Lake Ontario region is highly fragmented by agricultural lands, urbanization, and exurban development. The remaining and predominantly deciduous forest cover in this region ranges from 0.25 to 42.2% cover per 10×10 km squares (Fig. 6.1, Melles 2007). Here we use sampling units of 10×10 km squares to match the Atlas bird data (see below). In the rest of southern Ontario, south of the Canadian Shield, mixed deciduous and coniferous forests occur, fragmented by urban centers. This creates a gradient of forest cover in southern Ontario both from west to east and from south to north (Fig. 6.1).

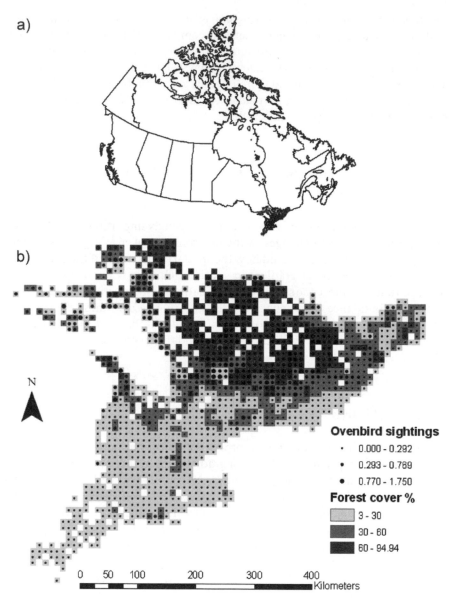

Fig. 6.1 (**a**) Map of Canada where the study area in southern Ontario is shown in *black*. (**b**) Study area extent showing the mean number of ovenbirds observed per point count (*proportional circles*) and percent forest cover (*grey shade squares*) per Atlas square 10×10 km (*n* = 1359)

6.3.2 Bird Data

Our model species, *S. aurocapilla*, is a neotropical migrant warbler that nests on the ground and has an insectivorous diet. This species was selected because it is a mature forest obligate that is sensitive to forest loss and forest fragmentation (e.g., Burke and Nol 1998). Ovenbirds breed in mature deciduous, mixed deciduous, and coniferous forests throughout eastern North America, ranging from the eastern deciduous forests of the United States to the Boreal forests of Canada. The habitat preferences of *S. aurocapilla* span all three of the ecoregions in southern Ontario.

Data reflecting the distribution and abundance of *S. aurocapilla* in southern Ontario were acquired from the Ontario Breeding Bird Atlas (OBBA) project (Ontario Breeding Bird Atlas Interim Database, Bird Studies Canada et al. 2006a). The OBBA was designed to assess the current distribution and abundance of bird species in Ontario at a 10×10 km spatial resolution. The Atlas is a volunteer-based, province-wide survey of breeding birds, conducted over a 5-year period (Cadman et al. 1987, Bird Studies Canada et al. 2006b). The first Ontario bird Atlas was initiated in 1981, covering the years 1981–1985, and the Atlas project is repeated every 20 years. We selected the most recent Atlas, covering the years 2001–2005 because these are the best available data to represent the current distribution and abundance of *S. aurocapilla*, and because corresponding land cover data were available for a roughly comparable time period (1991–1998). OBBA birds were surveyed aurally by volunteers for 5 min intervals at specified randomly selected points along road-side (and some off-road) locations within 10×10 km squares. Point count sampling was designed to attain a representative sample of the different habitat types present in each Atlas square. Both a limited radius (100 m) and an unlimited radius point count methodology were followed with a minimum interpoint distance of 250 m. We selected 100 m distance counts for relative abundance in this application because ovenbird song carries quite far and we wanted to reduce the potential of counting the same birds twice. In addition, the number of birds detected within 100 m should be less affected by observer skill and hearing than unlimited distance counts (Bird Studies Canada et al. 2006b). The target number of counts per square for the Atlas was set at 25 and in southern Ontario almost every 10×10 km square was covered with at least 20 h of volunteer survey effort (1678 squares). Following Atlas protocol (Bird Studies Canada et al. 2006b), only those squares with at least 10 point counts were included in our analysis ($n = 1359$) because mean abundance estimates per square did not change much with additional point counts. Point counts were conducted by volunteers during the early morning peak singing period in fair weather conditions (Bird Studies Canada et al. 2006b).

6.3.3 Forest Spatial Heterogeneity

Percent forest cover was determined from the most recent, publicly available landcover dataset for the province of Ontario derived from LANDSAT TM remotely sensed data (25 m grid resolution) recorded between 1986 and 1997

(OMNR 1991–1998). The majority of satellite data frames were recorded in the early 1990s (OMNR 1991–1998). Forested habitat included all coniferous, deciduous, and mixed woodland categories of the landcover data, but sparsely forested cover classes (<30% forested) were excluded because this type of forest is not suitable for ovenbirds.

There are some inherent problems with using landcover classes based on remotely sensed imagery to estimate forest cover. First, these data often have classification errors and uncertainties that are not accounted for. Landcover and other remotely sensed data are typically classified aspatially, resulting in image speckling and urban cells classified in the middle of forested areas. Classification errors may be particularly high in urban settings. Therefore, areas that are often changing most quickly due to development pressure could have the largest errors (Langford et al. 2006). Second, forest stands are interspersed with a matrix of urban development, agriculture, grasslands, and different types of land uses. Thus, the matrix has a heterogeneous composition as well and this suggests that all forest cover cells are not alike because the surrounding matrix habitat may affect bird–environment relationships. Third, increasing forest cover tends to be negatively correlated with forest fragmentation (i.e., isolation, configuration, edge effects), but not necessarily linearly so (Drolet et al. 1999). Forest configuration affects species dispersal behaviors with different species responding in dissimilar ways to the same forest configurations. Some species are more sensitive to forest fragmentation than others, demonstrating threshold like declines in population abundance and occurrence (Betts et al. 2006) Finally, the landcover data were determined 10 years or more prior to the Atlas data collection. This lack of temporal overlap is often inevitable. In this case, these are the only currently available forest cover data that cover the entire study area.

All of these aspects affect the estimation of forest loss and the relationship between forest cover and bird distribution, but in order to quantify these effects, we would need data at a much finer resolution than an Atlas square. The OBBA was designed to estimate the current distribution of *S. aurocapilla* within 10×10 km squares, thus it is at this resolution that we will consider deforestation in our analyses. We acknowledge that this is a simplification and that the insights made are limited because we do not have information on other relevant processes and sources of error. But for the purpose of demonstrating spatial analyses, they will help to characterize and quantify the spatial relationships between variables over the region.

6.3.4 Resolution and registration issues

Investigations into relationships between species and their environment on large spatial scales require large amounts of data summarizing the spatial processes of interest, often collected from differing sources and observed at disparate spatial resolutions. It is often difficult to match the scale of the response variable with the process of interest at an appropriate extent and resolution for analysis. Here, we are interested in the question of how the distribution of *S. aurocapilla* changes along a principal gradient of forest cover; however, bird and landcover data were available as point referenced and areal data, respectively. One solution to this spatial misalignment problem is to average the point-referenced data to the area level creating a common support for both variables. Because the bird Atlas project was designed to provide a representative sample of birds in Ontario within each 10×10 km square, the variable of interest was an estimate of mean relative abundance over all point counts within each square. The centroid of each Atlas square was used to represent its spatial position. Converting the point count to an areal estimate of relative abundance effectively averages out some of the variation in the underlying continuous spatial process represented by the original point count data. If we were interested in modeling the process represented by the point count data, a more correct approach would be to model the aggregation of points within a square using spatial integrals (e.g., block kriging; Journel and Huijbregts 1978). Block kriging is computationally more expensive, but provides a more accurate estimation of the spatial error.

The Universal Transverse Mercator (UTM) projection was used to map Atlas squares during the first and second Atlas. This projection and those like it (e.g., Lambert Conformal Conic projections in meters) provide a more accurate depiction of spatial distances on the ground than longitude and latitude coordinate systems that show relative position on the globe. The UTM projection divides the earth into zones; each zone is six degrees of longitude wide. Ontario, given its size, encompasses three UTM zones. A problem which arises from this is that Atlas squares on either side of zone lines have truncated triangle or 'sliver' shapes. The problem of sliver squares in this dataset was addressed by aligning and summarizing the two datasets to ensure that only one Atlas centroid occurred per landcover square (by averaging the occasional squares with two points). We acknowledge that potential smoothing errors were introduced at this stage. Yet a full uncertainty and error analysis is beyond the scope of this chapter. There are still disappointingly few investigations addressing how these types of registration issues affect inferences made in large scale ecological studies. This example highlights the need for researchers to be cognizant of practical issues such as map projections and datum during the design and analysis of data collected on large spatial scales and from disparate sources.

6.4 Spatial Exploratory Data Analysis

We perform two types of analyses to determine: (1) the degree of spatial autocorrelation (Section 6.4.1) and (2) the presence of spatially homogeneous subregions (Section 6.3.2).

6.4.1 Spatial Statistics

To determine the spatial properties of the data, global spatial autocorrelation can be computed using Moran's I coefficient (Fortin and Dale 2005). This coefficient is in essence the equivalent in space to Pearson's correlation coefficient where the relationship is not between two variables but rather between the values of a variable with itself at different sampling locations, spaced at a given spatial interval known as a distance lag or distance class "d". As with a correlation coefficient, Moran's I is standardized; hence positive spatial autocorrelation is indicated by positive values (usually ranging from 0 to ≥ 1), and negative spatial autocorrelation by negative ones (usually ranging from 0 to ≥ -1). In the absence of spatial autocorrelation, the expected value is slightly negative and close to zero ($E(I) = -(n-1)^{-1}$). For a given d distance lag, Moran's I is:

$$I(d) = \left(\frac{1}{W(d)}\right) \frac{\sum\limits_{\substack{i=1 \\ i \neq j}}^{n} \sum\limits_{\substack{j=1 \\ j \neq i}}^{n} w_{ij}(d)(x_i - \bar{x})(x_j - \bar{x})}{\frac{1}{n}\left(\sum\limits_{i=1}^{n} (x_i - \bar{x})^2\right)}, \tag{6.1}$$

where $w_{ij}(d)$ is the connectivity (or weight) matrix for the distance class d, which indicates whether a pair of sampling locations are in the same distance class; x_i and x_j are the values of the variable x at sampling location i and j; $\bar{x}$ is the mean value of the variable; $W(d)$ is the sum of the $w_{ij}(d)$, i.e., the number of pairs of sampling locations per distance class; and n is the number of sampling locations. A plot of Moran's I coefficient for each distance class is known as a correlogram and the shape of a correlogram can be informative of the type of spatial pattern (Fortin and Dale 2005). The term "global" is used to stress two properties of Moran's I: it is computed using all the sampling locations of the study under investigation, and therefore the value of each coefficient is an average value of spatial autocorrelation for the entire area (i.e., all sampling locations). This average implies that coefficient values do not represent a specific location in space but rather a statistical property of the data that is related to the spatial pattern. With our case study data, both forest cover and bird occurrence correlograms show a strong significant trend as the values range from highly positive at short distance intervals to negative at large distance intervals (Fig. 6.2a). Global and local (Section 6.4.2) Moran's I values were computed using PASSaGE (Pattern Analysis, Spatial Statistics and Geographic Exegesis) package software version 2 (developed by Dr. M. Rosenberg: http://www.passagesoftware.net).

The global degree of spatial autocorrelation can also be quantified using the semivariance function $\hat{\gamma}(h)$, which provides an estimate of how a variable changes in space. Unlike Moran's I, the semivariance is usually not standardized (but can be) and therefore has the same units as the variable of interest. Low values of semivariance (i.e., close to zero) indicate low spatial variance

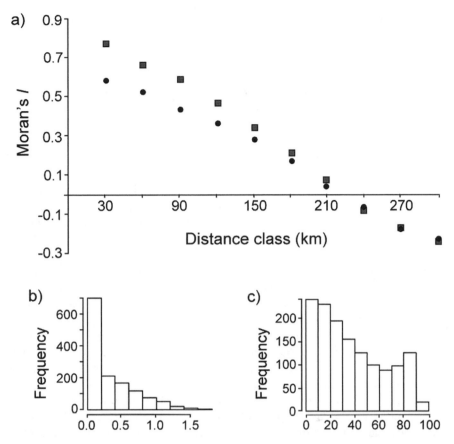

Fig. 6.2 (a) Global Moran's *I* of forest cover (*squares*) and mean number of ovenbirds (*circles*) in an Atlas square 10×10 km (*n* = 1359). Solid symbols indicate significant Moran's *I* values at $p \le 0.05$. (b) Histogram of mean number of ovenbirds in an Atlas square 10×10 km (*n* = 1359).

(i.e., a high degree of spatial autocorrelation), and as the distance increases between sampling points, the semivariance often increases as well. When the semivariance reaches a plateau, this indicates that the distance between sampling locations has no effect on the spatial variance. The distance at which this occurs is called the "range"; below the range, a variable is spatially autocorrelated, and beyond the range, values of the variable can be considered spatially independent (i.e., they lack spatial autocorrelation). A plot of semivariance values against distance is called a variogram and like the correlogram can provide insights into the spatial patterns in the data (Fortin and Dale 2005). Both Moran's *I* and semivariance values can be affected by the statistical distribution of the data (Fortin and Dale 2005). When the data are skewed, as is the case for both bird occurrences (Fig. 6.2b) and forest cover (Fig. 6.2c), a transformation of the data is often recommended to approximate a symmetric, Gaussian, distribution (Journel and Huijbregts 1978).

6.4.2 Delimiting Spatially Homogeneous Subregions

Ecologists who study species distributions over large spatial scales (e.g., states, provinces, continents) are often aware that the assumption of stationarity is unrealistic because dispersal patterns and the relationship between a species and its environment are likely to change with changes in ecotypes (e.g., eastern deciduous to boreal forests) and in land use (e.g., predominantly agricultural to predominantly forested). Yet there are few methods that can deal with non-stationarity (Osborne et al. 2007, see also Sections 6.3.4 and 6.4). This is why it is important to delineate spatially homogeneous subregions in terms of the major processes acting on them or subregions where parameter values (e.g., mean, variance) are more or less constant. Such homogeneous subregions can be delineated using edge and boundary detection algorithms, spatial cluster methods (Fortin and Drapeau 1995, Fortin and Dale 2005), or local spatial statistics (Philibert et al. 2008).

When the data are regularly spaced in a grid-like fashion, several edge detectors are available to determine boundaries (Jacquez et al. 2000, Csillag et al. 2001, Fortin and Dale 2005). The most effective ones are the multiscale detectors, such as wavelets (Csillag and Kabos 2002) or scale-space edge detectors (Faghih and Smith 2002). These techniques require complete spatial coverage, which is not the case in our study dataset due to the presence of lakes and incomplete sampling effort (Fig. 6.1). Therefore we used a lattice-wombling approach (Jacquez et al. 2000), which is a local edge detector that computes the magnitude of the rate of change among quantitative values of the four adjacent locations. Locational uncertainties can be taken into account using a randomization procedure that converts the rates of change into boundary membership values (Jacquez et al. 2000, Jordan et al. 2008).

To determine potential boundaries in our study area, we used all forest cover squares (i.e., forest cover ranging from 0 to 100%) to better identify the spatial properties of forest cover. Using this method, forest cover (Fig. 6.3a) showed two spatially homogeneous subregions of low boundary membership values (less than 0.153) in the southwest and in the northeast separated by a subregion with high boundary membership values (0.377–0.781). Boundary membership values for the ovenbird (Fig. 6.3b) resembled forest cover boundaries (0.401–0.797). Areas that lack any boundary membership value represent areas outside the study area (i.e., in the Great lakes) or areas that were excluded from the analysis (Fig. 6.1). The lattice-wombling and boundary membership values were computed using BoundarySeer (developed by TerraSeer Inc.: http://www.terraseer.com).

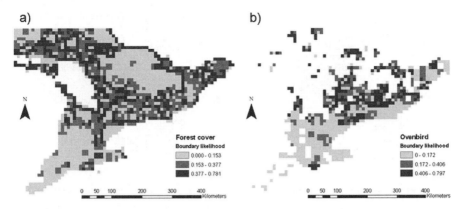

Fig. 6.3 Boundary likelihood probabilities based on lattice-Womble rates of change (**a**) percent forest cover and (**b**) mean number of ovenbirds by Atlas square 10×10 km ($n = 1359$)

Another way to identify spatial variability in the data is to use local spatial statistics (Anselin 1995, Fortin and Dale 2005, Jetz et al. 2005, Philibert et al. 2008). Local spatial statistics can be used as exploratory tools to statistically characterize the local spatial variability of a pattern (Boots 2002). Unlike global Moran's I, local Moran's I (Anselin 1995) computes spatial autocorrelation at each sampling location based on neighbouring values within a local neighbourhood search. Significance can be determined using randomization procedures or by transforming the local Moran's I values into their standardized corresponding z values (for mathematical detail, see Anselin 1995, Fortin and Dale 2005). Local Moran's I should be interpreted in a slightly different way than global Moran's I: positive z values can be due to values around i, and at i, that are larger (positive deviation) or smaller (negative deviation) than the average; negative z values indicate deviation from the average as well, but with negative local Moran's I the value at location i is of a different sign than its neighbours; and values close to zero imply that local deviation from the global average is very small.

The maps of local Moran's z values for each Atlas square were based on a local neighbourhood search of radius 30 km (Fig. 6.4). These local Moran's z value maps for both the ovenbird and forest cover data depict patterns that concur with the regionalization found in the boundary analysis (Fig. 6.3). Two subregions with high positive z values (≥ 1.96) were identified, separated by a subregion with few significant values (i.e., z values between -1.96 and 1.96). There were very few negative z values. The local Moran's I values were computed using PASSaGE package software version 2.

The local Getis statistic, G, is another local spatial statistic that can be used to determine spatially homogeneous areas (Getis and Ord 1996, Boots 2002, Philibert et al. 2008). As with local Moran's I, local Getis measures the degree of spatial association for each sampling location as the ratio of local averages

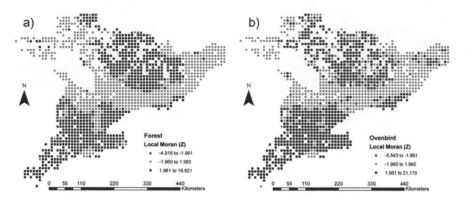

Fig. 6.4 Local Moran's *I* (reported as *z*-scores) computed using a 30 km radius search window on (**a**) percent forest cover and (**b**) mean number of ovenbirds by Atlas square 10×10 km ($n = 1359$)

(both at and around a location) to the global average of the variable of interest for the entire study area. In essence, local Getis calculates a local spatial moving average allowing the detection of local spatial clusters of high (hot spots) and low values (cold spots).

Standardized local *G* values are shown in Fig. 6.5. It is now apparent that the two subregions in our case study, which were also detected using boundary statistics and local Moran's *I* (Figs. 6.3 and 6.4) have different local *G* properties: the southern subregion is a 'cold spot' (with low *G* values) and the northern subregion is a 'hot spot' (with high *G* values). The local *G* values were computed using the spdep package for R (Bivand 2006).

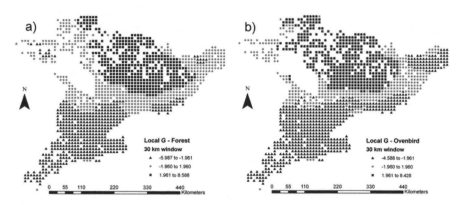

Fig. 6.5 Local Getis (reported as *z*-scores) computed using a 30 km radius search window (**a**) on percent forest cover and (**b**) mean number of ovenbirds by Atlas square 10×10 km ($n = 1359$)

6.5 Spatially Explicit Regression Methods

In this section, we outline and compare the performance of ordinary least squares regression and spatially explicit methods (i.e., regression kriging and geographical weighted regression) by analyzing model residuals for the presence of spatial pattern. The ordinary least squares regression was conducted in R (R Development Core Team 2007). The gstat package for R (Pebesma 2004) was used to perform RK, and the GWR 3 package was used for GWR (Charlton and Fotheringham 2004).

In ordinary least squares regression, OLS, the relationship between the dependent variable Y and the independent explanatory variables X can be assessed using the following equation, written in matrix notation:

$$Y = \beta X + \varepsilon, \tag{6.2}$$

where β is a vector of parameter coefficients for intercept and explanatory variables X to be estimated, and ε is a random error term assumed to be normally distributed.

OLS regression approaches (including multiple linear regression) assume that the errors are independent and normally distributed, which is often not the case when the dependent variable is based on species count data. Species count data are often, but not necessarily, highly skewed, and the distribution will be truncated at zero (as with our bird data Fig. 6.2b) because there are no negative counts. This can result in non-normally distributed residuals. One solution to this problem is to transform the dependent variable logarithmically $(\ln(y + 1)$ and to fit the regression model using the transformed variable. Often there is evidence of remaining spatial autocorrelation in the residuals of such a model even after data transformations (Henebry 1995). For our case study, the OLS regression was performed on the ln(mean overbirds + 1) data. Forest cover was logit transformed $(\ln(p/1-p))$, where p is the proportion of cover. The overall R^2 for this OLS regression was 0.43.

There are several ways to compute residuals. The most commonly used residuals with linear models are: (1) raw residuals (observed minus expected), (2) Pearson's residuals, and (3) deviance residuals (Myers et al. 2002). In some cases, i.e., with grouped or count data in Logistic or Poisson regression, the raw residuals are technically not appropriate because the variance of y_i depends on the number of observations within each group (Myers et al. 2002). If residuals are to be examined in further statistical analysis, then care must be taken to ensure that the statistical properties of those residuals meet the assumptions of further analyses. Whether or not it is appropriate, for example, to transform residuals in order to meet normality assumptions in complementary analyses is an open question and an important area for future research.

Residual analysis (e.g., mapping the residuals and computing global Moran's I statistics) can provide insight into model lack of fit or violations of

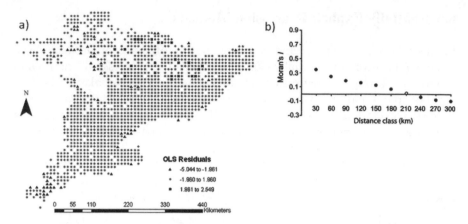

Fig. 6.6 (**a**) Standardized OLS regression residuals (sometimes known as internally studen-tized residuals) and (**b**) associated global Moran's *I* values for residuals of the regression between *ln*(mean overbirds + 1) and forest cover, (ln(*p*/1–*p*)), where *p* is proportion cover. *Solid circles* indicate significant Moran's *I* values

assumptions. This is definitely the case for our data as the map of the OLS residuals (Fig. 6.6a) revealed that high negative residuals were clustered in the southern part of Ontario where forest cover was lowest, and high positive residuals were predominant in the northern part of the study area where forest cover was highest. The spatial pattern remaining in the residuals was confirmed by the presence of positive spatial autocorrelation in the residuals up to 210 m, as indicated by the correlogram of Moran's *I* (Fig. 6.6). Spatially structured residuals may be related to four main factors:

1. missing an important environmental predictor in the model;
2. unaccounted for biotic processes or predictor variables;
3. incorrect model specification;
4. smoothing effects resulting from changes in scale (in terms of both grain and spatial extent).

In our case, the most likely reason for residual spatial pattern is that the study area contains sub-regions, which were identified by the spatial partitioning methods highlighted earlier (Figs. 6.3, 6.4, and 6.5). To explicitly account for this, the data can be partitioned and analyzed separately within each subregion. But how do spatially explicit regression methods, such as RK and GWR, deal with these kinds of data?

Regression kriging is known as a linear mixed spatial modeling technique (Zhang et al. 2005, Hengl 2007) because it combines regression with spatial interpolation of the regression residuals (Hengl et al. 2004, Hengl et al. 2007). Some authors reserve the term mixed model for models that include a mixture of both fixed and random effects in addition to the mean and error terms (Searle 1997), and thus RK may be more accurately described as a hybrid modeling approach because it includes only fixed linear effects and a spatially autocorrelated error term. In any case, a combination of techniques has been shown to result in better predictions than either approach alone (e.g., Zhang et al. 2005). Another strong point of RK is that the method apparently allows for the analysis of more complex forms of regression (e.g., Poisson regression, logistic regression, and the use of categorical or continuous data) along with separate modeling of the residual as a stationary random spatial function (Hengl et al. 2007). Moreover, residual interpolation does not require that the analyst have prior knowledge or hypotheses about the underlying causes of autocorrelation, which is often the case in models of species–environment relationships on large spatial scales. Thus, RK accepts incomplete knowledge of the underlying residual processes and models uncertainty in the system by means of a spatially autocorrelated stochastic residual (Hengl et al. 2007).

In RK, the underlying model is initially the same as for OLS, ($Y = X\beta + \varepsilon$), but predictions are made by modeling remaining spatial variation in the residual error term ε using an isotropic (i.e., equal in all directions) or anisotropic variogram model with zero mean and a covariance function that depends only on the distance between pairs of points. Typical forms used to model the residual variogram are Gaussian, spheroid, and exponential models (these models are also used to estimate covariance functions or correlation functions; Wackernagel 2003, Dormann et al. 2007). Interpolation of regression residuals is then performed by finding the kriging weights λ as determined by the semivariance or variogram model of the residuals:

$$\hat{Y} = X^T \hat{\beta} + \lambda^T \varepsilon, \tag{6.3}$$

The regression parameter estimates $\hat{\beta}_p$ are estimated from the sample by OLS or preferably by generalized least squares (GLS) estimation techniques, which account for residual spatial correlation (Hengl et al. 2004):

$$\hat{\beta}_{gls} = (X^T \Sigma^{-1} X)^{-1} X^T \Sigma^{-1} Y, \tag{6.4}$$

where Σ (sigma) is the $n \times n$ variance–covariance matrix of the residuals. It is important to realize that estimation of GLS residuals is an iterative process that first involves parameter estimation by OLS because before the regression is performed, the residuals are unknown. To find the OLS parameter estimates, the identity matrix (I, with ones on the diagonal and zeroes elsewhere in the matrix) is used in place of Σ. After estimation of parameter estimates by OLS, the covariance function of the residuals is used to obtain new GLS parameter estimates (Fig. 6.7). Arguably, these estimates should then be used to recompute the residuals, and so forth until convergence (Hengl et al. 2007). However Kitanidis (1993) showed that a single iteration to compute the residual variogram may be adequate (Hengl et al. 2007). Even so, the conditions under which this is true have not been fully explored. For a more detailed description of RK

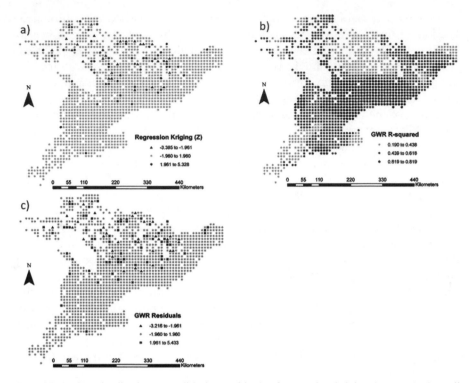

Fig. 6.7 (**a**) Standardized cross-validation residuals of regression kriging (*z*-scores). Overall adjusted R^2 for this hybrid spatial method was 0.68, an improvement over the OLS R^2 of 0.43. (**b**) Geographical weighted regression R^2 and (**c**) Standardized geographical weighted regression residuals (internally studentized residuals)

and its mathematical similarities to other methods like universal kriging or kriging with a trend model, see Hengl et al. (2004). The overall R^2 for our ovenbird and forest cover case study using RK was 0.68 (Fig. 6.7a), which was an improvement over the non-spatial OLS approach (Fig. 6.6, $R^2 = 0.43$). Moreover, the spatial distribution of the standardized cross-validation residuals for RK (*z*-scores) were more evenly distributed than standardized OLS residuals in the northern part of the study area.

Geographically weighted regression is one of the few currently available modeling approaches that was designed to account for a species–environment relationship that varies with location (Fotheringham et al. 2002, Shi et al. 2006, Osborne et al. 2007). GWR permits the parameter estimates to vary locally as the distance between points is taken into account during the estimation of regression coefficients (Hengl 2007):

$$\widehat{\beta}_{gls} = (X^T W X)^{-1} X^T W Y, \qquad (6.5)$$

W is a matrix of weights determined using some form of distance decay, usually a Gaussian kernel using either a fixed or adaptive bandwidth (i.e., local neighbourhood search radius; Shi et al. 2006, Hengl 2007). For our case study, GWR was computed using a Gaussian kernel and a fixed bandwidth of 30 km. The resultant R^2 values ranged from 0.190 to 0.819 according to location. The map of GWR residuals was similar to the corresponding map for RK, and residuals were sparsely distributed in the northern part of the study area (Fig. 6.7b). Note the similarities between $\hat{\beta}_{gls}$ parameter estimation by generalized least squares in RK and estimation of the weighted regression coefficients in GWR (Equations 6.4 and 6.5). Indeed, GLS estimation of the regression parameters in RK is a special case of geographically weighted regression, where the weights are determined more objectively by accounting for residual spatial correlation (Hengl 2007). This explains the similarities in the mapping of residuals between the results of our case study for RK and GWR (Figs. 6.7a and 6.7c).

The advantage of RK is that it provides a single set of regression parameters for the entire region, which are calculated by taking the marginal autocorrelation of the data into account. The advantage of GWR is that it helps to visualize different sub-regions in the data, and GWR models these subregions with different sets of parameter values. Overall, GWR is a good way to explore spatially regionalized data in search of missing spatial variables or an inappropriate functional model form, while RK is more appropriate for hypothesis testing across the region. This is because with RK, the degrees of freedom available for testing model significance do not vary over the study area as they do with GWR. Conversely, the advantage in RK of having a single set of regression parameters can also be considered a disadvantage because the results of GWR indicated that our study area had subregions with different parameter values, which was also suggested by our local spatial analyses of the area (Section 6.4.2).

6.6 Conclusion

Analysis of ecological data over large regions requires the exploration of not only the statistical, but also the spatial properties of the data. With information from preliminary global spatial analyses, researchers can better determine which methods they should use to evaluate species–environment relationships. For this case study, we limit our consideration to a few non-spatial and spatial methods which we feel are versatile and robust enough to be of use in a variety of cases. Regardless of the analytical method selected, we suggest that residual analysis is a key step that can improve our understanding of mechanisms for unexplained spatial autocorrelation. Residual spatial autocorrelation can highlight where our knowledge about the system is uncertain. The information provided by analysis of the residuals is why we caution against detrending data when insufficient knowledge of the underlying processes is available. Removing a spatial pattern in these cases without understanding the underlying mechanisms can result in a loss of essential information about the processes being studied. It is better to use spatially explicit regression methods that can model the marginal autocorrelation directly.

In our case study, residual analysis of the OLS regression showed significant spatial autocorrelation. Hence this non-spatial regression method was not able to fully capture the spatial pattern in ovenbird abundance, indicating that factors other than forest cover play a role in determining the species' spatial distribution. Spatially explicit methods were better able to account for the spatial pattern in ovenbird distribution over the entire region in different ways. With RK, residual spatial autocorrelation was modeled with a single (spherical) variogram and sampling locations within the range of autocorrelation were predominantly used to estimate the spatial component. Given that RK assumes the residual variogram is constant across the entire study area of southern Ontario, this method is more similar to other global statistical methods such as regression. Using GWR, the spatial pattern was modeled by allowing the regression coefficients to vary locally. In this case, a constant kernel was used and therefore a similar number of neighbors were included for local parameter estimation across the entire study area. A varying kernel bandwidth is also possible with GWR (Shi et al. 2006), which could be more appropriated for spatial exploration purposes. Recall that RK resulted in an overall R^2 of 0.68, indicating that 68 percent of the variation in mean ovenbird abundance per Atlas square was explained by this hybrid model; in contrast, GWR explained more variation in ovenbird abundance in some areas of the study area ($R^2 = 0.82$), but not in others ($R^2 = 0.19$); and the increase in fit was at the expense of having several local equations.

What insights can be gleaned from these results? The standardized cross-validation residuals of RK were spatially not that different from standardized GWR residuals (Fig. 6.7), but the varying R^2 values in GWR may provide some insight into areas where broad scale forest cover is not the most important predictor variable. Local fragmentation effects or other environmental variables may be playing a role here. This could justify initiating a more detailed field study in these areas to test some more direct hypotheses.

Both spatially explicit regression methods require data rich sample distributions with adequate spatial coverage. Indeed, in order to fit regression coefficients that vary in space (GWR), there must be an adequate number of samples within the GWR kernel. As such GWR may be prone to overfitting the data and it is not clear how many degrees of freedom are available for testing model significance. Regression coefficients calculated using GWR depend on position and for this reason GWR may be more useful as an exploratory tool rather than for hypothesis testing (Dormann et al. 2007). While RK may be more appropriate for hypothesis testing, it assumes a constant variance–covariance structure of residuals across the entire region, which may not be a realistic assumption in many large scale ecological studies.

Finally, regardless of the method used, determining the relative importance of biotic processes responsible for spatial autocorrelation requires data collected on multiple spatial scales. Without such data it is difficult, if not impossible, to determine if residual structure is the result of missing a spatially dependent environmental predictor or if residual structure is related to a spatially autocorrelated biotic process. Data on relevant biotic processes include information on species behavior, competition, and predation, which are available only at fine spatial scales over small extents.

Acknowledgments Ontario Provincial Landcover Data has been supplied under license by Members of The Ontario Geospatial Data Exchange. Thanks to the official sponsors of the Ontario Breeding Bird Atlas (Bird Studies Canada, Canadian Wildlife Service, Federation of Ontario Naturalists, Ontario Field Ornithologists, and Ontario Ministry of Natural Resources) for supplying Atlas data, and to the thousands of volunteer participants who gathered data for the project. We gratefully acknowledge the assistance of Trevor Middel for reading and editing previous versions of this chapter. A very special thanks to Randy McVeigh for discussions, analytical support, and assistance with preparing figures. Thanks also to colleagues at Wageningen and Utrecht University for discussions that helped to shape some of the ideas presented herein. Finally we would like to acknowledge the help of the four anonymous reviewers whose comments greatly improved the quality of this chapter.

References

Anselin, L. 1995. Local indicators of spatial association – LISA. Geographical Analysis **27**:93–115.

Austin, M. 2007. Species distribution models and ecological theory: A critical assessment and some possible new approaches. Ecological Modeling **200**:1–19.

Bélisle, M. 2005. Measuring landscape connectivity: The challenge of behavioral landscape ecology. Ecology **86**:1988–1995.

Bélisle, M., A. Desrochers, and M.-J. Fortin. 2001. Influence of forest cover on the movements of forest birds: A homing experiment. Ecology **82**:1893–1904.

Bellier, E., P. Monestiez, J.-P. Durbec, and J.-N. Candau. 2007. Identifying spatial relationships at multiple scales: Principal coordinates of neighbour matrices (PCNM) and geostatistical approaches. Ecography **30**:385–399.

Betts, M., G. J., Forbes, A. W. Diamond, and P.D. Taylor. 2006. Independent effects of fragmentation on forest songbirds: An organism-based approach. Ecological Applications **16**:1076–1089.

Bird Studies Canada, Canadian Wildlife Service, Ontario Nature, Ontario Field Ornithologists and Ontario Ministry of Natural Resources. 2006a. Ontario Breeding Bird Atlas Interim Database, 31 July 2006.

Bird Studies Canada, Canadian Wildlife Service, Ontario Nature, Ontario Field Ornithologists and Ontario Ministry of Natural Resources. 2006b. Ontario Breeding Bird Atlas Website. http://www.birdsontario.org/atlas/atlasmain.html.

Bivand, R. 2006. Implementing spatial data analysis software tool in R. Geography Analysis **38**:23–40.

Boots, B. 2002. Local measures of spatial association. Écoscience **9**:168–176.

Bowman, J., G. L. Holloway, J. R. Malcolm, K. R. Middel, and P. J. Wilson. 2005. Northern range boundary dynamics of southern flying squirrels: Evidence of an energetic bottleneck. Canadian Journal of Zoology **83**:1486–1494.

Borcard, D., and P. Legendre. 2002. All-scale spatial analysis of ecological data by means of principal coordinates of neighbour matrices. Ecological Modeling **153**:51–68.

Borcard, D., P. Legendre, C. Avois-Jacquet, and H. Tuomisto. 2004. Dissecting the spatial structure of ecological data at multiple scales. Ecology 85:1826–1832.

Burke, D. M., and E.Nol, E. 1998. Influence of food abundance, nest-site habitat, and forest fragmentation on breeding ovenbirds. Auk 115:96–104.

Cadman, M. D., P. F. J. Eagles, F. M. Helleiner, Federation of Ontario Naturalists, and Long Point Bird Observatory. 1987. Atlas of the Breeding Birds of Ontario. University of Waterloo Press, Waterloo, Ontario.

Charlton, M. and S. Fotheringham. 2004. National Center for Geocomputation, National University of Ireland. Maynooth, Ireland.

Clark, J. S. and A. Gelfand (editors). 2006. Hierarchical Modeling for the Environmental Sciences: Statistical Methods and Applications. Oxford University Press, New York.

Csillag, F., and S. Kabos. 2002. Wavelets, boundaries, and the spatial analysis of landscape pattern. Écoscience 9:177–190.

Csillag, F., B. Boots, M.-J. Fortin, K. Lowell, and F. Potvin. 2001. Multiscale characterization of boundaries and landscape ecological patterns. Geomatica 55:291–307.

Dale, M. R. T., and M.-J. Fortin. 2002. Spatial autocorrelation and statistical tests in ecology. Écoscience 9:162–167.

De'Ath, G. 2002. Multivariate regression trees: A new technique for modeling species-environment relationships. Ecology 83:1105–1117.

Diggle, P., and L. Søren. 2006. Bayesian geostatistical design. Scandinavian Journal of Statistics 33:53–64.

Dormann, C. F., J. M. McPherson, M. B. Araújo, R. Bivand, J. Bolliger, G. Carl, R. G. Davies, A. Hirzel, W. Jetz, W. D. Kissling, I. Kühn, R. Ohlemüller, P. R. Peres-Neto, B. Reineking, B. Schröder, F. M. Schurr, and R. Wilson. 2007. Methods to account for spatial autocorrelation in the analysis of species distributional data: A review. Ecography 30:609–628.

Dray, S., P. Legendre, and P. Peres-Neto. 2006. Spatial modeling: A comprehensive framework for principal coordinate analysis of neighbour matrices (PCNM). Ecological Modeling 196:483–493.

Drolet, B., A. Desrochers, and M.-J. Fortin. 1999. Effects of landscape structure on nesting songbird distribution in a harvested boreal forest. The Condor 101:699–704.

Dungan, J. L., S. Citron-Pousty, M. Dale, M.-J. Fortin, A. Jakomulska, P. Legendre, M. Miriti, and M. Rosenberg, 2002. A balanced view of scaling in spatial statistical analysis. Ecography 25:626–640.

Elith, J., C. H. Graham, R. P. Anderson, M. Dudik, S. Ferrier, A. Guisan, R. J. Hijmans, F. Huettmann, J. R. Leathwick, A. Lehmann, J. Li, L. G. Lohmann, B. A. Loiselle, G. Manion, C. Moritz, M. Nakamura, Y. Nakazawa, J. M. Overton, A. T. Peterson, S. J.Phillips, K. Richardson, R. Scachetti-Pereira, R. E. Schapire, J. Soberon, S. Williams, M. S. Wisz, and N. E. Zimmermann. 2006. Novel methods improve prediction of species' distributions from occurrence data. Ecography 29:129–151.

Ewers, R. M., and R. K. Didham. 2006. Confounding factors in the detection of species responses to habitat fragmentation. Biological Reviews 81:117–142.

Faghih, F., and M. Smith. 2002. Combining spatial and scale-space techniques for edge detection to provide a spatially adaptive wavelet-based noise filtering algorithm. IEEE Transactions on Image Processing 11:1069–1071.

Fahrig, L. 2003. Effects of habitat fragmentation on biodiversity. Annual Review of Ecology Evolution and Systematics 34:487–515.

Fischer, J., and D. B. Lindenmayer. 2007. Landscape modification and habitat fragmentation: A synthesis. Global Ecology and Biogeography 16:265–280.

Forman, R. T. T., D. Sperling, J. A. Bissonette, A. P. Clevenger, C. D. Cutshall, V. H. Dale, L. Fahrig, R. France, C. R. Goldman, K. Heanue, J. A. Jones, F. J. Swanson, T. Turrentine, and T. C. Winter. 2003. Road Ecology: Science and Solutions. Island Press, Washington, D.C.

Fortin, M.-J., and M. R. T. Dale. 2005. Spatial Analysis: A Guide for Ecologists. Cambridge University Press, Cambridge.

Fortin, M.-J., and P. Drapeau. 1995. Delineation of ecological boundaries: Comparison of approaches and significance tests. Oikos **72**:323–332.

Fortin, M.-J., and J. Gurevitch. 2001. Mantel tests: spatial structure in field experiments. In: Scheiner, S.M., and J. Gurevitch (editors). Design and Analysis of Ecological Experiments. Second edition. Oxford University Press, New York.

Fortin, M.-J., T. H. Keitt, B. A. Maurer, M. L. Taper, D. M. Kaufman, and T. M. Blackburn. 2005. Species ranges and distributional limits: Pattern analysis and statistical issues. Oikos **108**:7–17.

Fotheringham, A. S., C. Brunsdon, and M. E. Charlton. 2002. Geographically Weighted Regression: The Analysis of Spatially Varying Relationships. Wiley, Chichester.

Gaston, K. J. 2003. The Structure and Dynamics of Geographic Ranges. Oxford, Oxford University Press.

Getis, A., and J. K. Ord. 1996. Local spatial statistics: An overview. In Longley, P., and M. Batty (editors). Spatial Analysis: Modeling in a GIS Environment. pp. 261–277. Cambridge, GeoInformation International.

Griffith, D. A., and P. R. Peres-Neto. 2006. Spatial modeling in ecology: The flexibility of eigenfunction spatial analyses in exploiting relative location information. Ecology **87**:2603–2613.

Guisan, A., and W. Thuiller. 2005. Predicting species distribution: Offering more than simple habitat models. Ecology Letters **8**:993–1009.

Hastie, T., R. Tibshirani, and J. Friedman. 2001. The Elements of Statistical Learning: Data Mining, Inference, and Prediction. Springer.

Henebry, G. M. 1995. Spatial model error analysis using autocorrelation indexes. Ecological Modeling **82**:75–91.

Hengl, T. 2007. A practical guide to geostatistical mapping of environmental variables. JRC Scientific and Technical Reports. European Commission, Joint Research Centre, Luxemburg.

Hengl, T., G. B. M. Heuvelink, and D. G. Rossiter. 2007. About regression-kriging: From equations to case studies. Computers and Geosciences **33**:1301–1315.

Hengl, T., G. B. M. Heuvelink, and A. Stein. 2004. A generic framework for spatial prediction of soil variables based on regression-kriging. Geoderma **120**:75–93.

Jacquez, G. M., S. L. Maruca, and M.-J. Fortin. 2000. From fields to objects: A review of geographic boundary analysis. Journal of Geographical Systems **2**:221–241.

Jetz, W., and C. Rahbek. 2002. Geographic range size and determinants of avian species richness. Science **297**:1548–1551.

Jetz, W., C. Rahbek, and J. W. Lichstein. 2005. Local and global approaches to spatial data analysis in ecology. Global Ecology and Biogeography **14**:97–98.

Jordan, G. J., M.-J. Fortin, and K. P. Lertzman. 2008. Spatial pattern and persistence of historical fire boundaries in southern interior British Columbia. Environmental and Ecological Statistics (online).

Journel, A., and C. Huijbregts. 1978. Mining geostatistics. Academic Press, London.

Keitt, T. H., and D. L. Urban. 2005. Scale-specific inference using wavelet. Ecology **86**:2497–2504.

Keitt, T. H., O. N. Bjørnstad, P. M. Dixon, and S. Citron-Pousty. 2002. Accounting for spatial pattern when modeling organism-environment interactions. Ecography **25**:616–625.

Kerr, J. T., and J. Cihlar. 2004. Patterns and causes of species endangerment in Canada. Ecological Applications **14**:743–753.

Kitanidis, P. 1993. Generalized covariance functions in estimation. Mathematical Geology **25**:525–540.

Langford, W., S. E. Gergel, T. G. Dietterich, and W. Cohen. 2006. Map misclassification ca cause large errors in landscape pattern indices: Examples from habitat fragmentation. Ecosystems **9**:474–488.

Latimer, A. M., S. Wu, A. E. Gelfand, and J. A. Silander Jr. 2006. Building statistical models to analyze species distributions. Ecological Applications. **16**:33–50.

Legendre, P., and L. Legendre. 1998. Numerical Ecology. 2nd English edition., Elsevier Science BV, Amsterdam.

Lichstein, J. W., T. R. Simons, S. A. Shriner, and K. E. Franzreb. 2002. Spatial autocorrelation and autoregressive models in ecology. Ecological Monographs **72**:445–463.

Melles, S. J. 2007. Effects of forest connectivity, habitat availability, and intraspecific biotic processes on range expansion: Hooded warbler (*Wilsonia citrina*) as a model species. Ph.D. Thesis, Department of Ecology and Evolution, University of Toronto.

Melles, S., S. Glenn, and K. Martin. 2003. Urban bird diversity and landscape complexity: Species–environment associations along a multiscale habitat gradient. Conservation Ecology **7**(1):5. [online] URL: http://www.consecol.org/vol7/iss1/art5/.

Myers, R. H., D. C. Montgomery, and G. G. Vining. 2002. Generalized Linear Models. New York: John Wiley & Sons, Inc.

Odeh, I. O. A., A. B. McBratney, and D. J. Chittleborough. 1995. Further results on prediction of soil properties from terrain attributes: Heterotopic cokriging and regression-kriging. Geoderma **67**:215–226.

Olden, J. D. 2006. Biotic homogenization: A new research agenda for conservation biogeography. Journal of Biogeography **33**:2027–2039.

Ontario Ministry of Natural Resources (1991–1998) Ontario Provincial Landcover Data Base.

Osborne, P. E., G. M. Foody, and S. Suarez-Seoane. 2007. Non-stationarity and local approaches to modeling the distributions of wildlife. Diversity and Distribution **13**:313–323.

Osborne, P. E., and S. Suarez-Seoane. 2002. Should data be partitioned spatially before building large-scale distribution models? Ecological Modeling **157**:249–259.

Parmesan, C. 1996. Climate and species range. Nature **382**:765–766.

Pebesma, E. J. 2004. Multivariable geostatistics in S: The gstat package. Computers and Geosciences **30**:683–691.

Philibert, M. D., M.-J. Fortin, and F. Csillag. 2008. Spatial structure effects on the detection of patches boundaries using local operators. Environmental and Ecological Statistics (online).

Pyke, C. R. 2004. Habitat loss confounds climate change impacts. Frontiers in Ecology and the Environment **2**:178–182.

R Development Core Team. 2007. R: A language and environment for statistical computing. R Foundation for Statistical Computing, Vienna, Austria. ISBN 3-900051-07-0, URL http://www.R-project.org.

Searle, S. R. 1997. Linear Models. New York: John Wiley & Sons Inc.

Shi, H., E. J. Laurent, J. LeBouton, L. Racevskis, K. R. Hall, M. Donovan, R. V. Doepker, M. B. Walters, F. Lupi, and J. Liu. 2006. Local spatial modeling of white-tailed deer distribution. Ecological Modeling **190**:171–189.

Thomas, C. D., A. Cameron, R. E. Green, M. Bakkenes, L. J. Beaumont, Y. C. Collingham, B. F. N. Erasmus, M. Ferreira de Siqueira, A. Grainger, L. Hannah, L. Hughes, B. Huntley, A. S. van Jaarsveld, G. F. Midgley, L. Miles, M. A. Ortega-Huerta, A. T. Peterson, O. L. Phillips, and S. E. Williams. 2004. Extinction risk from climate change. Nature **427**(6970):145–148.

Wackernagel, H. 2003. Multivariate Geostatistics. Kindle Edition.

Wagner, H. H., and M.-J. Fortin. 2005. Spatial analysis of landscapes: concepts and statistics. Ecology **86**:1975–1987.

Zhang, L., J. H. Gove, and L. S. Heath. 2005. Spatial residual analysis of six modeling techniques. Ecological Modeling **186**:154–177.

Chapter 7
The Role of Paleoecology in Whole-Ecosystem Science

Suzanne McGowan and Peter R. Leavitt

7.1 Introduction

Paleoecological analyses provide exceptionally long time series of population abundance, community composition, environmental variability, ecosystem subsidies, and temporal variability that can uniquely inform and guide ecosystem managers and scientists conducting whole-ecosystem experiments (Battarbee 1999, Smol and Cumming 2000, Battarbee et al. 2005, Willis and Birks 2006). Although retrospective time series can be selective or biased representations of the underlying ecological or environmental processes, such biases are often predictable (Cuddington and Leavitt 1999), such that fossil records retain many essential time series characteristics of the original processes (e.g., rate of change, mode of variation, diversity, non-linear events, etc.). In general, paleoecological processes are most informative when investigators seek to quantify temporal variability and its controls within and among ecosystems. In particular, retrospective analyses can quantify baseline conditions, rates of departure from expected ecosystem states, threshold or tipping-point transitions, the presence of alternative stable states, and changes in the nature of variability (CV, range, etc.) – in essence, any measure of temporal variability which would otherwise be deemed as deviation from 'natural' state of the ecosystem (Cottingham et al. 2000, Scully et al. 2000, McGowan et al. 2005a). Because many changes in ecosystem properties develop over decades to centuries due to slow variation in direct (e.g., flux of energy and water to lake surface) or indirect effects of climate (e.g., mass transfer among ecosystems), paleoecological analyses are unique in their ability to place natural and anthropogenic disturbance into a context of baseline ecosystem variability, both through considered selection of contrasting ecosystems (e.g., lakes in clear-cut and undisturbed forests; Scully et al. 2000) and through state-of-the-art time series analyses (e.g., variance partitioning analysis, spectral analysis, etc.; Cumming et al. 2002).

S. McGowan (✉)
School of Geography, University of Nottingham, University Park, Nottingham,
NG7 2RD, UK
e-mail: suzanne.mcgowan@nottingham.ac.uk

S. Miao et al. (eds.), *Real World Ecology*, DOI 10.1007/978-0-387-77942-3_7,
© Springer Science+Business Media, LLC 2009

   This chapter illustrates the potential of retrospective studies to inform, guide, and refine ecosystem experimentation and management using case studies derived from paleoecological studies of freshwater lakes. The transfer of energy, water, and dissolved substances to the lake ensures that these sites integrate ecological and environmental information derived from aquatic, terrestrial and atmospheric sources. In particular, through the action of gravity, mass accumulates in an orderly fashion in lake sediments, usually following the *law of superposition* (deepest is oldest). Paleolimnoligsts recover, isolate, quantify and analyse all or part of these sediments to yield biased but interpretable time series lasting years to millennia, potentially with annual or better temporal resolution, and with many of the statistical properties most desired for numeric analysis or modelling (e.g., Stewart-Oaten and Murdoch 1986). Although similar archives exist for other ecosystems such as tree rings, ancient food caches, soil profiles, geological stratigraphy, ice cores, etc., we focus here on three case histories from lake ecology that illustrate unique insights derived from paleoecology.
   Paleoecological analysis of lake ecosystems has been advanced in the past 30 years by critical developments in analytical chemistry (e.g., mass spectroscopy and spectrometry, liquid chromatography), species taxonomy (e.g., diatoms), and statistics (e.g., multivariate analyses, weighed averaging models, time series analyses, Bayesian statistics, neural networks), each of which has improved investigators' ability to identify the causes and consequences of temporal variability arising from continuous or sudden forcing by climate and humans. For example, stable isotopes are now commonly used to identify change in the flux and cycling of nitrogen (N) and carbon (C) (Leavitt et al. 2006), while improved understanding of algal and invertebrate taxonomy, as well as decompositional processes forming fossil records (taphonomy), allow quantitative reconstruction of past fish populations and vertebrate predation regimes from sedimentary records of invertebrate remains (Jeppesen et al. 2001). Similarly, weighted-averaging models have allowed increasingly accurate reconstruction of the mean and variance of key regulatory variables, including nutrients that control eutrophication (Bennion et al. 1996), ionic content that reflects drought histories (Fritz et al. 1991, Laird et al. 2003), and dissolved organic matter content and C flux (Philibert et al. 2003), while new neural network approaches seem poised to increase the accuracy of retrospective studies (Racca et al. 2007). In all instances, effective use of lake sediments depends on accurate determination of sediment age (chronology), and explicit recognition of the potential limitations of both age determinations and temporal resolution of deposits (due to pre- and post-depositional mixing). The case studies that follow will demonstrate how many issues that limit the use and interpretation of whole-lake experiments can be resolved by the use of paleolimnological approaches, often in combination with modern monitoring or mass-balance studies, and by careful site selection and application of advanced numerical analyses (variance partitioning analysis, synchrony analysis and time series analysis).

7.2 Case History 1 – Sources of Temporal Variability in Greenlandic Lake Ecosystems

The development of lakes through time (lake ontogeny) is driven by diverse agents (e.g., climate, geology, ecological processes, humans) acting through both direct forcing (receipt of energy or mass by the lake) and indirect mechanisms that alter the interaction between lakes and their watersheds (e.g., vegetation development, depletion of geological and edaphic resources, groundwater hydrology) (O'Sullivan 2004), as well as ecological processes operating within individual lakes. Implicit in this idea is that external processes dictate or constrain the pathway of ecosystem ontogeny, with additional development arising from lake infilling that changes the presence and proportion of biological habitats, biogeochemical cycles, and species persistence. Unfortunately, to date, the links between external drivers (and their interactions) and lake trajectory through time have not been well defined (Engstrom et al. 2000), in part due to difficulties in disentangling the effects of multiple forcing mechanisms, many of which take centuries to millennia to develop and interact in unpredictable ways with landscape disturbance (e.g., extreme events). In this study, we aimed to overcome these challenges by using sediment sequences covering the past 6800 years in lakes from an area of southwest Greenland that is devoid of anthropogenic disturbance. In selecting undisturbed sites and using a combination of numerical analyses, our objectives were to quantify the natural mechanisms underlying lake ontogeny in the absence of human effects.

The area around Kangerlussuaq (Danish name Søndre Strømfjord; 66°59 N; 51°06 W; Fig. 7.1) has a geologically homogeneous terrain composed of hills of gentle relief (relative relief ca. 200 m, maximum altitude ca. 450 m). The terrestrial vegetation is scrub and heathland including *Betula nana*, *Salix glauca*, *Vaccinium spp.* and *Empetrum nigrum* (Böcher 1949). Among the numerous lakes in this area are a number of closed-basin sites (no surface outflow) with elevated salinity (conductivity 1500–4000 µS cm^{-1}) that are sensitive to present climatic variation (Anderson et al. 2001), and which lie in catchments that are thought to have been influenced by changes in temperature and precipitation over the past millennia (McGowan et al. 2003). Regional summers are short and temperate (mean ~10.5°C), winters are long and cold (mean –18°C), and there are low levels of precipitation (150 mm year^{-1}) (Hasholt and Søgaard 1978). In most cases, elevated lake-water salinity arises because annual evaporation (and sublimation) exceeds precipitation, and lakes are maintained as a result of runoff that also transfers dissolved substances to the lake basin. However, superimposed over this general trend, lake salinity fluctuates among years, with higher ionic content during warmer and drier periods, and lower conductivity during humid intervals. Lake ontogeny in this region, therefore, is expected to be mediated by variations in the flux of energy, water and dissolved salts that have both long-term, directional trends (baseline), and shorter-term cyclic, threshold or aperiodic variability (disturbance). Because these chemical and

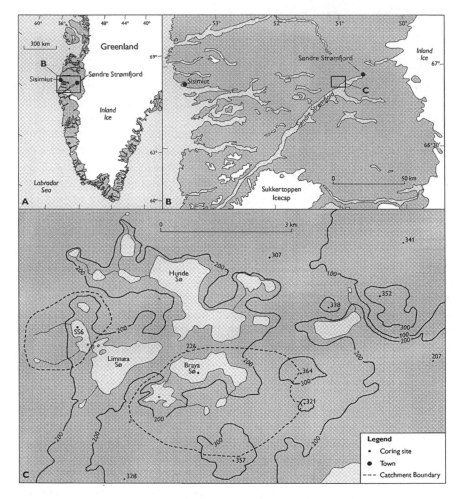

Fig. 7.1 Map of (A) West Greenland, (B) the Søndre Strømfjord region and (C) the area in which Braya Sø and Lake SS6 are located (the dashed line indicates the limits of the lake catchments). Source: McGowan et al. (2008) with permission from the Journal of Paleolimnology

thermal changes are known to strongly regulate species composition and abundance in high-latitude lakes, analysis of historical changes in the fossil record of algal community composition and production using pigment biomarkers (Leavitt and Hodgson 2001) should provide important insights into the nature and mechanism of ecosystem change.

Modern limnological studies demonstrate that the biota of saline lakes respond to changes in climate via a number of mechanisms. At present, several lakes exhibit chemical stratification by depth (termed meromixis) in which dense saline waters are overlain by more ion-poor freshwaters (Anderson et al. 2000). Photosynthetic biota are usually limited to the well-lit surface waters, although several species of phototrophic sulphur bacteria predominate along the strong salinity gradient (e.g., a 25% change in specific conductivity in Braya Sø over 1 m depth) separating upper and lower water bodies. Changes in the influx of freshwaters are thought to enhance the separation between saline deep and fresh surface waters by increasing the density difference between them. Second, because changes in water balance affect lake depth and the degree of inundation of nearshore land, changes in climate also affect biotic composition and production by altering the proportion of benthic and planktonic (suspended) habitats (Stone and Fritz 2006). Third, changes in atmospheric temperature alter the timing and extent of ice cover, as well as the phenology of species replacement during the short growing seasons. Ice cover may affect biota in saline Greenlandic lakes by preventing turbulent mixing (suspension of heavy biota), light penetration (habitat for algal and bacterial growth), and near-surface water chemistry by exclusion of salts from ice (osmotic environment). Unfortunately, while all mechanisms undoubtedly influence lake biota, little is known of their importance in regulating lake ontogeny, or if the relative importance of individual controls change over millennia.

To address these issues, we used paleoecological analysis of algal species (diatoms) and biochemical fossils (pigments) to quantify the relative importance of temperature, salinity and the passage of time on the abundance and composition of lake autotrophs. We focused on salinity because modern limnological surveys and experiments suggest that it is the principle factor explaining algal variability among lakes (Ryves et al. 2002), temperature because of its role in regulating ice cover and because of our interest in the effects of global warming, and time because we anticipated that some slowly developing but immeasurable control mechanisms (e.g., mineralization of geological deposits to supply nutrients and dissolved ions) may vary predictably over centennial or millennial timescales.

7.2.1 Methodological Approach

We used basic and well-tested paleolimnological techniques to quantify the unique and interactive effects of salinity, temperature and the passage of time on lake ontogeny, but paid particular attention to the sampling design and subsequent numerical analyses. Sediment cores covering the last >6800 years (entire Holocene sequence is ca. 8000 years in this region) were recovered from lakes in the Kangerlussuaq region (McGowan et al. 2003, McGowan, et al. 2008). At all sites, accelerator mass-spectrometric (AMS) analysis of ^{14}C activity in the organic matter of lake sediments was used to establish the tentative age of deposits at a number of depth intervals within the sediment cores (McGowan et al. 2003). In addition, historical variation in autotrophic production was estimated by using high performance liquid chromatography to quantify changes in the sedimentary concentrations of chlorophyll and carotenoid pigments from diverse groups of algae and phototrophic bacteria (Fig. 7.2)

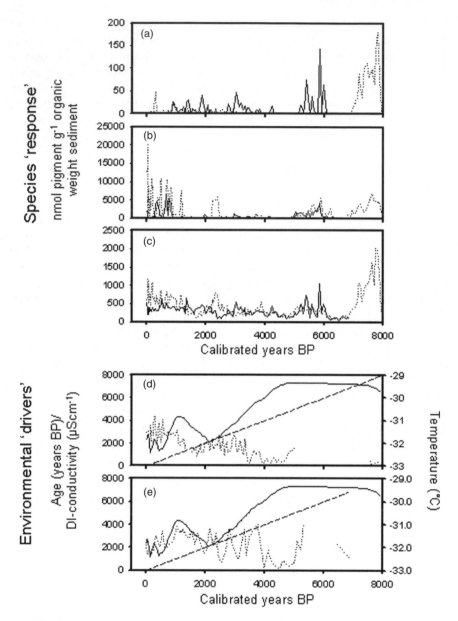

Fig. 7.2 Time series of response variables (pigments **a**, **b**, **c**) and environmental variables (**d**, **e**). Pigments shown include those from filamentous cyanobacteria (myxoxanthophyll, a), purple sulphur bacteria (okenone, b) and all algae (β-carotene, c) in Braya Sø (*dotted line*) and Lake SS6 (*solid line*). Environmental variables for the Braya Sø (d) and Lake SS6 (e) analyses include published records of lake diatom-inferred conductivity (dotted line; McGowan et al. 2003), temperatures from published GRIP records (*solid line*; Dahl-Jensen et al. 1998) and lake age (*dashed line*)

(Leavitt and Hodgson 2001). Changes in the species composition of fossil diatoms were used to infer changes in the salinity of lake water within the well-lit waters and benthos of each lake (Ryves et al. 2002, McGowan et al. 2003).

Because our aim was to quantify patterns of lake ontogeny in response to temperature, salinity and time, rather than to test differences among individual lake basins, we selected similar sites (chemistry, biology, morphology, catchment) that would be expected to respond synchonously to common forcing mechanisms. Specifically, we selected Braya Sø (66°59.3′ N; 51°02.8′ W; 170m a.s.l.) and previously-unnamed Lake SS6 (66°59.8′ N; 51°06.6′ W; 175m a.s.l.). These basins are located at a similar altitude, are meromictic, and have a similar size and maximum depth (~ 50 ha, ~20 m). In addition, our study, sites were located within < 3 km of each other, to ensure that climatic influences on them and their catchments had been as similar as possible.

7.2.1.1 Synchrony Analysis

Our first approach to partition the effects of lake development and climate (salinity and temperature) was to conduct an analysis of lake synchrony both with and without the effects of the passage of time. The assumption underlying this analysis is that historical changes in algal production in the absence of human influence will be strongly correlated only with common strong climatic forcing or to the slow, directional effects of the passage of time (Magnuson and Kratz 2000). By comparing measures of synchrony calculated with and without temporal autocorrelation arising from long-term trends, we also sought to separate non-directional climate variability from directional processes well correlated to the passage of time. If historical variation of primary production in the lakes was highly correlated, then we assumed that climate or the passage of time regulated this aspect of lake ontogeny. By fitting linear regressions to production time series, removing linear trends, and retesting for lake synchrony, we sought to quantify the unique effects of climate. Conversely, if lake algal production varied asynchronously, then climate and the passage of time must not have common effects on lake ontogeny. Finally, a significant but negative correlation among pigment time series indicated similar forcing by either climate or time, but modified by lake-specific processes (Blenckner 2005).

Before synchrony among lakes could be quantified, we needed to first harmonize time series from the two lakes to account for errors in age estimates (chronology), unequal temporal resolution among sites, and variable intervals between samples within a site (Patoine and Leavitt 2006). Analysis of synchrony is more straightforward in modern monitoring and experimental settings than in paleolimnological studies, as sampling is often sufficiently contemporaneous to allow errors to be ignored (George et al. 2000, McGowan et al. 2005b). In contrast, harmonization of paleoecology time series must account for both analytical errors, as well as true differences in the mechanisms by which materials accumulate at the lake bottom (e.g. Cuddington and Leavitt 1999).

Four main steps are required to harmonize sedimentary time series. First, the tentative estimates of sediment age must be corrected for the effects of isotopically depleted ('old') carbon and incomplete mixing of isotope pools. Arctic lakes commonly contain substantial stores of old organic carbon – radio-isotopically inactive C derived from terrestrial sources that dilute ^{14}C activity and lead to overestimates of sediment age (McGowan et al. 2003). Similarly, dissolved inorganic carbon in high latitude sites with prolonged ice cover may be in disequilibrium with the atmosphere sources (the 'reservoir effect'), particularly in stratified or meromictic systems with limited water circulation. Under disequilibrium conditions carbon that has undergone radioactive decay in the bottom waters of the lake and thus is depleted in ^{14}C may be used by algae in photosynthesis, and subsequently deposited at lake bottom. Dates derived from this organic material may be erroneously old. Fortunately, anomalously old dates observed at Braya Sø could be corrected using accelerator mass spectrometric (AMS) analysis of individual wood fragments entombed in the lake deposits at the same stratigraphic level as the bulk sediment age determinations (Oldfield et al. 1997, McGowan et al. 2003). Second, error estimates must be produced for the individual age determinations. Expected radiocarbon activity, and hence sediment age, depends on a variable rate of cosmogenic production of ^{14}C, as well as the more predictable rate of decay to daughter isotopes. As a result, statistical models with associated errors are needed to convert apparent ^{14}C-derived sediment age (termed radiocarbon years) into more familiar calendar year intervals. For example, the Stuiver et al. (1998) model is commonly used to derive calendar years from radiocarbon dates. This model was applied to our Greenlandic cores and produced dating errors (standard deviations) that restricted statistical inferences of more highly resolved time series. Third, we needed to minimize dating errors from fitting regression models to the relationship between burial depth and calendar year of the deposit. The age-depth model was based on six dates in Braya Sø and four dates in Lake SS6, each constrained by an addition analysis of ^{210}Pb activity in sediments <100 years old. Rather than simply extrapolating between dates, we used a regression approach which gave greater weighting to dates with low errors (Heegaard et al. 2005) and allowed us to approximate ages and errors for other, undated depth intervals within each core. Taken together, estimated ages for given intervals varied by 30–270 years in Braya Sø and 68–681 years in Lake SS6. Finally, running averages of 70 and 300 years duration (the upper and lower limits of the error ranges of dates) were applied to each core to harmonize lake time series by exactly matching sampling dates, eliminating the effects of high frequency variation and quantifying the role of multi-century variation on lake synchrony.

Measures of lake synchrony were calculated as the simple Pearson correlation coefficient of the harmonized time series (Patoine and Leavitt 2006) of individual algal and bacterial fossil pigments from Braya Sø and Lake SS6 (Table 7.1). In all cases, raw pigment concentrations were $\log_{10}(x + 1)$ transformed to normalize variance. In general, correlations based on time series

Table 7.1 Synchrony of individual chlorophyll and carotenoid pigments from Braya Sø and Lake SS6 as determined by correlation analysis using Pearson's *r*. Prior to analysis, pigments were smoothed to (a) 70 year intervals and (b) 300 year intervals and detrended using the residuals of a linear regression against age and log $(x + 1)$-transformed. Significant correlations at the $p<0.05$ level are underlined

(a) 70-year smoother

Algal group	Pigment	r	P	r (detrended)	p (detrended)
Colonial cyanophytes	Myxoxanthophyll	0.092	0.450	0.034	0.784
Purple-sulphur bacteria	Okenone	<u>0.737</u>	<u>0.000</u>	<u>0.697</u>	<u><0.001</u>
All algae	β-Carotene	<u>0.744</u>	<u>0.000</u>	0.277	0.021

(b) 300-year smoother

Algal group	Pigment	r	P	r (detrended)	p (detrended)
Colonial cyanobacteria	Myxoxanthophyll	0.117	0.594	0.114	0.606
Purple-sulphur bacteria	Okenone	0.365	0.086	0.369	0.083
All algae	β-Carotene	0.089	0.688	−0.098	0.657

smoothed over 70 years were both greater than those derived from 300-year moving averages, and more statistically significant, particularly for ubiquitous pigments indicating total algal production (β-carotene). In contrast, those from colonial cyanobacteria (myxoxanthophyll) were never synchronous, suggesting that the controls of the ontogeny of total algal production differed from those of the component taxonomic groups. Consistent with this view, historical variation in pigments from purple sulphur bacteria (okenone) were significantly and positively synchronous when some sub-centennial variance was retained (70-year moving average), but not when 300-year filters were applied to the time series (Table 7.1). Overall, there was no evidence that multi-millennial trends (passage of time) contributed to lake synchrony, as the removal of linear trends by fitting regression analysis had little effect on any measure of time series synchrony. Together, these results suggest that the ontogeny of total algal production, and the presence of purple sulphur bacteria, is regulated by common, likely climatic, mechanisms, but these controls operate over relatively fine (~centennial) timescales. In contrast, the abundance of colonial cyanobacteria varied asynchronously between lakes, suggesting that lake-specific processes (e.g., N: P ratios, light penetration, UV radiation) were more important controls than was long-term climatic variability.

7.2.1.2 Variance Partitioning Analysis

Variance partitioning analysis (VPA) was used to independently assess the role of temperature, lake salinity (conductivity), and the passage of time in regulating changes in community composition of primary producers during lake ontogeny. Unlike analysis of synchrony, VPA uses a combination of constrained canonical analyses to quantify the unique and covarying statistical relationships between independently reconstructed environmental variables and a diverse suite of algal taxonomic groups (Borcard et al. 1992, Borcard and Legendre 1994). Furthermore, because analyses are conducted separately on each lake, VPA has fewer requirements for time series harmonization among sites, although it still requires identical sampling dates for predictor and response time series (Hall et al. 1999, Patoine and Leavitt 2006). In this approach, fossil pigments were used to reconstruct historical changes in the abundance of total algae (Chl *a*, pheophytin *a*, β-carotene), and component groups such as diatoms (diatoxanthin), siliceous algae (fucoxanthin), crypto-phytes (alloxanthin), total cyanobacteria (echinenone), colonial cyanobacteria (myxoxanthophyll, oscillaxanthin), chlorophyte algae (Chl *b*, pheophytin *b*), a combination of green algae and cyanobacteria (lutein-zeaxanthin), and purple sulphur bacteria (okenone). Changes in the absolute abundance of these response variables were then compared with time series of environmental data to evaluate the correlative relationships among predictors and responses in a multivariate context.

In principle, time series of predictor variables can be derived from direct observation (monitoring), modelling, or paleoecological reconstructions. Ide-ally, predictor time series will be derived from sources completely independent of those providing the response variable time series. However, while this is relatively easy to accomplish when quantifying environmental and sedimentary change during the past 100–300 years (e.g., Leavitt et al. 1999b, Patoine and Leavitt 2006), care must be taken when analyses involve predictor and response time series derived from reconstructions at the same site. In this latter case, changes in lake condition may alter deposition or preservation of fossil records independent of underlying causal relationships between predictor and response, leading to apparent inflation of the importance of these predictors. Similarly, because the variance partitioning approach can also be performed with multiple time series in each predictor category (e.g., nitrogen, phosphorus, silica, etc., in a single nutrient category), care must be take to eliminate redundant or collinear time series prior to conducting VPA. For further details on problems with inflation of explained variance, see Hall et al. (1999).

Regardless of the data source, VPA necessitates preparation of both pre-dictor and response time series prior to analysis. In the case of Greenlandic lakes, we used the published borehole measurements of the Greenland Ice Coring Project (GRIP) to infer historical changes in temperature (Dahl-Jensen et al. 1998), diatom-based reconstructions of lake-water salinity (as conductiv-ity) to measure effects of changes in net water balance (McGowan et al. 2003), and lake age from the ^{14}C measurements described above to estimate the effects of the passage of time. Because of dating errors in both ice-core and lake chonologies, we used a 300-year running average to remove high frequency variability, and to produce a harmonized temporal resolution in both predictor and response time series.

Holocene changes in lake-water conductivity were derived from analysis of sedimentary diatom species assemblages. Siliceous cell walls (frustules) were isolated from individual sediment samples by oxidative digestion of organic and fine mineral matter, and cleaned diatoms were enumerated to species-level taxonomic resolution using differential-interference-contrast optical micro-scopy and detailed, region-specific, taxonomic sources (Ryves et al. 2002). Diatom-inferred conductivity was reconstructed in both lakes using statistical models developed from the contemporary distribution of diatom taxa in prox-imate geographic regions (Fritz et al. 1991, Ryves et al. 2002). These models, referred to as transfer functions, relate the species abundance and composition of recently deposited diatom assemblages to the monitored chemical conditions of the overlying water. Many transfer functions eliminate rare or patchily distributed species, weight the influence of individual taxa according to their mean abundance, and down-weight species if they are ubiquitous or show little fidelity to the predominant chemical gradient among survey lakes. Further details on the inference of past limnological conditions from sedimentary diatom assemblages are provided by Birks (1998), Hall and Smol (1999), and Smol (2002).

To conduct VPA, we first had to determine whether species distribution along an environmental gradient (e.g., conductivity gradient) conformed to a linear or unimodal model. Here we used detrended correspondence analysis (DCA) to determine the length of environmental gradient associated with the main axis of variation in an ordination of $\log_{10}(x + 1)$-transformed pigment concentrations. For this and other calculations, we followed the procedures of Hall et al. (1999) and used the computer program CANOCO v. 4.0 (ter Braak and Šmilauer 1998). Species distributions along short gradients (<2 standard deviations) can be approximated by a linear function, and should be analysed using redundancy analysis (RDA) or other canonical techniques that incorpo-rate a linear model (Lepš and Šmilauer 2003), whereas unimodal species dis-tributions are best approximated using canonical correspondence analysis (CCA). In both cases, it is usually necessary to normalize variance in predictor time series through appropriate transformation.

In the case of Greenlandic lakes, VPA was performed using RDA to relate $\log_{10}(x + 1)$ pigment concentrations with diatom-inferred conductivity (Co), untransformed temperature (T) and sediment age (A). Variance partitioning analysis used a series of constrained and partially constrained redundancy analyses to determine the amount of variance in past pigment assemblages that was correlated to individual predictor (environmental) variables, in either a unique or 'shared' (covarying) manner. Initially both the total amount of variance explained by all predictors (Co + T + A) and the amount unexplained (100 – Co + T + A) were calculated as the sum of the canonical eigenvalues in an unconstrained RDA. Next, a series of partial canonical ordinations was used to calculate variance explained by the unique effects of each category (Co, T,

or A). In this step, ordinations of individual predictor categories were conducted with the remaining two categories as covariables. Partial canonical
ordinations were used also to calculate the proportion of variation in pigment
assemblages explained by the unique effects of predictors combined with covarying effects of pairs of predictors (e.g., Co + CoT, Co + CoA, T + CoT, T +
AT, A + CoA, A + AT). In each trial, one category of predictor was paired with
one of the remaining categories acting as a covariable. Next, we determined the
explanatory power of pairs of predictor variables (CoT, AT, CoA) by subtracting appropriate terms generated above (e.g., CoA = [Co + CoA] − Co). Finally,
the second-order interaction (CoTA) was calculated as the difference between
100 percent and the sum of variance captured in the first, second and fourth
steps (CoTA = 100 - Co -T - A - CoT - CoA - AT - unexplained).

VPA suggested that time-dependent effects of temperature were most highly
correlated with changes in the abundance and composition of primary producers within Greenlandic lakes during the most recent 6800 years (Table 7.2).
Consistent with the analysis of synchrony described above, the passage of time
alone (A) explained only 6–8% of historical variance in primary producers.
However, the unique effects of temperature (T) and lake-water salinity (Co)
were also unexpectedly slight, with their variance explaining only 1–6% of past
changes in algal and bacterial communities. Instead, age-dependent effects of
temperature (Lake SS6) or the age-dependent covarying effects of temperature
and conductivity (Braya Sø) explained nearly 40% of past variation in the
assemblage of fossil pigments. In agreement with this analysis, colonial cyanobacteria (Fig. 7.2 a) generally declined and purple-sulphur bacterial abundance
and total algae (Fig. 7.2 b and c) increased through time, but shorter-term
variance in abundance of each pigment was correlated either positively or
negatively with changes in conductivity and temperature (Fig. 7.2 d and e).
Extremely slow, long-term effects of temperature are consistent with the

Table 7.2 Proportion of variance in fossil pigment assemblages explained by changes in
temperature (GRIP ice cores, Dahl-Jensen et al 1998), diatom-inferred lake-water salinity
(as conductivity), and the passage of time (age) using variance partitioning analysis of
sediment time series from 6800 years before present to modern day

| | % Variance explained | |
Component	Braya Sø	Lake SS6
Total	58.6	59.6
Age	7.5	5.6
Temperature	5.7	1.0
Conductivity	1.7	4.7
Age-temperature	1.5	40.1
Age-conductivity	3.1	0.2
Temperature-conductivity	0.1	0.3
Age-temperature-conductivity	39	7.7
Unexplained	41.4	40.4

documented role of changes in solar irradiance arising from variations in Earth's orbital position (Berger 1978), while the covarying effects of temperature and conductivity suggest that changes in thermal regime may partly underlie historical variation in past lake-water salinity. Unexpectedly, there were no substantial, unique or age-dependent effects of conductivity alone on total algal composition or production. This suggests that while diatoms might be highly responsive to historical variation in the ionic content of lakes, other algal or cyanobacterial groups might be less responsive to changes in conductivity. However, this result may be partly explained by the poor preservation of the diatom record (ca. 7000–5200 years BP) during a period of elevated conductivity which led to this part of the record being excluded from the analysis (McGowan et al. 2003). This exclusion of high conductivity values may have reduced the degree of correlation with pigment assemblages by removing the periods of most extreme change and reducing the overall variance in the conductivity time series. In contrast, variation in the temperature regime through time appears to be an important control of lake production and primary producer composition during ecosystem ontogeny at these high latitudes, consistent with the suggestions of the IPCC (2007) on the consequences of global warming.

7.2.2 Statistical Issues, Caveats, and Future Directions

Paleoecological analyses provide useful insights into ecosystem function when they are applied to mechanisms whose expected effects are expressed over timeframes that are not amenable to monitoring or direct experimentation (e.g., lake ontogeny, development of terrestrial vegetation, urbanization, climate change, alteration of global biogeochemical cycles, etc.). A paleoecological approach can provide particularly powerful information when it is applied to multiple sites and, in general, is more cost-effective than replicated ecosystem experiments. For example, using a combination of simple (synchrony) and multivariate (RDA, VPA) statistical approaches at two similar sites, we could infer that climate, as historical changes in global temperature, appeared to explain ca. 40% of historical changes in lake production and algal community composition during the past 6800 years, but the effects of temperature vary through time. This insight provides experimental scientists with a clear testable hypothesis for future lake manipulations, as well as a warning that non-linear or time-dependent effects of climate may be expected. Although our approaches are potentially sensitive to errors in chronology, and demonstrate correlation rather than prove causation, both statistical analyses can be applied easily to many of the common time series produced by aquatic and terrestrial paleoecologists (e.g., tree-rings, ice cores, geothermal profiles of soils, lithostratigraphy, stable isotopes, etc.). In particular, the variance partitioning approach is proving itself an extremely powerful tool to quantify and contrast the relative

importance of climate and human activities on ecosystem structure and function (Patoine and Leavitt 2006), as well as identifying precise mechanisms
underlying ecosystem change observed in both monitoring (Hall et al. 1999)
and experimental situations (Leavitt et al. 1999a).

As with all multivariate approaches, the success of synchrony, VPA and
other time series approaches rely, in part, on recognition of underlying assumptions and limitations. In the case of synchrony analysis, investigators must
realize that such a correlation approach does not account for lagged ecosystem
response to forcing (e.g., soil development following temperature increases),
although this lag can be addressed by calculating the cross-correlation among
time series to identify potential lags (Verleyen et al. 2004). Similarly, inferences
concerning possible causal relationships are obscured in ecosystems responding
to simultaneous change in climate and land use, as each can synchonize lake
variability over large spatial scales (George et al. 2000, Patoine and Leavitt
2006). The output of VPA is also highly sensitive to the use of inappropriate
data transformations, lurking (non-causal) correlations between predictor and
response variables (e.g., both derived from a single core), colinearity of predictor time series (misidentification of mechanisms) and weak statistical models
which underlie transfer functions (species distribution along environmental
gradients). Fortunately, there is a rapidly expanding literature concerning
recent advances in these multivariate analyses, including the development
of neural network applications (Racca et al. 2003, 2004, 2007), guidelines for
minimum transfer function requirements (Telford et al. 2004, Telford and Birks
2005) and Bayesian statistics and time-series modelling approaches, such as
dynamic linear models (Cottingham et al. 2000) and multivariate autoregressive
models (Ives et al. 2003).

7.3 Case History 2 – Sockeye Salmon Ecology and Management in Bristol Bay, Alaska

The Bristol Bay region of southwest Alaska (Fig. 7.3) supports the largest
sockeye salmon fishery in the world and is of critical importance to the econ-
omy, culture and ecology of Alaska (Naiman et al. 2002). These anadromous
fishes accumulate most of their biomass as marine predators before returning to
coastal freshwater ecosystems to spawn and die (Groot and Margolis 1991).
Adult sockeye that are not harvested by the marine commercial fishery (termed
'escapement') can contribute over 30% of ecosystem nitrogen (N) and up to
90% of ecosystem phosphorus (P) to nursery lakes, attributable to the fishes'
large size (2–20 kg), high nutrient content (0.36% P, 3.0% N) and elevated
abundance in natal lakes (2000–20000 fish km^{-2}) (Naiman et al. 2002, Brock
et al. 2006). However, because Alaskan escapement levels have declined
over 50% during the 20th century, there is mounting concern that loss of
marine-derived nutrients (MDN) may reduce the primary and secondary production in nursery lakes (Naiman et al. 2002, Ruckelshaus et al. 2002).

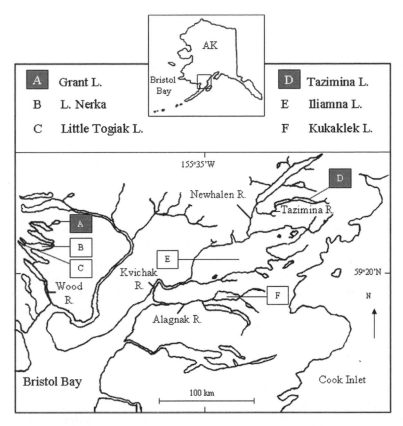

Fig. 7.3 Map of the Bristol Bay, Alaska, showing the location of the sampling sites. Reference sites without salmon are indicated in *shaded grey*. Reproduced from Brock et al. (2007) with permission from Limnology and Oceanography

Reductions in salmon populations in Alaska may arise from climate variations, commercial harvest and natural cycles of population abundance, either alone or in association with other forcing mechanisms. For example, the abundance of salmon within Bristol Bay varies in concert with the Pacific Decadal Oscillation (PDO), a coupled atmospheric-oceanic system that alters coastal production and survival of juvenile salmon entering the Bering Sea over cycles of 20 to 70 years (Mantua et al. 1997). However, despite such large-scale coherence, salmon production appears to vary asynchronously among the individual river districts that compose the Bristol Bay fishery (Hilborn et al. 2003), with striking declines in some stocks leading to closure of local fisheries (Fair 2003) while other districts have exhibited apparently unprecedented increases in escapement during the past decade (Clark 2005). Unfortunately, because monitoring records rarely exceed 50 years in this region, little is known of the magnitude, variability or production potential of unregulated salmon populations prior to the development of the regional fisheries (ca. 1910), or of the relative role of climate, commercial harvest and natural population cycles in leading to recent declines in overall salmon abundance.

Paleoecological analysis of coastal lakes was undertaken to reconstruct past population densities of sockeye salmon, quantify how fish populations have varied in the absence of commercial fishing, evaluate the relative effects of climate and MDN on nursery lake production and determine whether primary production regulated salmon production. Because adult salmon are predators in the marine ecosystem, their tissues are enriched in ratios of stable isotopes of N (δ^{15}N ~11%) relative to other components of the marine food web. Furthermore, because the stable isotope signature of coastal lakes is only ~2% in the absence of anadromous fishes, this marine-derived N can enrich δ^{15}N signatures of nursery lakes and their sediments in direct proportion to the escapement density of salmon (spawning sockeye km^{-2}) (Finney et al. 2000, Brock et al. 2006). Application of simple mass-balance models of N flux in nursery and reference lakes can then be used to reconstruct past changes in salmon population density from sedimentary profiles of N stable isotopes. Similarly, pigments from many primary producers preserved in lake sediments can be isolated, quantified and used to reconstruct historical variation in past algal abundance and gross community composition in both nursery and reference lakes (reviewed in Leavitt and Hodgson 2001).

Changes in past fish populations were estimated by comparing 450-year time series of δ^{15}N from nursery and reference lakes allowing us to identify the effects of commercial fishing on sockeye stocks. Application of spectral analyses to nursery lakes isotope records and population reconstructions further allowed us to quantify the presence of intrinsic cycles of salmon population abundance. Finally, comparisons of sedimentary pigment and δ^{15}N time series were used to determine how historical changes in salmon abundance affected primary production in nursery lakes, and whether there was a subsequent feedback between algal and salmon production. In most cases, these questions could not be addressed through surveys, modelling, or ecosystem experiments.

7.3.1 Methodological Approach

The Bristol Bay region has three drainage systems including the Wood, Kvichak, and Alagnak River catchments (Fig. 7.3). Lakes along this system have low nutrient concentrations, high transparency, chronic P limitation, and low algal production but differ in catchment area, proportion of basin with forest, and intensity of thermal stratification. Climate is maritime and the vegetation is mixed coastal tundra and boreal forest with moist shrubby lowlands. Two lakes, Tazimana (Newhalen-Tazimana River) and Grant Lake (Wood River), were chosen as reference sites because they have waterfalls at their outlets that are inaccessible to migratory fish. Nerka, Little Togiak, Iliamna, and Kukaklek lakes were selected as representative nursery lakes for sockeye salmon because of long histories of monitoring and escapement estimates (e.g., Hilborn et al. 2003).

Cores were sampled from each lake and sectioned into 1.7-mm thick slices representing 2–4 years of sediment accumulation. Chronology of the cores was established using gamma spectrometric analysis of ^{210}Pb activity and by application of the constant-rate-of-supply (CRS) calculation (Appleby and Oldfield 1978). In all lakes, declines in ^{210}Pb activity were near-exponential with depth in the sediment core, revealing that sediments were deposited in an orderly fashion, that post-depositional mixing was limited, and that sediment age could be estimated with good accuracy (Brock et al. 2007). Because ^{210}Pb activity was undetectable in sediments deposited before ca. 1850, linear plots of cumulative mass vs. burial depth were used to approximate the ages of older sediments, while corrected AMS ^{14}C activity (see Case 1 above) was used to estimate sediment age and validate extrapolated ^{210}Pb chronologies at some sites. Stable isotope ratios were estimated on freeze-dried whole sediments using an isotope ratio mass spectrometer and were expressed as the deviation of the ^{15}N:^{14}N ratio from a standard atmospheric sample (δ^{15}N). Sockeye salmon escapement was monitored by the Alaska Department of Fish and Game and the University of Washington since the 1950s and MDN influx was estimated based on observed escapement and average nutrient content of adult salmon ($\sim$82 g N, $\sim$10 g P). Since 1956, the annual average nutrient contribution from salmon has ranged from 103–1080 kg N km^{-2} and 13–132 kg P km^{-2}. Finally, historical changes in abundance and gross community composition of primary producers were estimated by high performance liquid chromatographic (HPLC) analysis of sediments from each lake, as detailed in Case 1 above and in Leavitt and Hodgson (2001).

7.3.1.1 Natural Variation in Sockeye Salmon Abundance

A two-box mixing model was used to translate changes in sedimentary δ^{15}N into estimates of sockeye salmon escapement (E_t) in Bristol Bay nursery lakes (Schindler et al. 2005, 2006). In this approach, changes in E_t are estimated using a combination of observed salmon escapement and historical changes in fossil δ^{15}N signatures of both nursery and reference lakes. Specifically, the mixing model is expressed as:

$$E_t = N_{ws} R_t + e_t, \qquad (7.1)$$

where

$$R_t = \left(\frac{\delta^{15}N_{sed,t} - \delta^{15}N_{ws,t}}{\delta^{15}N_{salm} - \delta^{15}N_{sed,t}} \right). \qquad (7.2)$$

In these calculations, N_{ws} is the total mass N exported from the watershed to the lake, $\delta^{15}N_{sed,t}$ is the δ^{15}N signature of the nursery lake sediments at time

t (a mixture of N derived from salmon and land), $\delta^{15}N_{ws,t}$ is the $\delta^{15}N$ value of terrestrial N sources at time t, $\delta^{15}N_{salm}$ is the $\delta^{15}N$ of adult sockeye salmon tissue (assumed constant at 11.2%), and e_t is random normally distributed errors with mean equal to zero. N_{ws} is the only unknown term and can be estimated by regressing E_t against R_t for the period (1955–2002) when escapement to a specific nursery lake was enumerated directly by the Alaska Department of Fish and Game (Clark 2005). Because the values of R_t are obtained from sediment samples that include more than one year of time and because sediment mixing causes some blurring of the salmon-derived nitrogen signature in the sedimentary record, we smoothed the enumerated escapement data with a 5-year running mean before estimating N_{ws}. Values of E_t from the smoothed series were selected to match the average age of each of the 25 sediment samples from Kukaklek Lake between 1955–2003.

Four model scenarios were used to account for the effects of variation in the isotopic signature of terrestrially derived N ($\delta^{15}N_{ws}$) on the quantification of R_t and to better constrain historical estimates of salmon population abundance. Two scenarios assumed that the N isotopic signature of watershed N was constant through time and used a core-wide average value derived from sediments of each of two proximal reference systems. In addition, we also assumed that $\delta^{15}N_{ws}$ could vary through time, and so used historical changes in individual sedimentary $\delta^{15}N$ signatures of each reference lake to capture this variation. As described in Case 1, core chronology was first harmonized among sites by fitting a running-average to each N isotope profile and interpolating dates from nursery and reference lakes to develop time series with common temporal resolution and sampling intervals. Overall, predictions of sockeye salmon abundance fit observed escapement data accurately, with best-fit model explaining $\sim70\%$ of variation in monitored fish abundance (Schindler et al. 2006).

Morlet wavelet transformations (Torrence and Compo 1998) were used to quantify the presence of cycles in historical time series of sockeye abundance in nursery lakes. First, we created a single continuous time series by combining enumerated salmon abundances (catch + escapement) from 1955 to the present with sedimentary reconstructions during 1507–1955. This integrated time series was then interpolated to obtain abundance values on a regular 4-year interval, as required by the time series analysis. Finally, a power spectrum was calculated following Torrence and Compo (1998) to identify the presence of statistically significant cycles within the 450-year record. Software for this wavelet analysis is available at http://paos.colorado.edu/research/wavelets/.

Salmon populations were reconstructed at Kukaklek Lake, one of the critical nursery lakes within the Alagnak River drainage basin, to determine whether recent unexpected increases in sockeye production and escapement were a result of reduced fishing pressure to conserve stock (Clark 2005) or whether they were part of long-term population variability arising from as-yet-undocumented climatic, lacustrine, or oceanic processes (Schindler et al. 2006). Comparison among sites immediately revealed that both the absolute magnitude and historical variability of sedimentary $\delta^{15}N$ were much greater in the nursery lake than in either Grant or Tazimina lake reference sites (Fig. 7.4a).

When converted to estimates of past fish abundance, analysis of sedimentary N isotopes demonstrated that sockeye salmon populations had varied by over 10-fold prior to the onset of commercial fishing, that population maxima occurred every ~100 years, and that recent documented increases in escapement coincided with the most recent population maximum (Fig. 7.4b). In general, population estimates were not sensitive to model assumptions concerning the

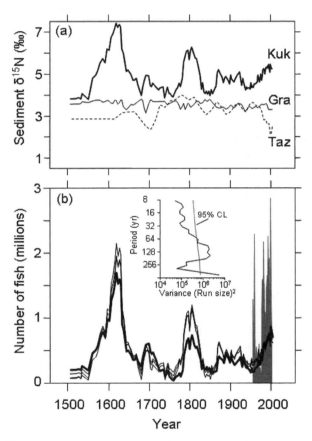

Fig. 7.4 Raw $\delta^{15}N$ signature of sediments (%) from Kukaklek Lake and reference lakes Grant and Tazimina (**a**) and reconstructed time series of sockeye salmon abundance in the Alagnak River from four different baseline scenarios with the best fit (Tazimina Lake as a reference) indicated as a heavy line (**b**). The *shaded region* represents the total enumerated annual run size to the Alagnak River from 1955–2002. The enumerated run size exceeds reconstructed escapement because of removal by fishing. Inset represents results of wavelet spectral analysis of the time series of reconstructed salmon populations for Kukaklek Lake, with significant periodicity recorded for values to the right of the 95% confidence interval (*vertical dashed line*). Reproduced from Schindler et al. (2006) with permission from the Canadian Journal of Fisheries and Aquatic Sciences

δ^{15}N signature of terrestrially derived N. Instead, comparison of the present population maximum with other historical events suggested that modern salmon production had been matched or exceeded twice previously, but the duration of these maxima rarely exceeded 50 years. Wavelet analysis confirmed the presence of multi-decadal cycles (see inset of Fig. 7.4b), some of which were similar to the longer mode of PDO variance (ca. 70 years duration). Taken together, these findings suggest that record escapements were not simply the result of improved conservation measures, but rather reflect cyclic variation in past salmon populations arising from climate and other natural forcing mechanisms. Furthermore, because these peaks were both transient in nature and appeared to arise regularly from intervals of very low population abundance, they suggest that present harvest rates could be maintained without detrimental consequence for the fishery.

7.3.1.2 Effects of Fishing on Algal and Salmon Production

Although reconstruction of salmon populations in Kukaklek Lake revealed little unambiguous evidence of the effects of commercial fishing, comparison of sedimentary δ^{15}N profiles among a wider suite of nursery lakes in southwestern Alaska suggests that fisheries exploitation since 1900 has reduced the density of spawning sockeye by up to 50% relative to values reconstructed for the previous century (Finney et al. 2000, Schindler et al. 2005). For example, reconstructed densities of sockeye salmon in Nerka Lake (Wood River drainage basin) declined from 7000 fish km^{-2} to 2500 fish km^{-2} after $\sim$1900 (Fig. 7.4b) corresponding to substantial declines in sedimentary δ^{15}N relative to those of the Grant Lake reference system (Fig. 7.4a). Similar changes in fossil N isotope ratios have been recorded for other lakes of the Wood River drainage (Brock et al. 2007), as well as for sites located on nearby Kodiak Island (Finney et al. 2000). Interestingly, analysis of time-series synchrony during the past 300 years (conducted as in Case 1) suggests that variations in salmon production are not strongly synchronous among nursery lakes in the absence of commercial harvest (Brock et al. 2007), a result which is consistent with analyses of modern escapement records (Hilborn et al. 2003). Thus although fishing has apparently reduced salmon stocks across much of Bristol Bay, the region remains characterized by high population variance among individual nursery lakes.

Fossil analyses also revealed that commercial fishing has reduced the influx of marine-derived nutrients, leading to substantial declines in primary production in lakes where salmon import a substantial proportion of total ecosystem nutrients (Schindler et al. 2005, Brock et al. 2007). For example, algal abundance estimated from fossil concentration of pigments from diatoms (diatoxanthin), chlorophytes (lutein-zeaxanthin), and total algae (β-carotene) declined concomitantly with reductions in sedimentary δ^{15}N in both Nerka (Fig. 7.5c) and Little Togiak nursery lakes (Brock et al. 2007). In contrast, inferred primary production was relatively constant during the past 300 years in both Grant and Tazimina lake reference ecosystems (Fig. 7.5d and Brock et al. 2007).

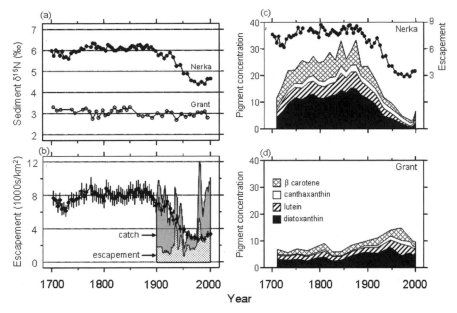

Fig. 7.5 Historical variation of sedimentary $\delta^{15}N$ (**a**) and reconstructed escapement densities of sockeye salmon in Lake Nerka (**b**) generated from the mixing model using the data in (**a**). The *solid circles* in (b) are the mean ± standard deviation of values generated from Monte Carlo simulations of the mixing model. Historical fishery catch and escapement records from Lake Nerka from 1958 to 2000 have been smoothed with a 5-year running mean and are overlaid on the sediment-derived escapement estimates (b). Algal abundance from four dominant fossil pigments (nmole pigment g^{-1} dry sediment) compared with the reconstructed sockeye salmon escapement in Lake Nerka (**c**). Time series of algal abundance in the reference Grant Lake (**d**). Pigments in (c) and (d) include β-carotene (indicating abundance of all algae), canthaxanthin (colonial cyanobacteria), lutein-zeaxanthin (chlorophytes, cyanobacteria), and diatoxanthin (mainly diatoms). Source: Schindler et al. (2005) with permission from Ecology

Controversially, these findings suggest that 50 years of diversion of MDN influx by commercial fishing has had no obvious effect on the total production of adult salmon (Fig. 7.5b) contrary to observations from whole-lake experiments which demonstrate that the production of young-of-the-year salmon is increased by fertilization of nursery lakes (Hyatt 1985, Stockner 2003). In fact, when catch and escapement estimates for Nerka Lake were combined for the monitoring period, modern production of sockeye salmon was indistinguishable from that reconstructed from sedimentary archives for the pre-fishing period of 1700–1900. Taken together, these findings suggest that while primary production in nursery lakes can be regulated by the influx of marine-derived nutrients in salmon, the survival of Alaskan fish to a harvestable size is regulated by some mechanism other than the productivity of nursery lakes (e.g., estuary habitat quality, oceanic production, marine predation regimes, mortality during migration, etc.).

7.3.1.3 Relative Effects of Salmon and Climate on Lake Ecosystems

One of the most significant challenges to the use of paleoecology to study ecosystem processes is the difficulty in demonstrating the precise mechanism underlying observed historical variation. As with ecosystem experimentation, this issue is addressed best through the use of replication and reference ecosystems, particularly when the agent causing environmental variability can contrast strongly between impacted and reference ecosystems (e.g., effects of agriculture, forestry, urbanization, fisheries, etc.). However, in contrast to determinations of the effects of humans on ecosystem processes, it is now recognized that regional climatic variation may have different effects on individual lakes depending on how energy and mass are transferred to the lake from the catchment (Blenckner 2005). In particular, localized variation in the timing and form of precipitation (e.g., intensity, duration, and seasonality of snow or rain) may interact with unique features of individual catchments (elevation, slope, exposure, vegetation, soil development, groundwater, etc.) to reduce the spatial synchrony of lake response to climate (Pham et al. 2008). Because catchment runoff also regulates nutrient flux to nursery lakes, temporal and spatial variability in climate effects on lakes may confound reconstruction of past salmon abundance from sedimentary deposits, obscure the role of marine-derived nutrients in regulating lake production, and alter the spatial synchrony of algal and fish production.

To determine how climatic variability might interact with changes in salmon abundance to regulate nursery lake production, we first developed a predictive conceptual model of how the effects of salmon should vary among nursery lakes in the absence of climatic forcing. From first principles, we argued that the influx of MDN in salmon should have the greatest effect on lake production in instances where nutrients from fish constitute a substantial proportion of total nutrient influx to the natal lake. Specifically, we hypothesized that the correlation between sedimentary $\delta^{15}N$ and biomarker indices of past algal abundance should vary as a direct function of the proportion of ecosystem nutrients derived from salmon, with the highest Pearson correlation coefficients between pigments and $\delta^{15}N$ observed in lakes with the greatest escapement (salmon km^{-2}) relative to background N influx. Although modern limnological analysis demonstrates that algal abundance is limited mainly by the influx of P in Alaskan nursery lakes (cited in Brock et al. 2007), the elementary ratio of N:P is relatively constant in adult salmon, such that changes in $\delta^{15}N$ and salmon densities is expected to predict influxes of both N and P from marine sources. Using a combination of observed escapement for each of six lakes and the mass balance approach described above (equations 1 and 2), we estimated the proportion of total N in each lake derived from salmon during the past 50 or 100 years and compared that with the correlation between fossil pigments from diatoms (diatoxanthin) or total algae (β-carotene) and sedimentary $\delta^{15}N$ for the appropriate lake and time interval.

As predicted, algal abundance was regulated by influx of MDN associated with salmon. For example, diatom and total algal abundance (both $\log_{10}(x+1)$

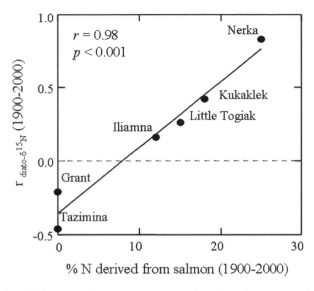

Fig. 7.6 Relationship between the percentage proportion of total ecosystem nitrogen derived from sockeye salmon during the twentieth century and the correlation between concentration of fossil pigments from diatoms ($\log 10(x + 1)$–transformed diatoxanthin) and untransformed sedimentary δ^{15}N over the same time interval. Regression line is highly significant, $r = -0.98$, $p < 0.001$. Note that pigment–isotope correlations are negative in salmon-free reference lakes and positive in sockeye salmon nursery lakes. Source: Brock et al. (2007) with permission from Limnology and Oceanography

transformed) during the twentieth century was significantly correlated ($p < 0.05$) with changes in sedimentary δ^{15}N for all study lakes, with positive correlations observed in nursery lakes (Nerka, Iliamna, Little Togiak, Kukaklek) and negative correlations in reference lakes (Grant, Tazimina) (Fig. 7.6). Overall, the strength of fossil pigment- isotope correlations during 1900–2000 was linearly correlated ($r > 0.97$, $p < 0.001$) to both the fraction of ecosystem N derived from salmon since 1900 (Fig. 7.6) and the mean documented escapement during the past 50 years (Brock et al. 2007). The incidence of negative correlation coefficients for lakes lacking salmon was unexpected, but is proposed to reflect the role of watershed runoff which both increases the export of N from land to lake and reduces the δ^{15}N signature of terrestrially derived N by preventing slow microbial transformation of terrestrial N (e.g., denitrification) which increases the ^{15}N content of terrestrial N pools (Brock et al. 2007). However, regardless of the precise mechanism, comparison of nursery and reference lakes suggested that the sign of the correlation between fossil pigments and stable isotopes could be used to distinguish between lakes or periods of time in which MDN controlled lake production (positive correlation) and those in which climatic processes related to hydrology regulated algal abundance (negative correlation).

Periods in which climate exerted a strong effect on lake production were identified for nursery lakes by quantifying historical changes in the correlation between sedimentary pigments and isotope ratios in overlapping 50-year intervals throughout the past 300 years (Brock et al. 2007). In this approach, we first harmonized the time series of each lake to decadal resolution by either averaging multiple observations to the nearest central year of the decade (surface sediments), using the nearest observation ($\pm$ 3 year), or by interpolating linearly among dated intervals (deepest sediments). Pearson correlations were then calculated between $\delta^{15}N$ and $\log_{10}(x+1)$ transformed pigment concentrations for overlapping consecutive 50-year intervals throughout each core (e.g., $r_{2000-1950}$, $r_{1990-1940} \cdots r_{1750-1700}$). This procedure resulted in a new time series of pigment-isotope correlations for each site that could be compared among lakes to determine whether the effects of salmon (positive correlation) and climate (negative correlation) varied synchronously among regional nursery lakes.

Analysis of time series of pigment- isotope correlations revealed that climatic events occasionally overwhelmed the effects of MDN on algal production in all nursery lakes, but the precise interval of climatic control differed among sites (Fig. 7.7). For example, although historical changes in algal abundance and MDN flux (as $\delta^{15}N$) were positively correlated in all nursery lakes (e.g., Fig. 7.6), fossil pigment concentrations were negatively correlated with $\delta^{15}N$ within Nerka Lake sediments during the both early-1700s and mid-1800s (Fig. 7.7a). Similar significant ($p < 0.05$) periods of negative pigment–isotope correlation were also recorded for all other nursery lakes, although there was no evidence of synchrony among all sites even during the last 50 years when errors in sediment chronology were minimal. We infer that the negative correlation arose from non-salmonid (climatic) mechanisms because negative correlations were recorded in some lakes during periods when monitoring revealed substantial salmon populations (e.g., $\sim$1960 in Iliamna Lake). Absence of strong spatial synchrony in climate effects likely arises both because salmon production and escapement vary asynchronously among nursery lakes (Hilborn et al. 2003) and because hydrologic inputs vary greatly among lakes (Pham et al. 2008) due to differences in catchment characteristics that regulate the transfer of water and dissolved substances from land to lakes (e.g., slope, aspect, vegetation, soil, etc.) (Blenckner 2005).

7.3.2 Management Insights, Caveats and Future Directions

Paleoecological analyses provided several novel insights into the ecology and management of sockeye salmon that reinforce, refine, and guide knowledge derived from ecological monitoring, modelling, and ecosystem experimentation. First, reconstruction of past salmon populations allowed us to establish that natural population variability must be incorporated into management for

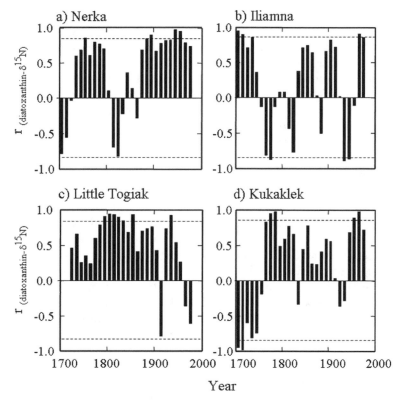

Fig. 7.7 Time series of Pearson correlation coefficients (r) for the relationship between sedimentary diatom pigments and $\delta^{15}N$ during overlapping consecutive 50-year intervals between 1700–2000 in Nerka (**a**), Iliamna (**b**), Little Togiak (**c**) and Kukaklek (**d**) Lake sediment cores. All time series were converted to decadal resolution correlation analysis. Positive correlations indicate that lake production is controlled mainly by changes in the influx of marine derived nutrients associated with spawning salmon, whereas negative correlations suggest that climate regulates changes in algal abundance during that interval. Individual histograms greater than the *dashed line* are significant at $p < 0.05$. Reproduced with permission from Limnology and Oceanography (Brock et al. 2007)

sustainable yield of fisheries both to determine the upper limit to ecosystem production and sustainable harvest and to take advantage of apparently unprecedented upsurges in fish production (e.g., Fig. 7.4). Such retrospective analyses also suggest that the primary production of nursery lakes does not have a powerful effect on the regulation of salmon escapement or recruitment to harvestable size (Fig. 7.4), although further research is required to determine how this finding may vary along continental gradients (Stockner 2003). Similarly, comparison of centennial time series from sediment cores confirmed the findings of annual and decadal studies (Hilborn et al. 2003) that salmon production varies asynchronously among lakes within a fisheries district. These fossil analyses further demonstrated that the timing of effects of MDN and climate on lake production varies greatly among lakes. Thus although the precise mechanisms underlying this temporal and spatial variability have yet to be demonstrated conclusively, these results caution investigators against extrapolation of short-term or site-specific studies to regional landscapes.

Application of the paleoecological approach along continental margins bordering the north Pacific Ocean may provide important insights on the mechanisms by which large climate systems (e.g., PDO, ENSO, etc.) regulate salmon production and the structure and function of near-shore lake ecosystems. In this regard, Finney et al. (2002) have used similar retrospective analyses to those described above (e.g., $\delta^{15}N$, fossil diatoms) to suggest that long-term reorganization of oceanic circulation may lead to synchronous but out-of-phase changes in coastal fish production in the Alaskan and Californian current regions. Climatic variation over decades to centuries may be expected to produce similar coherent variation in sockeye salmon populations located in eastern (North America) and western (Russia) parts of their range. Finally, additional paleoecological analyses are needed within constrained geographic regions to further refine how natural population variability interacts with effects of commercial harvest to regulate the abundance and synchrony of stocks within a given fishing district. Such a study would help management agencies determine how to best conserve stocks (e.g., by lake, catchment, or district).

Several challenges remain in the application of paleoecological analyses to the management of salmonid populations. First, the mass balance and synchrony studies described above suggest that further improvement in the estimation of past salmon densities will require a site-specific correction for temporal variation in the influx of nutrients to individual nursery lakes. Although the use of reference ecosystems appears to correct for pronounced regional variation in climatic regimes (and N isotope export to lakes), further refinement of salmon population estimates will require investigators to correct for site-specific differences in local hydrology (e.g., Fig. 7.7). In this regard, we believe that application of advanced time series statistics (e.g., dynamic linear models, multivariate autoregressive models, cross spectral analysis) may prove beneficial in identifying, and perhaps correcting for, changes in the statistical relationships among time series. As with ecosystem experiments, choice of an appropriate local reference ecosystem with similar catchment filters of climate (*sensu* Blenckner 2005) may prove instrumental in improving population estimates. Similarly, a more predictive understanding is required for additional factors that may constrain the accuracy of population reconstructions; for example, how do the relationships between $\delta^{15}N$ and escapement, sedimentary $\delta^{15}N$ and pigments, and climate and N influx vary as a function of N versus P limitation of primary production (e.g., Brahney et al. 2006, Hobbs and Wolfe 2007)? Finally, new paleoecological indicators could be developed to independently quantify historical variation in hydrology (e.g., compound-specific analysis of $\delta^{2}H$, $\delta^{18}O$), input of terrestrial organic matter (e.g., lignin phenols;

higher plant biomarkers), or independent metrics of past fish abundance (e.g., $\delta^{35}S$, Strontium). By improving our predictive understanding of the conditions that constrain the accuracy of retrospective studies, paleoecological studies will be better capable of improving ecological insights that are essential to robust management and conservation strategies.

7.4 Case History 3 – Water Quality Loss in Continental Lakes

Eutrophication of lakes by point and diffuse nutrient influx remains a serious global problem despite decades of intense research to determine factors that regulate water quality (Carpenter et al. 1998, Schindler 2006). Mechanisms that control eutrophication have been studied extensively for stratified temperate lakes, but remain poorly understood for polymictic hardwater lakes of central North America (the Prairies). The inability to identify and regulate the mechanisms that cause eutrophication continues to hamper management initiatives for prairie lakes (e.g., Davies 2006) and likely reflects the confounding effects of simultaneous change in climate, land use, and urbanization as well as the naturally high concentrations of phosphorus (P) which may lead to conditions in which nitrogen (N) influx regulates lake production (Hall et al. 1999, Bunting et al. 2007).

   Water quality in prairie lakes is impacted simultaneously by a highly variable climate, intensive resource-use, and social change leading to urbanization (Hall et al. 1999). For example, most of the northern Great Plains of Canada lie in sub-humid to semi-arid climates where moisture deficits (annual precipitation – potential evaporation) range from 20 to 60 cm $year^{-1}$, and alterations in energy and water flux can concentrate nutrients and salts (Pham et al. 2008). Furthermore, because mean annual temperatures are near $0°C$, this region is highly sensitive to global warming, and ice cover has both decreased by over 40 days since ca. 1860 and remains highly variable among years within a given decade (~30 days) (Hall et al. 1999). Changes in water-column mixing associated with ice melting regulate the depletion of oxygen, release of N and P from sediments, build-up of toxic ammonia and hydrogen sulphide, and fish kills during spring and winter (Hammer 1973). At the same time, this region has undergone profound changes in aquatic and terrestrial resource-use during the past century. Over 95% of the arable grassland has been converted to crop production and animal husbandry using intensive agricultural practices that increase soil erosion and nutrient export (Bennett et al. 1999), while management of fisheries and catchment hydrology have altered food-web composition and hydrologic supply of nutrients to many lakes (Hall et al. 1999). Finally, cultural changes resulting from profound regional drought (1930s) and international conflicts (1940s) have lead to a shift from a widely distributed agrarian population during the early twentieth century to a 10-fold increase in urban populations and associated pollution (Hall et al. 1999, Leavitt et al. 2006). Although all factors are likely to contribute to the eutrophication of prairie lakes, managers can still improve water quality substantially by establishing policies based on cost-effectiveness and the magnitude of expected improvement.

Eutrophication arising from N influx may be common in lakes of the northern Great Plains, sites which often exhibit naturally high nutrient content, low N: P ratios, surface blooms of N_2-fixing cyanobacteria, and periodic fish kills (Patoine et al. 2006). In many prairie catchments, supply of P from glacial tills and soils is great (Allen and Kenney 1978) leading to poor predictive relationships between P influx and algal abundance (Campbell and Prepas 1986) which contrast those derived from boreal lake districts (Smith et al. 2003, Schindler 2006). Further, it is increasingly evident that long-term agricultural practices can saturate soils with P (Bennett et al. 1999, Foy et al. 2003), increase P export to surface waters (Bennett et al. 2001), and create conditions in which diffuse and point sources of N may degrade water quality (Bunting et al. 2007). Unfortunately, relatively few studies have quantified the role of N pollution in the eutrophication of freshwaters (Kilinc and Moss 2002, James et al. 2003), possibly because whole-lake experiments and lake surveys suggest that biological fixation of atmospheric N_2 by cyanobacteria is an important but uncontrollable source of N to biota (reviewed in Schindler 2006). As a result, managers can be hesitant to invest in expensive or controversial policies that seemingly contradict conventional wisdom concerning the mechanisms of water quality control (i.e., N vs. P).

In this section, we use a combination of paleoecology, multivariate statistics, long-term ecological monitoring, stable isotopes, and whole-lake mass balances to identify the main causes of water quality change during the twentieth century, prioritize management strategies, and quantify the role of N in regulating lake production and water quality degradation. We focused on lakes of the Qu'Appelle River drainage basin because this system supplies water to nearly one-third of the population of the Canadian Prairies, and because lakes within the drainage are characterized by poor water quality, surface blooms of toxic cyanobacteria (*Aphanizomenon, Anabaena, Microcystis*), excessive macrophyte growth, and kills of commercially important fishes (reviewed in Hall et al. 1999). Rather than focus intensely on a single sentinel site, we explicitly selected a suite of lakes with similar food-web structure (McGowan et al. 2005b), but contrasting exposure to climate, land use practises, and urbanization, to better identify the causes of ecosystem change. Conceptually, this comparative approach is similar to that of whole-ecosystem experiments in that both types of studies maximize contrasts among ecosystem states through application of known causal mechanisms (e.g., sites upstream and below wastewater outfall).

7.4.1 Methodological Approach

The Qu'Appelle River drains ~52,000 km^2 of mixed-grass Prairie in southern Saskatchewan (SK), Canada (Fig. 7.8). Land use within the drainage basin is

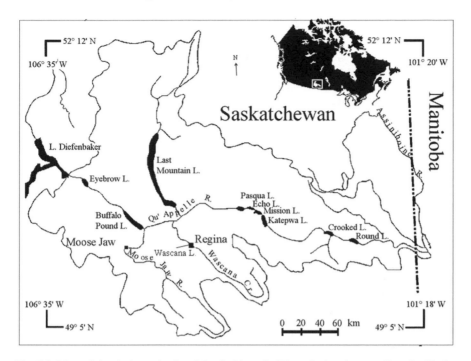

Fig. 7.8 Map of the drainage basin of the Qu'Appelle River, Saskatchewan, Canada. Under natural conditions, flow arises near Eyebrow Lake and passes through Buffalo Pound, Pasqua, Echo, Mission, Katepwa, Crooked, and Round lakes. Last Mountain Lake and Wascana Lake drain into the Qu'Appelle River via Last Mountain and Wascana creeks, respectively. In addition, water is transferred into the catchment from the Lake Diefenbaker reservoir and is occasionally forced into Last Mountain Lake during the spring discharge maximum. The City of Regina (population ~200,000) discharges tertiary-treated wastewater into Wascana Creek downstream of Wascana Lake, while urban wastes from Moose Jaw (population ~45,000) are diverted entirely from surface waters during most years. Reproduced with permission from Limnology and Oceanography (Leavitt et al. 2006)

composed of cereal cultivation (wheat, barley, canola), pastures (cattle, swine), and natural mixed grasslands (~12%), while the cities of Regina and Moose Jaw are the main urban centres (Hall et al. 1999). Ten connected water bodies (0.5–552 km^2) lie within the central Qu'Appelle catchment, including three head-water reservoirs and seven natural lakes (Table 7.3). Reservoirs were created by impoundment of the South Saskatchewan River (Lake Diefenbaker), Wascana Creek (Wascana Lake), and the outflow of Buffalo Pound Lake. Under natural conditions, the Qu'Appelle River flows from west to east through Buffalo Pound Lake, a central chain of four basins (Pasqua, Echo, Mission, Katepwa lakes), and two downstream lakes (Crooked, Round). Water from Last Mountain Lake and Wascana Lake enters the Qu'Appelle River via Last Mountain Creek and Wascana Creek, respectively. Limnological monitoring since 1993 reveals pronounced gradients of lake characteristics from west to east, with

S. McGowan and P.R. Leavitt

Table 7.3 Morphometric and limnological characteristics of study lakes within the Qu'Appelle drainage basin. Limnological variables means of summer sampling period (DOY 130–248). Data from Hall et al. (1999a), Quinlan et al. (2002), Patoine et al. (2006) and unpublished data. See Fig. 6.8 for lake location

Lake	Longitude (°W)	Area (km²)	Volume (m³×10⁶)	Mean depth (m)	Conductivity (µS cm⁻¹)	pH	Dissolved N (µg L⁻¹)	Dissolved P (µg L⁻¹)	N:P (dissolved, by mass)	Chl a (mg L⁻¹)	DOC (mg L⁻¹)	Secchi depth (m)	Water residence (year)
Diefenbaker	106.63	430.0	9400	21.9	510	8.4	331	28.2	29.7	5.0	5.9	3.4	1.5
Buffalo Pound	105.30	29.1	87.5	3.0	602	8.5	524	33.9	20.3	29.5	7.3	1.1	0.7
Last Mountain	105.14	226.6	1807.2	7.9	1844	8.7	874	57.8	31.5	13.3	12.5	2.2	12.6
Wascana	104.67	0.5	0.7	1.4	929	9.0	1449	360.0	6.5	42.4	17.7	0.9	0.05
Pasqua	104.00	20.2	120.8	6.0	913	8.9	1012	147.6	6.9	28.5	10.6	1.1	0.7
Echo	103.49	12.5	122.1	9.8									0.7
Mission	103.44	7.7	62.9	8.2									0.4
Katepwa	103.39	16.2	233.2	14.4	1128	8.7	911	166.3	7.7	23.0	11.0	1.8	1.3
Crooked	102.44	15.0	120.9	8.1	1189	8.7	881	117.0	12.1	28.4	10.8	1.5	0.5
Round	102.22	10.9	83.9	7.7									

increased concentrations of total dissolved phosphorus (P) and chlorophyll *a*
(Chl *a*), and declines in secchi-depth transparency and dissolved N:P ratios in
lakes not directly receiving urban wastewaters (Soranno et al. 1999). Although
blooms of cyanobacteria are common in all lakes by late summer (McGowan
et al. 2005b), strong landscape gradients of N:P favour particularly intense
biological fixation of atmospheric N_2 in eastern lakes (Patoine et al. 2006).

Sediment cores were collected from nine lakes within the Qu'Appelle River
catchment to quantify changes in sediment geochemistry and stable isotope
content (Leavitt et al. 2006), algal production and gross community composition (Hall et al. 1999, Patoine and Leavitt 2006), diatom species composition
(Dixit et al. 2000), and the abundance and taxonomic composition of benthic
invertebrate communities (Quinlan et al. 2002). Sediments were obtained using
freeze-coring procedures (Leavitt and Hodgson 2001) that allowed annually
resolved samples to be recovered from a continuous sediment column that
spanned key events during the past 200 years, including the onset of European-style agriculture (ca. 1890), intensive urbanization (ca. 1930), and installation of tertiary sewage treatment (1977). Sediments were sectioned into 5-mm
intervals that correspond to the mean annual deposition of bulk sediments. The
chronology of each sediment core was established as described in Case 2 but
using alpha spectrometric analysis of ^{210}Pb activities (Appleby and Oldfield
1978). In addition, ^{210}Pb-derived chronologies were validated by comparison
against independent gamma spectrometric determinations of changes in ^{137}Cs
activities that signal the onset (ca. 1955) and maximum (ca. 1964) of atmospheric nuclear weapons tests. As described in Case 2, best-fit models of the
relationship of mass accumulation rates versus burial depth were used to
approximate ages of sediments deposited prior to ca. 1875. Because our intent
was to use variance partitioning analysis to statistically compare changes in
each fossil time series with documented records of climate, land use and urba-
nizations available since 1911, we were especially careful to avoid contamina-
tion among finely resolved samples (see Hall et al. 1999, Leavitt and Hodgson
2001).

As much as possible, all fossil analyses were conducted on the same homo-
genized sediment sample, although the mode of sediment preparation varied
among fossil metrics. Aliquots of samples for analysis of stable isotopes of
carbon (C) and N were freeze-dried completely before combustion in an ele-
mental analyser and the introduction of gases into an isotope ratio mass
spectrometer (IRMS) (Savage et al. 2004). As described in Case 2, stable isotope
ratios were reported in the conventional δ notation with respect to atmospheric
N ($\delta^{15}N$) or standard C reference material ($\delta^{13}C$).

Sedimentary pigments were extracted immediately from freeze-dried material, filtered, and dried under N_2 gas following Leavitt and Hodgson (2001).
Carotenoid, chlorophyll, and derivative concentrations were quantified using
reversed-phase high-performance liquid chromatography (RP- HPLC) calibrated with both internal standards and against authentic reference pigments.
Whole wet sediments were digested in a series of oxidation steps (see Case 1)

to isolate diatom frustules that could be identified to species using interference light microscopy. Finally, chironomid remains were isolated from entire sediments after deflocculation in warm KOH, screening onto 95-μm Nitex (mesh, preliminary visual sorting at 50x magnification, mounting onto permanent glass slides, and identification to genus using light microscopy. Further analytical details are available in our publications on stable isotopes (Leavitt et al. 2006), pigments (Hall et al. 1999), diatoms (Dixit et al. 2000), and fossil invertebrates (Quinlan et al. 2002). Such a multi-proxy approach is increasingly common in paleoecological analyses and takes advantage of the fact that individual stressors often influence geochemical, algal, and invertebrate parameters in different but well documented ways. Interpretation of fossil assemblage change is more reliable in instances where changes in indicator groups are both synchronous and are each consistent with their responses predicted from whole-lake experiments, models, or monitoring programs.

7.4.1.1 Relative Effects of Climate, Land Use and Urbanization on Water Quality

Effective management of the Qu'Appelle lakes has been hampered by the absence of long-term monitoring programs during much of the twentieth century, lack of agreement on the relevance of predictive models derived from boreal lakes, and concomitant variation in several potential forcing mechanisms. Historical records of European colonists, as well as modern surveys over diverse geographic regions, suggest that many prairie lakes may be naturally productive, but provide little insight on whether modern conditions are degraded substantially relative to historical states. Therefore our first research objective was to generate reliable fossil time series of changes in algal abundance, diatom community composition, and fossil invertebrate assemblages, and to compare these statistically with instrumental records of climate change, resource exploitation, and urbanization using variance partitioning analysis (VPA) (see Case 1 above).

Continuous annual time series for 83 variables were collected to determine the impacts of environmental change on water quality (1890–1995) and were assigned to climate, resource-use, or urban categories (Hall et al. 1999). Climate time series included monthly, seasonal, and annual estimates of precipitation, minimum and maximum temperatures, and gross evaporation, as well as ice thaw and freeze dates, and river flow at several locations within the catchment. Resource-use time series included annual records of commercial fish harvest and stocking in individual lakes, and catchment-level estimates of cattle, swine and poultry biomass, field crop area, and total farmland area. Finally, the time course of urbanization was established from annual records of urban population growth in Regina and Moose Jaw, rural populations in the Qu'Appelle Valley, annual influx of total P (TP) and total N (TN) from Regina's sewage treatment plant (STP) to Wascana Creek (Fig. 7.7), and N:P ratios of total influx.

VPA analysis required environmental time series to be harmonized to a uniform period (1920–1993), mainly because the availability and reliability of individual monitoring records differed in the early twentieth century. Similarly, numerical analyses were restricted to fossil time series of relative (%) abundance of the most common diatom and chironomid taxa (>1% abundance) and $\log_{10}(x + 1)$-transformed concentrations of the 13 most common fossil pigments for 1920–1993. In general, VPA was conducted as detailed in Case 1, using a series of constrained and partial canonical ordinations based on redundancy analysis (RDA). In addition, all time series were first smoothed using a three-point running average to focus our investigation on the controls of long-term biological changes rather than inter-annual variability (Hall et al. 1999). Further, because VPA requires similar numbers of historical variables within each explanatory category to avoid bias arising from happenstance correlations, we selected only explanatory variables that accounted for significant amounts of variance in fossil data (critical $\alpha = 0.05$), based on a series of RDAs constrained to a single explanatory variable at a time. Significant variables were assigned to one of the explanatory categories. Finally, we performed a series of RDAs on each category to sequentially eliminate explanatory variables until all variance inflation factors were <20 indicating that none of the remaining environmental variables were co-varying (Hall et al. 1999). This procedure resulted in similar numbers of variables in each predictor category.

Analysis of fossil diatoms, pigments and invertebrates each revealed that all Qu'Appelle lakes were naturally eutrophic prior to agricultural and urban development. For example, fossil diatom assemblages in Pasqua Lake (Fig. 7.9) were composed almost exclusively of species characteristic of nutrient-enriched waters (e.g., *Stephanodiscus niagare*, *Aulaccoseira granulata*, *Fragilaria capucina/bidens*) prior to 1890, while pigments from colonial (myxoxanthophyll) and N_2-fixing cyanobacteria (aphanizophyll) were abundant during the pre-agricultural era, and benthic invertebrates were composed mainly of the chironomid genus *Chironomus* ('bloodworm') a reliable indicator of the oxygen-poor bottom waters common in eutrophic lakes. Although the precise species composition varied among lakes, fossils from eutrophic indicator taxa were abundant, species from nutrient poor lakes were absent, and pigments from cyanobacteria were common in pre-agricultural sediments of all Qu'Appelle lakes (Dixit et al. 2000, Quinlan et al. 2002). Together, these patterns demonstrate clearly that Qu'Appelle lakes should be managed for a naturally productive condition, rather than for a hypothetical oligotrophic status (Hall et al. 1999).

Although Qu'Appelle lakes are naturally productive, fossil pigment analyses demonstrated that total algal abundance increased by three-fold and colonial cyanobacterial abundance by five-fold during the twentieth century in Pasqua Lake, the first site downstream of the city of Regina (Fig. 7.9a). Similarly, sedimentary diatom assemblages changed to become composed mainly of *Stephanodiscus hantzchii*, one of the most reliable species indicators of highly eutrophic waters (Fig. 7.9b). Although the relative taxonomic composition of

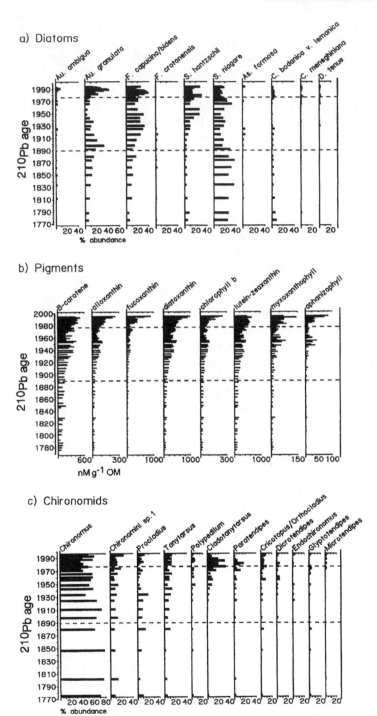

Fig. 7.9 Temporal changes in biological community composition as reflected in the sediment core from Pasqua Lake, based on: percent abundances of common diatom taxa

chironomid fossils remained composed mainly of *Chironomus* spp. (Fig. 7.9c), absolute concentrations of invertebrate headcapsules declined substantially (data not shown), a pattern that is also consist with intensification of deepwater anoxia during worsening eutrophication (Quinlan et al. 2002). Concomitant but less pronounced increases in production, diatom community composition, and benthic invertebrate abundance were also recorded immediately downstream in Echo, Mission, and Katepwa lakes (Hall et al. 1999, Dixit et al. 2000, Quinlan et al. 2002), but not at eastern sites near the base of the catchment, or in headwater lakes that do not receive wastewater from the Regina Sewage Treatment Plant (STP) (see below). Taken together, this geographically structured analysis of lake production and species composition demonstrates to managers that despite naturally high production algal biomass could be reduced by up to 300%, that potentially toxic cyanobacteria should decline to an even greater extent, but beneficial effects of improved management would be restricted to a short chain of four lakes that received wastewater from the City of Regina.

Installation of tertiary wastewater treatment by Regina in 1977 did not improve water quality in Pasqua Lake even after 20 years (Fig. 7.9), despite the fact that P in urban effluent accounted for >80% of P flux in the Qu'Appelle River during the 1960s and early 1970s, and that tertiary treatment reduced total P (TP) release from 120 tonnes year^{-1} to 20 tonnes year^{-1} by 1979 (Hall et al. 1999). Similarly, both diatom and invertebrate assemblages deposited since 1977 were characterized by high inter-annual variability, but showed no evidence of a return to baseline community composition when evaluated by similarity or principal components analyses (Hall et al. 1999). Instead, historical changes in the abundance of bloom-forming algae (as fossil pigments lutein-zeaxanthin) and total cyanobacteria (as echinenone) were linearly correlated with the mass of N released in Regina's effluent but not with TP in wastewaters (Leavitt et al. 2006). This result suggests that pollution with urban N is sustaining eutrophic conditions in Pasqua, Echo, Mission and Katepwa lakes, a finding that is consistent with biweekly nutrient-limitation bioassays during the past 14 years, in situ 2000-L mesocosm experiments, and with whole-lake mass balance studies which demonstrate that most N is exported to lake sediments rather than river outflow (Patoine and Leavitt 2006). In contrast, mass balances conducted during the 1980s revealed that P was predominantly

Fig. 7.9 (continued) (**a**), concentrations (nmole g^{-1} organic matter) of the most abundant chlorophyll and carotenoid pigments (**b**), and percent abundances of common chironomid taxa (**c**). Pigments (b) include β-carotene (all algae), alloxanthin (cryptophytes), fucoxanthin (mainly siliceous algae), diatoxanthin (mainly diatoms), chlorophyll *b* (chlorophytes), lutein-zeaxanthin (chlorophytes and cyanobacteria), myxoxanthophyll (filamentous cyanobacteria), and aphanizophyll (N$_2$-fixing cyanobacteria). *Upper and lower dashed lines* indicate the onset of European-style agriculture in 1890 and the onset of tertiary sewage treatment in Regina in 1977, respectively. Source: Hall et al. (1999), reproduced with permission from Limnology and Oceanography

exported to downstream lakes (Kenney 1990), suggesting that it is presently well in excess of algal demands even following improved wastewater treatment.

Variance partitioning analysis captured 71–97% of variation in fossil composition using only 10–14 of 84 potential predictors (Fig. 7.10) (Hall et al. 1999). In particular, resource use (cropland area, livestock biomass) and urbanization (N in wastewater) were stronger correlates of algal and chironomid community change in Pasqua Lake than were factors related to climate (temperature, evaporation, river discharge), although the explanatory power of individual categories varied with lake position in the landscape and with fossil group. For example, historical variation in regional climate explained a much higher proportion of change in chironomid assemblages in remote lakes than those that received urban wastewater N (Quinlan et al. 2002). Similarly, the proportion of variance in algal production (fossil pigments) explained by land use practices was greatest in headwater lakes and declined with distance downstream, whereas urban effects on algae were most apparent at sites immediately downstream from Regina (Hall et al. unpublished data). Consistent with results of univariate regression analysis, TN from Regina or N: P ratios of effluent were retained in the final set of urban variables in independent analyses of diatoms, algal pigments, and invertebrates, whereas TP was never a significant predictor of lake community change during the twentieth century (Hall et al. 1999). Taken together, these results again point to an important role of nitrogenous wastes from cities and farms as regulators of prairie lake eutrophication.

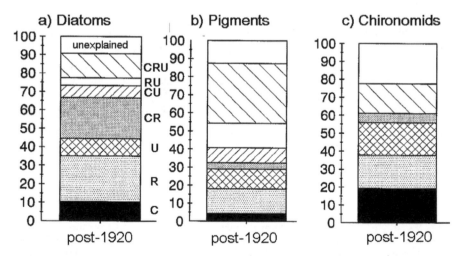

Fig. 7.10 Effects of climate (C), resource use (R) and urban factors (U) on fossil assemblages of diatoms (**a**), pigments (**b**) and chironomids (**c**) from Pasqua Lake determined using variance partitioning analyses with partial canonical ordination. Variance partitioning results for ca. 1920–1993, 1950–1993 and 1970–1993. Pairs of letters (e.g., CR) represent combined effects of factors (i.e., C + R). Source: Hall et al. (1999), reproduced with permission from Limnology and Oceanography

7.4.1.2 Quantification of Nitrogen Effects on Water Quality

Although paleoecological analysis can provide direct measurement of past lake conditions, quantify the direction and rate of ecosystem change, and set clear benchmarks against which lake recovery can be evaluated, many lake managers have been hesitant to base remediation strategies solely on retrospective studies. In particular, lack of familiarity with paleoecological techniques and multivariate analysis of time series can undermine acceptance of management options based on fossil analyses, especially in instances where proposed strategies contradict conventional wisdom (e.g., N vs. P regulation of eutrophication). In the case of the Qu'Appelle Valley lakes, removal of N from Regina wastewaters will require a $70 million upgrade of the Regina STP, a cost that must be borne by a small tax base (ca. 200,000 urban inhabitants); therefore, both government and managers require strong evidence that ecosystem N is derived mainly from urban sources. In addition, concern exists in both scientific and management communities that reduction in the N:P of effluent will promote development of potentially toxic N_2-fixing cyanobacteria, even though N:P ratios are already low in impacted lakes (<6:1) and >80% of all algal biomass is presently composed of cyanobacteria (Leavitt et al. 2006, Patoine et al. 2006). To address these issues, we combined stable isotope analysis of lake sediments with whole-lake mass balance models and catchment-scale isotopic tracer studies to quantify the unique role of urban N in regulating lake eutrophication.

Stable isotopes of N have proven value in identifying sources and distribution of urban wastewater (e.g., Wayland and Hobson 2001), although the precise isotopic signature of effluent depends on the sewage treatment process employed. For example, primary wastewater treatment removed only large particles and lipids; therefore, total solid concentrations remain high and final effluent is only moderately enriched with ^{15}N (3–8%) (deBruyn et al. 2003). In contrast, secondary (colloid removal, microbial processing) and tertiary (chemical flocculation of P) treatment favours sedimentation, ammonia volatilization, and denitrification in clarifying tanks, processes which both greatly enrich ^{15}N content of dissolved N and reduce particle and colloid loads. Under these conditions, $\delta^{15}N$ of final effluent can exceed 20% (Bedard-Haughn et al. 2003). However, because background $\delta^{15}N$ values vary greatly among recipient ecosystems, often due to agricultural activity (Anderson and Cabana 2005), unambiguous identification of urban N sources requires landscape-scale analyses of both impacted sites and those remote from urban wastewaters (Steffy and Kilham 2004).

Our main objectives were to quantify how urban N is transported through a chain of eight prairie lakes and to evaluate how this N regulates landscape patterns of algal production. We hypothesized that if urban N were regulating lake eutrophication, elevated $\delta^{15}N$ signatures characteristic of urban sources would be restricted to rivers and lakes downstream of the STP, and that only these sites should show substantial increases in lake production. These ideas were tested by directly measuring the stable isotope content of Regina's effluent at all stages of sewage processing during two years to estimate how $\delta^{15}N$ signatures may have varied with technological changes during the twentieth century (Leavitt et al. 2006). In addition, we quantified the $\delta^{15}N$ values of total dissolved N, benthic algae (periphyton), particulate organic matter (POM), and sediments collected biweekly from both headwater rivers, and those located downstream from the City of Regina to trace the flow path of urban N to Pasqua Lake and other downstream sites.

Historical changes in the $\delta^{15}N$ signature of sediments of all Qu'Appelle lakes were quantified using IRMS (see Case 2 above) analysis of cores previously collected for determination of fossil pigments, diatoms, and chironomids (Fig. 7.9). Time series of $\delta^{15}N$ change in headwater Buffalo Pound, Wascana and Last Mountain lakes were used to quantify the effects of agriculture and climate change in the absence of urban N influx, whereas Pasqua, Echo, Mission, Katepwa, Crooked, and Round lakes represented a gradient of increasing distance downstream from wastewater sources of N. Two-box mixing models were used to estimate the fraction of total N in Pasqua Lake derived from urban sources (Leavitt et al. 2006). Finally, lake-specific N mass-balance budgets were developed for all Qu'Appelle lakes to determine whether N derived from N_2 fixation and river inflow was stored in lake sediments (indicative of N limitation) or was exported to downstream lakes (indicative of N sufficient conditions). Further details of these analyses are presented in Patoine et al. (2006) and Leavitt et al. (2006).

Stable isotope analysis revealed that wastewater N was transported effectively from Regina to downstream Pasqua Lake where it accounted for ~70% of total ecosystem N (Leavitt et al. 2006). Wastewater N was highly enriched in ^{15}N ($16 \pm 2\%$) relative to dissolved or particulate N from upstream rivers or from sediments or algae from headwater lakes (~6.5%) (Leavitt et al. 2006, Patoine and Leavitt 2006). However, N isotope signatures of both dissolved and periphytic N in Wascana Creek increased seasonally from 8 to 15% at sites immediately downstream of the STP and remained highly enriched throughout the 175 km transit of river water to Pasqua Lake. In contrast, there was little (<2%) spatial or temporal variation in N isotope ratios at western reference sites in agricultural catchments (e.g., upper Qu'Appelle River, Moose Jaw Creek).

Once at Pasqua Lake, urban N was rapidly incorporated into algal biomass and deposited into the lake sediments (Leavitt et al. 2006), leading to a substantial increase in the $\delta^{15}N$ signature of surface deposits (12.8%) relative to both pre-agricultural sediments at that site and those observed in headwater reference lakes (~6.5%) (Fig. 7.11a). Similar but less pronounced historical changes were observed clearly in downstream Echo, Mission and Katepwa lakes, but not in eastern-most Round or Crooked lake (Fig. 7.11b). Application of a suite of two-box mixing models revealed that 41–80% of total N in Pasqua Lake was derived from urban sources (Leavitt et al. 2006), but this proportion declined with distance downstream due to the combined effects of sequestration

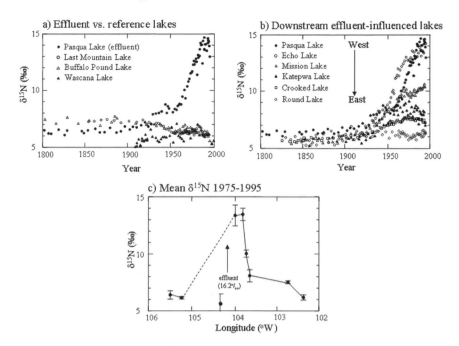

Fig. 7.11 Historical changes in $\delta^{15}N$ of bulk sediments from Pasqua Lake and its upstream reference ecosystems (**a**), and from Qu'Appelle lakes receiving urban wastewaters (**b**). Mean ± SD $\delta^{15}N$ values for sediments deposited during 1975–1995 are presented in panel (**c**). Reproduced with permission from Limnology and Oceanography (Leavitt et al. 2006)

in algae and lake sediments, and by the increasing dilution of ^{15}N signatures by the biological fixation of isotopically depleted (~0%) atmospheric N_2 (Patoine et al. 2006). Taken together, these analyses establish clearly that urban rather than agricultural N is the main component of lakes downstream from the City of Regina.

Several lines of evidence suggest that the eutrophication of Pasqua Lake was caused by influx of N from urban sources (Fig. 7.12). First, there was strong linear relationship ($r^2 = 0.84$) between total mass of N released from Regina's STP and the $\delta^{15}N$ signature of Pasqua Lake sediments during the twentieth century (Fig. 7.12a), consistent with both transportation pathway studies and mixing models based on stable isotopes. Second, both N flux from Regina and sedimentary $\delta^{15}N$ were highly correlated ($r^2 = 0.69$) to changes in concentrations of fossil pigments from both bloom-forming algae (lutein-zeaxanthin) and total cyanobacteria (echinenone) (Fig. 7.12b). In contrast, historical variations in fossil pigment concentration (Fig. 7.9b) were always uncorrelated ($r^2 < 0.1$) with changes in the total quantity of P released from Regina (Hall et al. 1999). Finally, comparison of landscape patterns of change in algal abundance (Fig. 7.12c) with those of sedimentary $\delta^{15}N$ (Fig. 7.11c) revealed that lake eutrophication was only evident at sites with unambiguous isotopic evidence

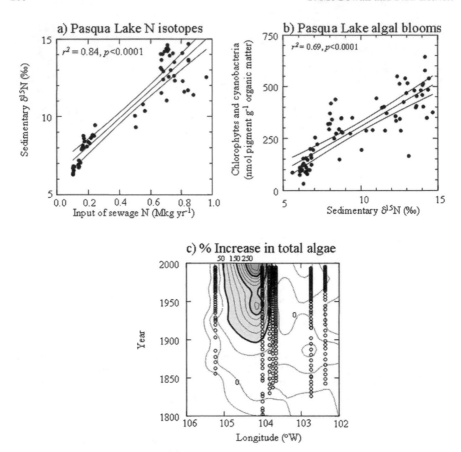

Fig. 7.12 Linear regression of sedimentary $\delta^{15}N$ from Pasqua Lake as a function of total N efflux from the City of Regina 1875–1995 (**a**), and as a predictor of past changes in the abundance of bloom-forming algae (chlorophytes, cyanobacteria; as nmol fossil lutein-zeaxanthin g^{-1} organic matter) during 1780–1995 (**b**). Panel (**c**) represents percentage increase in total algal abundance (between ~1800 and 1880) in all Qu'Appelle lakes (as ubiquitous fossil β-carotene) when standardized against basin-specific levels of fossil algae existing prior to the onset of agriculture (ca. 1890). Wastewater enters the Qu'Appelle immediately upstream of Pasqua Lake (~104°W longitude). Results from Wascana Lake are not included because the lake was deepened in 1931. Contour lines in (**c**) are interpolated between lakes using distance-weighted least-squares regression in SYSTAT v. 10. See Leavitt et al. (2006) for additional details. Source: Leavitt et al. (2006), reproduced with permission from Limnology and Oceanography

of urban N inputs, a pattern which eliminates diffuse nutrients as the cause of eutrophication. This result is consistent with whole-lake N mass balances that demonstrate 40–80% of total N inputs are sequestered in the sediments of successive lake basins and show that chains of freshwater lakes can serve as an effective filter to remove urban N pollution (Leavitt et al. 2006).

7.4.2 Management Insights, Caveats, and Future Directions

Insights from the combined use of stable isotope analyses, whole-lake mass balances, long-term monitoring, and paleoecology resulted in specific recommendations to government, regional managers, and urban engineers. First, we suggested that future improvements in water quality will be linked to reductions in N influx when lakes are replete with P from natural or anthropogenic sources. Although specific to the P-rich prairie lakes, we note that N influx may also regulate eutrophication of lakes in regions where agricultural management during the nineteenth and early twentieth centuries has increased soil P content to the point that diffuse runoff of P may no longer control algal growth (e.g., Bunting et al. 2007). Second, we proposed that sewage treatment plants should be upgraded to include biological nutrient removal (BNR) protocols that will remove over 90% of total N load through microbial denitrification. Declines in N flux are expected to reduce total cyanobacterial abundance by up to 500% (Fig. 7.9b), while restricting the high rates of toxin production which are common in rapidly growing cyanobacteria (Giani et al. 2005). At present, concentrations of the hepatotoxin microcystin are 3–10 fold greater in N-polluted Pasqua Lake than in reference lakes or sites well downstream from urban N sources (D. Donald and P.R. Leavitt, unpublished data). Third, we suggested that watershed managers clearly inform special-interest groups concerning the degree of water quality improvement expected for each lake. Although substantial improvements are expected for Pasqua, Echo, and Mission lakes, algal biomass should change relatively little at other sites (Fig. 7.9b, 7.12c). Finally, because mass-balances suggest that internal loading from sediments may supply 22–55% of water-column N, we suggested that complete lake recovery may take decades (c.f. Battarbee et al. 2005) and that public expectations must be scaled correctly by the rate of ecosystem response to nutrient diversion. However, it is important to note that slow ecosystem response should not be used as justification to avoid expensive management of N, as both monitoring and retrospective studies suggest that water quality will continue to decline in the future.

Despite improved understanding of the role of N in degrading surface water quality, there are several unresolved issues that continue to constrain development of effective management strategies in these and other P-rich lakes. First, relatively little is known of how the precise chemical form of N may interact with the availability of light to alter the effects of N on algal growth. To date, strong correlations between N influx and algal production have been demonstrated only in unstratified eutrophic lakes (e.g., Leavitt et al. 2006, Bunting et al. 2007), sites in which light energy may be insufficient to allow substantial biological fixation of atmospheric N_2 and where allochthonous N may comprise a high proportion of total N influx. At present, little is known of whether the same patterns would occur in thermally stratified lakes in which algae may remain suspended within the photic zone. Similarly, more information is needed on the relative effects of organic and inorganic forms of dissolved N on algal growth. For example, urea and other organic compounds are common farm fertilizers, a common component of total N in wastewaters, and may preferentially stimulate the growth of cyanobacteria and other toxic algae in some aquatic ecosystems (Glibert et al. 2004). Finally, improved understanding of the biogeochemistry of stable isotopes of C and N may help simplify interpretation of sedimentary time series and the causes of spatial variation among ecosystems.

7.5 Summary

In principle, paleoecological techniques can be applied to any issue in which the passage of time plays a central role in understanding the regulation of ecosystem structure or function. In our opinions, many successful retrospective studies exhibit several common features related to study design and analysis. First, we find that fossil analysis are more likely to yield important insights on ecosystem function if retrospective studies are conducted at multiple sites with contrasting geographic position relative to known or suspected forcing agents. In the same manner as experiments seek to gain insights from moving ecosystems among contrasting states (oligotrohic to eutrophic), paleoecological studies should be designed to maximize the contrast among sites arranged along well-defined environmental gradients. In this regard, use of reference systems ('controls') is as important in paleoecology as in any other branch of modern ecology, as is the requirement to base testable hypotheses on clearly defined mechanisms.

Second, we believe that accurate interpretation of fossil records requires sophisticated understanding of the mechanisms by which fossil records are formed, including the basic ecology, biogeochemistry, and energetic constraints of the system being studied. For example, simulation models of algal growth, degradation, and deposition have greatly improved the understanding of pigment biogeochemistry and allowed investigators to distinguish among confounding mechanisms that alter sedimentary time series (e.g., production vs. preservation) (Cuddington and Leavitt 1999).

Third, although evaluation of causal mechanisms is always difficult with retrospective analyses, we feel that more robust insights can be gained from studies that quantify a diverse suite of geochemical and biological time series. Because many organisms exhibit contrasting responses to common environmental forcing, the use of multiple proxies of environmental change can help avoid errors associated with happenstance variation in individual indicators or hidden causal mechanisms (lurking correlations). Similarly, recent advances in multivariate numeric techniques (e.g., variance partitioning analysis) have helped paleoecologists to better quantify the unique and covarying effects of multiple concomitant stressors (e.g., Case 3).

Finally, we suggest that future advances in paleoecology will arise from integrating the retrospective approach into a common theoretical framework and methodological toolkit, rather than by enforcing pre-existing disciplinary distinctions. In this regard, we believe that paleoecology provides unique temporal insights to ecology, but it would also benefit from a more explicit theoretical framework.

Acknowledgments This research was funded by an EU Marie Curie Fellowship (SMcG), NSERC Canada, the Canada Research Chair program, Saskatchewan Learning, the National Science Foundation (grant DEB-0553766 to the National Center for Ecological Analysis and Synthesis during P.R.L. sabbatical), the Betty and Gordon Moore Foundation, the University of Notttingham (UK) and University of Regina (Canada). The authors also gratefully acknowledge the diverse contributions of past and present members of the University of Regina Limnology Laboratory, as well as our colleagues at University of Washington (D. E. Schindler), University of Loughborough (N. J. Anderson), Geological Survey of Denmark and Greenland (R. K. Juhler), Queen's University (J. P. Smol, S. Dixit, A. Dixit), York University (R. Quinlan), McGill University (I. Gregory-Eaves) and University of Alaska Fairbanks (B. Finney). We thank the four anonymous reviewers for their insights in improving earlier versions of this chapter.

References

Allen, R. J., and B. C. Kenney 1978. Rehabilitation of eutrophic prairie lakes. Mitteilungen Internationale Vereinigung für theoretische und angewandte Limnologie **20:** 214–224.

Anderson, C., and G. Cabana 2005. $\delta^{15}N$ in riverine food webs: effects of N inputs from agricultural watersheds. Canadian Journal of Fisheries and Aquatic Sciences **62:** 333–340.

Anderson, N. J., R. Harriman, D. B. Ryves, and S. T. Patrick. 2001. Dominant factors controlling variability in the ionic composition of West Greenland lakes. Arctic, Antarctic and Alpine Research **33:** 418–425.

Anderson, N. J., A. Clarke, R. K. Juhler, S. McGowan, and I. Renberg. 2000. Coring of laminated sediments for pigment and mineral magnetic analyses, Søndre Strømfjord, West Greenland. Geology of Greenland Survey Bulletin **186:** 83–87.

Appleby, P. G., and F. Oldfield. 1978. The calculation of lead-210 dates assuming a constant rate of supply of unsupported 210Pb to the sediment. Cantena **5:** 1–8.

Battarbee, R. W., N.J. Anderson, E. Jeppesen, and P.R. Leavitt. 2005. Combining palaeolimnological and limnological approaches in assessing lake ecosystem response to nutrient reduction. Freshwater Biology **50:** 1772–1780.

Battarbee, R. W. 1999. The importance of palaeolimnology to lake restoration. Hydrobiologia **395/396:** 149–159.

Bedard-Haughn, A., J. W. van Groenigen, and C. van Kessel. 2003. Tracing N-15 through landscapes: potential uses and precautions. Journal of Hydrology **272:** 175–190.

Bennett, E. M., S. R. Carpenter, and N. F. Caraco. 2001. Human impact on erodable phosphorus and eutrophication: A global perspective. Biosciences **51:** 227–234.

Bennett, E. M., T. Reed-Anderson, J. N. Houser, J. R. Gabriel, and S. R. Carpenter. 1999. A phosphorus budget for the Lake Mendota watershed. Ecosystems **2:** 69–75.

Bennion, H., S. Juggins, and N. J. Anderson. 1996. Predicting epilimnetic phosphorus concentrations using an improved diatom-based transfer function and its application to lake eutrophication management. Environmental Science and Technology **30:** 2004–2007.

Berger, A. L. 1978. Long-term variations of calorific insolation resulting from the earth's orbital elements. Quaternary Research **9:** 139–67.

Birks, H. J. B. 1998. D. G. Frey & E. S. Deevey Review #1. Numerical tools in palaeolimnology-Progress, potentialities, and problems. Journal of Paleolimnology **20**: 307–332.

Blenckner, T. 2005. A conceptual model of climate-related effects on lake ecosystems. Hydrobiologia **533**: 1–14.

Borcard, D. and P. Legendre. 1994. Environmental control and spatial structure in ecological communities: an example using oribatid mites (Acari, Oribatei). Environmental and Ecological Statistics **1**: 37–61.

Borcard, D., P. Legendre and P. Drapeau. 1992. Partialling out the spatial component of ecological variation. Ecology **73**: 1045–1055.

Brahney, J., D. G. Bos, M. G. Pellatt, T. W. D. Edwards, and R. Routledge. 2006. The influence of nitrogen limitation on $\delta^{15}N$ and carbon: nitrogen ratios in sediments from sockeye salmon nursery lakes in British Columbia, Canada. Limnology and Oceanography **51**: 2333–2340.

Brock, C. S., P.R. Leavitt, D. E. Schindler, and P. D. Quay. 2007. Variable effects of marine-derived nutrients on algal production in salmon nursery lakes of Alaska during the past 300 years. Limnology and Oceanography **52**: 1588–1598.

Brock, C. S., P.R. Leavitt, D. E. Schindler, S. P. Johnson, and J. W. Moore. 2006. Spatial variability of stable isotopes and fossil pigments in surface sediments of Alaskan coastal lakes: constraints on quantitative estimates of past salmon abundance. Limnology and Oceanography **51**: 1637–1647.

Bunting, L., P.R. Leavitt, C. E. Gibson, E. J. McGee, and V. A. Hall. 2007. Degradation of water quality in Lough Neagh, Northern Ireland, by diffuse nitrogen flux from a phosphorus-rich catchment. Limnology and Oceanography **52**: 354–369.

Campbell, C. E., and E. E. Prepas. 1986. Evaluation of factors related to the unusually low chlorophyll levels in prairie saline lakes. Canadian Journal of Fisheries and Aquatic Sciences **43**: 846–854.

Carpenter, S. R., N.F. Caraco, D. L. Correll, R. W. Howarth, A. N. Sharpley and V. H. Smith. 1998. Nonpoint pollution of surface waters with phosphorus and nitrogen. Ecological Applications **8**: 559–568.

Clark, J. H. 2005. Abundance of sockeye salmon in the Alagnak River system of Bristol Bay Alaska. Anchorage: Alaska Department of Fish and Game.

Cottingham, K. L., J. A. Rusak, and P.R. Leavitt. 2000. Increased ecosystem variability and reduced predictability following fertilization: evidence from palaeolimnology. Ecology Letters **3**: 340–348.

Cuddington, K., and P.R. Leavitt. 1999. An individual-based model of pigment flux in lakes: implications for organic biogeochemistry and paleoecology. Canadian Journal of Fisheries and Aquatic Sciences **56**: 1964–1977.

Cumming, B. F., K. R. Laird, J. R. Bennet, J. P. Smol, and A. K. Salomon. 2002. Persistent millenial-scale shifts in moisture regimes in western Canada during the past six millenia. Proceedings of the National Academy of Sciences **99**: 16117–16121.

Dahl-Jensen, D., K. Mosegaard, N. Gundestrup, G. D. Clow, S. J. Johnsen, A. W. Hansen, and N. Balling. 1998. Past temperatures directly from the Greenland Ice sheet. Science. **282**: 268–271.

Davies, J.-M. 2006. Application and tests of the Canadian water quality index for assessing changes in water quality in lakes and rivers of central North America. Lake and Reservoir Management **22**: 308–320.

deBruyn, A. M. H., D. J. Marcogliese, and J. B. Rasmussen. 2003. The role of sewage in a large river food web. Canadian Journal of Fisheries and Aquatic Sciences **60**: 1332–1344.

Dixit, A. S., R. I. Hall, P.R. Leavitt, R. Quinlan, and J. P. Smol. 2000. Effects of sequential depositional basins on lake response to urban and agricultural pollution: a palaeoecological analysis of the Qu'Appelle Valley, Saskatchewan, Canada. Freshwater Biology **43**: 319–337.

Engstrom, D. R., S. C. Fritz, J. E. Almendinger, and S. Juggins. 2000. Chemical and biological trends during lake evolution in recently deglaciated terrain. Nature **408**: 161–166.

Fair, L. F. 2003. Critical elements of Kvichak River sockeye salmon management. Alaskan Fisheries Research Bulletin **10:** 95–103.

Finney, B. P., I. Gregory-Eaves, M. S. V. Douglas, and J. P. Smol. 2002. Fisheries productivity in the northeastern Pacific Ocean over the past 2,200 years. Nature **416:** 729–733.

Finney, B. P., I. Gregory-Eaves, J. Sweetman, M. S. V. Dougas, and J. P. Smol. 2000. Impacts of climatic change and fishing on Pacific salmon abundance over the past 300 years. Science **290:** 795–799.

Foy, R. H., S. D. Lennox, and C. E. Gibson. 2003. Changing perspectives on the importance of urban phosphorus inputs as the cause of nutrient enrichment in Lough Neagh. Science of the Total Environment **310:** 87–99.

Fritz, S. C., S. Juggins, R. W. Battarbee, and D. R. Engstrom. 1991. Reconstruction of past changes in salinity using a diatom-based transfer function. Nature **352:** 706–708.

George, D. G., J. F. Talling, and E. Rigg. 2000. Factors influencing the temporal coherence of five lakes in the English Lake District. Freshwater Biology **43:** 449–461.

Giani, A., D. F. Bird, Y. T. Prairie, and J. F. Lawrence. 2005. Empirical study of cyanobacterial toxicity along a trophic gradient of lakes. Canadian Journal of Fisheries and Aquatic Sciences **62:** 2100–2109.

Glibert, P. M., C. A. Heil, D. Hollander, M. Revilla, A. Hoare, J. Alexander, and S. Murasko. 2004. Evidence for dissolved organic nitrogen and phosphorus uptake during a cyanobacterial bloom in Florida Bay. Marine Ecology Progress Series **280:** 73–83.

Groot, C., and L. Margolis. 1991. Pacific salmon life histories. UBC Press.

Hall, R. I., P. R. Leavitt, R. Quinlan, A. S. Dixit, and J. P. Smol. 1999. Effects of agriculture, urbanization and climate on water quality in the Northern Great Plains. Limnology and Oceanography **44:** 739–756.

Hall, R. I., and J. P. Smol. 1999. Diatoms as indicators of lake eutrophication. In: The diatoms: applications for the environmental and earth sciences, ed. E. F. Stoermer and J. P. Smol, pp. 128–169. Cambridge: Cambridge University Press.

Hammer, U. T. 1973. Eutrophication and its alleviation in the upper Qu'Appelle River system, Saskatchewan, Canada. Hydrobiologia **37:** 473–507.

Hasholt, B., and H. Søgaard. 1978. Et forsøg på en klimatisk-hydrologisk regionsinddeling af Holsteinsborg Kommune (Sisimiut). Geografisk Tidsskrift **77:** 72–92.

Heegaard, E., H. J. B. Birks and R. J. Telford. 2005. Relationships between calibrated ages and depth in stratigraphical sequences: an estimation procedure by mixed-effect regression. Holocene **15:** 612–618.

Hilborn, R., T. P. Quinn, D. E. Schindler, and D. E. Rogers. 2003. Biocomplexity and fisheries sustainability. Proceedings of the National Academy of Sciences **100:** 6564–6568.

Hobbs, W. O., and A. P. Wolfe. 2007. Caveats on the use of paleolimnology to infer Pacific salmon returns. Limnology and Oceanography **52:** 2053–2061.

Hyatt, K. D. 1985. Responses of sockeye salmon (*Oncorhynchus nerka*) to fertilization of British Columbia coastal lakes. Canadian Journal of Fisheries and Aquatic Sciences **42:** 320–331.

IPCC. (2007). Summary for policymakers. In Climate Change 2007: the physical science basis. Contribution of working group I to the Fourth Assessment Report of the Intergovernmental Panel on Climate Change, ed. S. Solomon, D. Qin, M. Manning, Z. Chen, M. Marquis, K. B. Averyt, M. Tignor and H. L. Miller. Cambridge: Cambridge University Press.

Ives, A. R., B. Dennis, K. L. Cottingham, and S. R. Carpenter. 2003. Estimating community stability and ecological interactions from time-series data. Ecological Monographs **73:** 301–330.

James, C., J. Fisher, and B. Moss. 2003. Nitrogen-driven lakes: the Shropshire and Cheshire Meres? Archiv für Hydrobiologie **158:** 249–266.

Jeppesen, E., P.R. Leavitt, L. De Meester, and J. P. Jensen. 2001. Functional ecology and palaeolimnology: Using cladoceran remains to reconstruct anthropogenic impact. Trends in Ecology and Evolution **16:** 191–198.

Kenney, B. C. 1990. Lake dynamics and the effects of flooding on total phosphorus. Canadian Journal of Fisheries and Aquatic Sciences **47:** 480–485.

Kilinc, S., and B. Moss. 2002. Whitemere, a lake that defies some conventions about nutrients. Freshwater Biology **47:** 207–218.

Laird, K. R., B. F. Cumming, S. Wunsam, J. Rusak, R. J. Oglesby, S. C. Fritz, and P. R. Leavitt. 2003. Lake sediments record large-scale shifts in moisture regimes across the northern prairies of North America during the past two millenia. Proceedings of the National Academy of Sciences **100:** 2438–2488.

Leavitt, P. R., C. S. Brock, C. Ebel, and A. Patoine. 2006. Landscape-scale effects of urban nitrogen on a chain of freshwater lakes in central North America. Limnology and Oceanography **51:** 2262–2277.

Leavitt, P. R., and D. A. Hodgson. 2001. Sedimentary pigments. In Tracking environmental changes using lake sediments, vol. 3: Terrestrial, algal and siliceous indicators, eds. J. P. Smol, H. J. B. Birks and W. M. Last. Dordrecht: Kluwer Academic Publishers.

Leavitt, P. R., D. L. Findlay, R. I. Hall, and J. P. Smol. 1999a. Algal responses to dissolved organic carbon loss and pH decline during whole-lake acidification: evidence from paleolimnology. Limnology and Oceanography **44:** 757–773.

Leavitt, P. R., R. D. Vinebrooke, R. I. Hall, S. E. Wilson, J. P. Smol, R. E. Vance, and W. M. Last. 1999b. Multiproxy record of prairie lake response to climatic change and human activity: Clearwater Lake, Saskatchewan. Bulletin of the Geological Survey of Canada **534:** 125–138.

Lepš, J., and P. Šmilauer. 2003. Multivariate analysis of ecological data using CANOCO. Cambridge: Cambridge University Press.

Magnuson, J. J., and T. K. Kratz. 2000. Lakes in the landscape: approaches to regional limnology. Verhandlungen Internationale Vereinigung für theoretische und angewandte Limnologie **27:** 74–87.

Mantua, N. J., S. R. Hare, Y. Zhang, J. M. Wallace, and R. C. Francis. 1997. A Pacific interdecadal climate oscillation with impacts on salmon production. Bulletin of the American Meteorological Society **78:** 1069–1079.

McGowan, S., R. K. Juhler, and N.J. Anderson. 2008. Autotrophic response to lake age, conductivity and temperature in two West Greenland lakes. Journal of Paleolimnology **39:** 301–317.

McGowan, S., P. R. Leavitt, R. I. Hall, N. J. Anderson, E. Jeppesen, and B. V. Odgaard. 2005a. Controls of algal abundance and community composition during ecosystem state change. Ecology **86:** 2200–2211.

McGowan, S., A. Patoine, M. D. Graham, and P. R. Leavitt. 2005b. Intrinsic and extrinsic controls on lake phytoplankton synchrony. Verhandlungen Internationale Vereinigung für theoretische und angewandte Limnologie **29:** 794–798.

McGowan, S., D. B. Ryves, and N.J. Anderson. 2003. Holocene records of effective precipitation in West Greenland. Holocene **13:** 239–249.

Naiman, R. J., R. E. Bilby, D. E. Schindler, and J. M. Helfield. 2002. Pacific salmon, nutrients, and the dynamics of freshwater and riparian ecosystems. Ecosystems **5:** 399–417.

Oldfield, F., P. R. J. Crooks, D. D. Harkness, and G. Petterson. 1997. AMS radiocarbon dating of organic fractions from varved lake sediments: an empirical test of reliability. Journal of Paleolimnology **18:** 87–91.

O'Sullivan, P. E. 2004. Palaeolimnology. In The lakes handbook: Limnology and limentic ecology, Volume 1, eds. P. E. O'Sullivan and C. S. Reynolds. Oxford: Blackwell Publishing.

Patoine, A., M. D. Graham, and P. R. Leavitt, P. R. 2006. Spatial variation of nitrogen fixation in lakes of the northern Great Plains. Limnology and Oceanography **51:** 1665–1677.

Patoine, A. and P. R. Leavitt. 2006. Century-long synchrony of algal fossil pigments in a chain of Canadian prairie lakes. Ecology **87:** 1710–1721.

Pham, S. V., P. R. Leavitt, S. McGowan, and P. Peres-Neto. 2008. Spatial variability of climate and land-use effects on lakes of the northern Great Plains. Limnology and Oceanography **53:** 728–742.

Philibert, A., Y. T. Prairie, and C. Carcaillet. 2003. 1200 years of fire impact on biogeochemistry as inferred from high resolution diatom analysis in a kettle lake from the Picea mariana-moss domain (Quebec, Canada). Journal of Paleolimnology **30:** 167–181.

Quinlan, R., P. R. Leavitt, A. S. Dixit, R. I. Hall, and J. P. Smol. 2002. Landscape effects of climate, agriculture, and urbanization on benthic invertebrate communities of Canadian prairie lakes. Limnology and Oceanography **47:** 378–391.

Racca, J. M. J., R. Racca, R. Pienitz, and Y. T. Prairie. 2007. PaleoNet: new software for building, evaluating and applying neural network based transfer functions in paleoecology. Journal of Paleolimnology **38:** 467–472.

Racca, J. M. J., I. Gregory-Eaves, R. Pienitz, and Y. T. Prairie. 2004. Tailoring palaeolimnological diatom-based transfer functions. Canadian Journal of Fisheries and Aquatic Sciences **61:** 2440–2454.

Racca, J. M. J., M. Wild, H. J. B. Birks, and Y. T. Prairie. 2003. Separating wheat from chaff: Diatom taxon selection using an artificial neural network pruning algorithm. Journal of Paleolimnology **29:** 123–133.

Ruckelshaus, M. H., P. Levin, J. B. Johnson, and P. M. Kareiva. 2002. The Pacific salmon wars: What science brings to the challenge of recovering species. Annual Review of Ecology and Systematics **33:** 665–706.

Ryves, D. B., S. McGowan, and N.J. Anderson. 2002. Development and evaluation of a diatom-conductivity model from lakes in West Greenland. Freshwater Biology **47:** 995–1014.

Savage, C., P. R. Leavitt, and R. Elmgren. 2004. Distribution and retention of effluent nitrogen in surface sediments of a coastal bay. Limnology and Oceanography **49:** 1503–1511.

Schindler, D. E., P. R. Leavitt, S. P. Johnson, and C. S. Brock. 2006. A 500-year context for the recent surge in sockeye salmon (Oncorhynchus nerka) abundance in the Alagnak River, Alaska. Canadian Journal of Fisheries and Aquatic Sciences **63:** 1439–1444.

Schindler, D. E., P. R. Leavitt, C. S. Brock, S.P Johnson, and P. D. Quay. 2005. Marine-derived nutrients, commercial fisheries, and production of salmon and lake algae in Alaska. Ecology **86:** 3225–3231.

Schindler, D. W. 2006. Recent advances in the management and understanding of eutrophication. Limnology and Oceanography **51:** 356–363.

Scully, N. M., P. R. Leavitt, and S. R. Carpenter. 2000. Century-long effects of forest harvest on the physical structure and autotrophic community of a small temperate lake. Canadian Journal of Fisheries and Aquatic Sciences **57:** 50–59.

Smith, D. W., E. E. Prepas, G. Putz, J. M. Burke, W. L. Meyer, and I. Whitson. 2003. The Forest Watershed and Riparian Disturbance study: a multi-discipline initiative to evaluate and manage watershed disturbance on the Boreal Plain of Canada. Journal of Environmental and Engineering Science **2:** S1–S13 Suppl. 1.

Smol, J. P. 2002. Pollution of lakes. A paleoenvironmental perspective. London: Arnold.

Smol, J. P., and B. F. Cumming. 2000. Tracking long-term changes in climate using algal indicators in lake sediments. Journal of Phycology **36:** 986–1011.

Soranno, P. A., K. E. Webster, J. L. Riera, T. K. Kratz, J. S. Baron, P. A. Bukaveckas, G. W. Kling, D. S. White, N. Caine, R. C. Lathrop, and P. R. Leavitt. 1999. Spatial variation among lakes within landscapes: Ecological organization along lake chains. Ecosystems **2:** 395–410.

Steffy, L. Y., and S. S. Kilham. 2004. Elevated $\delta^{15}N$ in stream biota in areas with septic tank systems in an urban watershed. Ecological Applications **14:** 637–641.

Stewart-Oaten, A. and W. W. Murdoch. 1986. Environmental impact assessment: "Pseudoreplication in time?". Ecology **67:** 929–940.

Stockner, J. G. 2003. Nutrients in salmonid ecosystems: Sustaining production and biodiversity. American Fisheries Society Symposium 34.

Stone, J. R., and S. C. Fritz. 2006. Three-dimensional modeling of lacustrine diatom habitat areas: Improving paleolimnological interpretation of planktic: benthic ratios. Limnology and Oceaongraphy **49**: 1540–1548.

Stuiver, M., P. J. Reimer, E. Bard, J. W. Beck, G. S. Burr, K. S. Hughen, B. Kromer, G. McCormac, J. Van der Plicht, and M. Spurk. 1998. INTCAL98 Radiocarbon age calibration, 24,000-0 cal BP. Radiocarbon **40**: 1041–1083.

Telford, R. J., and H. J. B. Birks. 2005. The secret assumption of transfer functions: problems with spatial autocorrelation in evaluating model performance. Quaternary Science Reviews **24**: 2173–2179.

Telford, R. J., C. Andersson, H. J. B. Birks, and S. Juggins. 2004. Biases in the estimation of transfer function prediction errors. Paleoceanography **19**: Art. No. PA4014.

ter Braak, C. J. F., and P. Smilauer. 1998. CANOCO Reference Manual and User's Guide to Canoco for Windows: Software for Canonical Ordination (version 4). Ithaca: Microcomputer Power.

Torrence, C., and G. P. Compo. 1998. A practical guide to wavelet analysis. Bulletin of the American Meteorological Society **79**: 61–78.

Verleyen, E., D. A. Hodgson, P. R. Leavitt, K. Sabbe, and W. Vyverman. 2004. Quantifying habitat-specific diatom production: A critical assessment using morphological and biogeochemical markers in Antarctic marine and lake sediments. Limnology and Oceanography **49**: 1528–1539.

Wayland, M., and K. A. Hobson. 2001. Stable carbon, nitrogen, and sulfur isotope ratios in riparian food webs on rivers receiving sewage and pulp-mill effluents. Canadian Journal of Zoology **79**: 5–15.

Willis, K. J., and H. J. B. Birks. 2006. What is natural? The need for a long-term perspective in biodiversity conservation. Science. **314**: 1261–1265.

Chapter 8
A Spatially Explicit, Mass-Balance Analysis of Watershed-Scale Controls on Lake Chemistry

Charles D. Canham and Michael L. Pace

Abbreviations AIC – Aikaike's Information Criterion, ALSC – Adirondack Lakes Survey Corporation, ANC – acid neutralizing capacity, DEM – digital elevation model, DOC – dissolved organic carbon, EPA – Environmental Protection Agency, Fe – iron, GIS – Geographic Information System, Hg – mercury, N – nitrogen, P – phosphorus, RMSE – root mean squared error, SA – lake surface area, SAI – surface area input

8.1 Introduction

Most ecological research approaches are best suited to small scale experiments and single system models. This situation partly reflects the complexity of ecological problems but is also a function of orientation. Ecologists are primarily trained to work at scales that can be most easily encompassed by traditional field methods (e.g., experimentation, observation). Many scientific as well as management problems, however, need to be confronted on larger scales. One tactic is to scale-up traditional experiments and modeling efforts. Another tactic is to think differently and develop alternate methods. Here, we present a method for evaluating the relationship between watersheds and lake chemistry. The approach requires large numbers of watershed-lake systems to implement, and hence the approach is an inherently large-scale, multi-watershed method.

The subject of chemical constituents loading in lakes has a long history of research that is of enormous practical and theoretical significance. The loading concept provides an explicit focus on the transfer of vital (or harmful) materials from a watershed to a lake. The study of this transfer across ecosystem boundaries was the foundation of successful modeling and management of problems like lake eutrophication and acidification. Traditionally, nutrient loading in lakes is characterized using either highly mechanistic models based on detailed

C.D. Canham (✉)
Cary Institute of Ecosystem Studies, Box AB, Millbrook, NY 12545, USA
e-mail: ccanham@ecostudies.org

S. Miao et al. (eds.), *Real World Ecology*, DOI 10.1007/978-0-387-77942-3_8,
© Springer Science+Business Media, LLC 2009

case studies of individual watersheds, or much more descriptive regression models based on comparative analyses of a large number of lakes (Vollenweider 1976, Reckhow and Chapra 1983, Boyer et al. 1996, Soranno et al. 1996). Detailed hydrologic models are difficult to parameterize and implement across many watersheds, and thus have limitations for use in regional analyses. Regression models typically lack spatial detail and use a wide array of variables (such as mean slope of the watershed) that are clearly just surrogates for other, more implicitly causal, parameters. In the simplest models, the movement of solutes from terrestrial to aquatic ecosystems is often characterized in terms of area-weighted exports from watersheds to receiving waters. These models traditionally represent inputs for a year either as an average value for the entire watershed area, or as some function of the relative abundance of land-cover types (Reckhow and Simpson 1980, Soranno et al. 1996). Most empirical regression models (e.g., Rasmussen et al. 1989, Kortelainen 1993, Houle et al. 1995, D'Arcy and Carignan 1997) have not considered spatially explicit data, leaving open questions about how different cover types contribute to and how exports from different areas are affected by distance from the receiving waters (Gergel et al. 1999). Many current models do not utilize mass balance but instead transform inputs via "retention" factors to predict concentration (e.g., empirical phosphorus models, see Reckhow and Chapra 1983). This approach does not explicitly consider washout, burial in sediments, or chemical and biological processes that degrade or transform materials in the lake.

We have developed an intermediate approach to analysis of the linkages between watersheds and lakes (Canham et al. 2004). The analysis is based on a simple spatially explicit, watershed-scale model of lake chemistry. The model is based on mass balance and predicts the steady-state mid-summer concentration of any specific nutrient in a lake. Our approach maintains a mechanistic foundation while combining the power of data from a very large number of watersheds. The approach is specifically designed to exploit the growing availability of spatially explicit data from watersheds mapped over large regions. Specific hypotheses are tested by comparing alternate forms of the model, using model comparison methods and information theory statistics. Many of the most important results, however, come directly from the maximum likelihood estimates of the parameters of the most parsimonious model.

In this chapter, we illustrate this approach using our previous studies of the effects of loading from different source areas within watersheds on regional variation in lake dissolved organic carbon (DOC *See* dissolved organic carbon, sample size $\sim$ 500 watersheds) and iron (Fe *See* iron, sample size $\sim$ 100 watersheds) (Canham et al. 2004; Maranger et al. 2006). Our current research is focused on the effects of regional atmospheric pollution on loading of nitrogen (N), and the effects of lakeshore development on loading of phosphorus (P), against the backdrop of the natural variation in loading of N and P to individual lakes as a function of variation in watershed and lake basin properties. Our analyses provide quantitative estimates of the loading of a chemical constituent to a lake from any given source area within a watershed as a function of (1) export in relation to land cover characteristics (e.g., vegetation type, land-use history, wetland hydrologic regime) and (2) loss in relation to the length and nature of the flowpath to the lake. The concentration of the constituent in the lake is then estimated as a function of total estimated loading and in-lake processes (e.g., sedimentation), and watershed-scale hydrologic features (e.g., flushing rate, lake volume, position of a lake in a chain of lakes). We illustrate how the analysis provides robust empirical estimates of nutrient loadings in lakes. The analysis can also provide quantitative answers to questions about (1) the relative importance of different land cover types as a source of loading, (2) how losses along flow paths are only significant for a few cover types, and (3) how forests quantitatively dominate inputs while wetland contributions explain much of the among lake variation in nutrient inputs. Overall, the approach provides a means to test questions on regional scales using the power of large numbers of watersheds that enable robust parameter estimates and comparison of models. One of the distinct advantages of our approach is that it allows assessment of the cumulative impacts of proposed changes in the distribution of cover types and land management within watersheds across a region (Canham et al. 2004).

8.2 The Study Region

Our research has been focused on lakes in New York State's Adirondack Park, which covers 2.4 million ha and consists of both public and private lands (Fig. 8.1). There are over 2,750 lakes greater than 0.2 ha in size within the Park with a total surface area of approximately 100,000 ha (Kretser et al. 1989). Regional surveys have highlighted the striking diversity in lake biological and chemical properties within the Adirondacks (Linthurst et al. 1986, Baker et al. 1990, Driscoll et al. 1994). Approximately 40% of the forests within the Park are protected as "forever wild" by the state constitution, while a comparable proportion is managed as commercial forestland by private owners. The Park has the largest concentration of oldgrowth temperate deciduous forests remaining anywhere in the world, and there is growing evidence that these oldgrowth forests are distinctively different than second growth forests in their loadings of both nitrogen and organic carbon to lakes (Goodale and Aber 2001). Many different types of wetland ecosystems occupy roughly 10% of the landscape and are critical habitat for a significant percentage of the area's biodiversity (Curran 1990). There is very little agricultural land within the Park, but lakeshore residential development has accelerated in recent years, and is now considered a significant issue for water quality in the Park. There are very few lakes with significant point sources of pollutants. However, the entire region has been subject to significant rates of atmospheric deposition of a variety of pollutants (particularly different forms of N, S, and Hg) for decades (Weathers et al. 2006).

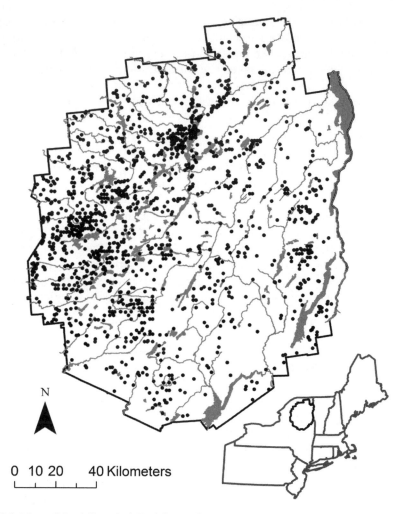

Fig. 8.1 Map of the Adirondack Park in northern New York state. Individual *dots* show the locations of the 1469 lakes sampled by the Adirondack Lakes Survey Corporation (ALSC) in the 1980s. Inset map shows the northeastern United States

8.3 A Spatially Explicit, Mass-Balance Analysis of Lake Chemistry

Our approach is based on the principles of mass-balance, in which variation in the amount of a chemical constituent Y at time t can be understood as a balance between total inputs to the lake, primarily from the surrounding watershed, and net losses, primarily as a result of in-lake degradation and output in lake discharge (Canham et al. 2004). Expressed as a difference equation:

$$Y_{t+1} = Y_t + \text{Inputs}_{t \to t+1} - \text{Degradation}_{t \to t+1} - \text{Discharge}_{t \to t+1} \quad (8.1)$$

where Y is the total amount of the nutrient in the lake, and inputs and losses are scaled to a predefined time interval (e.g., a year). Inputs to the lake are assumed to be independent of in-lake concentration, while losses are assumed to be proportional to in-lake concentration. This results in a predicted steady-state when lake concentration $[Y]$ (in g/m^3) reaches a level where losses balance inputs. Our analysis is designed to predict average mid-summer concentrations within individual lakes.

8.3.1 Inputs

There are typically three major allochthonous inputs of chemical constituents to lakes: (a) atmospheric deposition to the surface of the lake, (b) inflowing stream water and groundwater from wetlands and upland areas within the immediate watershed, and (c) input via streams that carry material exported from upstream lakes and their associated watersheds. In addition, there can be within-lake "production" of many chemical constituents (i.e., in-lake N fixation, autochthonous production of DOC). For the purposes of our analyses, we assume that both within-lake production and atmospheric deposition directly to the lake are linearly proportional to lake surface area (SA), so we combine these two sources into a single, net lake surface area input (SAI).

We consider the watershed of a given lake as a grid of source areas of fixed size (typically 10 × 10 m), in which each source area is classified as a discrete cover type based on vegetation and potentially other features such as drainage and land-use. Inputs to the lake originating from a given grid cell move along flow paths that conceptually include both overland and groundwater flow, until they reach surface water (either the lake shore or streams feeding into the lakes). Our analysis typically does not discriminate between overland vs. groundwater flow, but instead lumps this as "ground" flow, as distinct from "stream" flow inputs to the lake. We originally made this distinction because of the potential for very different processing and loss rates along surface water versus terrestrial flowpaths.

In the simplest analysis of a headwater lake, total annual input (g) to the lake is specified by:

$$\text{Input} = (\text{SAI} \times \text{SA}) + \sum_{i=1}^{N} \text{Export}_c \ e^{-\alpha_c D_i^{\beta_c}} \tag{8.2}$$

Export$_c$ is the export (in g) of the ith grid cell (0.01 ha) of a given cover type c within the immediate watershed. The fraction of the export that reaches the lake (i.e., loading) is specified by an exponential loss as a function of the flow path distance (D_i) from the grid cell to the lake. The loss function is flexible enough to accommodate a wide range of shapes of the distance decay function, according to the estimated parameters α and β, both of which are estimated separately for each landscape cover type. The specific processes that cause loss along a flow path will vary depending on the chemical constituent. For DOC, loss is assumed to occur because of a several processes, including (a) decomposition, (b) sedimentation and mineral complexing in soils and sediments along the flowpath, and (c) loss to deep groundwater. For nitrogen, the term for loss along the flow path would include denitrification.

For lakes that are downstream from other lakes, the analysis is recursive and calculates the total discharge from the headwater lakes first, and then estimates the fraction of that amount that reaches the next downstream lake, and so on down the lake chain:

$$\text{Inputs} = (\text{SAI} \times \text{SA}) + \sum_{j=1}^{M} \lambda \times \text{ULE}_j + \sum_{i=1}^{N} \text{Export}_c \ e^{-\alpha_c D_i^{\beta_c}}, (8.3)$$

where ULE is the upstream lake export (in g) from $j = 1 \ldots M$ immediately upstream lakes, and λ is the average proportion of upstream lake export that is not lost through processing within a stream before it reaches the next downstream lake. In our models to date, we have assumed that λ is independent of stream length, but this assumption could be easily relaxed to estimate λ as a function of the stream distance from the upstream lake. Note that since the analysis is recursive, input from lakes located further up a lake chain are already taken into account when calculating the discharge from the immediately upstream lake.

Equations 2 and 3 are simple additive models of non-point inputs in which each unit area of the watershed is a potential source, and the amount of a chemical constituent from each source area that reaches the lake is a declining function of the distance of the source area from the lake. In this simplest model, loss along a flow path that originated from an upslope source area does not depend on the nature of the cover type that the constituent has to pass through on its way to the lake.

8.3.2 Losses

Losses of a chemical constituent from the lake are conceptually separated into (1) lake discharge and (2) within-lake losses. Loss via lake discharge is estimated from flushing rates based on data on runoff from within the immediate watershed, lake morphometry, and discharge from upstream lakes. The exact processes that lead to within-lake losses vary depending on the chemical constituent. Degradation of DOC in aquatic systems is typically an amalgamation of processes that include direct photodecay, microbial degradation, and flocculation/sedimentation (Wetzel 2001, Molot and Dillon 1997). Within-lake losses of iron occur primarily as sedimentation whereby Fe hydroxides formed in the surface waters rapidly settle to the lake bottom. Regardless of the specific processes involved, we combine their effects into a single decay constant:

$$\text{Degradation} = k \times \text{volume} \times C \qquad (8.4)$$

where C is concentration in the lake.

We have also considered alternative formulations of within-lake losses that were related to a variety of other factors. For DOC, these included (a) lake depth (Rasmussen et al. 1989, Dillon and Molot 1997), which could be expected to reduce decay, (b) the proportion of watershed DOC loading from wetlands, which could be expected to increase decay because of higher loading of more labile DOC from wetlands (Engstrom 1987), and (c) lake acid neutralizing capacity (ANC), which could be expected to increase decay (Reche et al. 1999). Thus, for the analysis of DOC (Canham et al. 2004), we tested alternate models in which k in Eq. (8.4) was replaced by one of the following equations:

$$k = k' \exp^{(-A \times \text{depth})} \qquad (8.4a)$$

$$k = k' + A \times \%\text{Wet land Loading} \qquad (8.4b)$$

$$k = k' + A \times \text{ANC} \qquad (8.4c)$$

Combining Eqs. (8.1, 8.2, 8.3, and 8.4), at steady state:

$$[Y] = \frac{(\text{SAI} \times \text{SA}) + \sum_{j=1}^{M} \lambda \times \text{ULE}_j + \sum_{i=1}^{N} \text{Export}_c e^{-\alpha_c D_i^{\beta_c}}}{\text{volume}(k + \text{ flushing rate})} \qquad (8.5)$$

8.3.3 Interannual Variability in Watershed Loading

When the lake chemistry data is collected in different years, it is necessary to incorporate terms to account for year-to-year variation in rainfall and hydrology. For example, the 1469 lakes sampled by the Adirondack Lakes Survey Corporation (ALSC) (Kretser et al. 1989) were sampled in mid-summer of one of 4 years (1984–1987). In each year, the sampled lakes were widely distributed across the region and well stratified across watershed characteristics such as lake size and flushing rate. Nonetheless, lakes sampled in 1986 had a significantly higher DOC concentration than lakes sampled in the other 3 years (Canham et al. 2004). In a separate study, Pace and Cole (2002) examined temporal variation of DOC in a set of Michigan lakes and found a high degree of synchrony across lakes within a region. Years with high mid-summer DOC concentrations were associated with higher-than-normal runoff in spring and early summer. Strictly from the perspective of algebra, years with high runoff could alter lake chemistry because of higher inputs (changes to the numerator in Eq. 8.5) or higher flushing rates (in the denominator of Eq. 8.5). In the case of the ALSC data, the year with the highest rainfall (1986) also had the highest average lake DOC concentrations, indicating that the net effect of higher runoff was to increase loading (outweighing an increase in the flushing of DOC out of the lake). On this basis, we incorporated a very simple term in the analysis to allow for interannual variation in total DOC loading from within the watershed. 1984 was set as a benchmark, and the analysis then estimated the variation in total within-watershed loading for the 3 other years (1985–1987) needed to account for the observed interannual variation in lake DOC concentration (Canham et al. 2004).

8.3.4 Data Sources

There are a number of features that make the Adirondacks a useful setting for our research. One of the most significant is the availability of extensive regional datasets on watershed physical and biological characteristics. The wetlands for 10 of the major river drainages in the Park – including: the $\sim$ 400,000 ha Oswegatchie and Black River drainages, the 695,000 ha of the Upper Hudson and Sacanadaga River drainages, and the 647,000 ha in the watersheds of four rivers that drain into the St. Lawrence River (the Grass, Racquette, St. Regis, and Salmon Rivers) – have been mapped and classified by the Adirondack Park Agency (APA) with funding from EPA's State Wetlands Protection Program. As a companion to their wetlands mapping program, APA has assembled an extensive set of GIS-referenced data layers on the physical and biological characteristics of the watersheds in those drainages, including mapping of upland vegetation within the park from remote sensing (Roy et al. 1997, Primack et al. 2000). There have also been a number of regional surveys of lake chemistry. The most important of these was a study of 1469 lakes between 1984 and 1987 by the ALSC (Kretser et al. 1989). Besides collecting extensive data on the chemistry and biology of the lakes, ALSC determined the volume of each lake and estimated flushing rates based on regional analyses of precipitation and stream flow. Given either the original data on lake chemistry from lakes within the ALSC sample (Canham et al. 2004) or resampling of a subset of the ALSC lakes for either current lake concentrations (Maranger et al. 2006) or for constituents like total N that were not previously available (our current research), the ALSC datasets and the GIS datalayers assembled by the Park Agency together provide all of the independent data needed for the analyses (Box 8.1)

   The most obvious challenge in the use of our approach is that it is extremely data-intensive. It requires detailed spatial information on the composition and physical structure of each watershed. Remotely sensed data on wetland and

Box 8.1 Adirondack watershed data sources. Many of these have been compiled by the Adirondack Park Agency (APA). See Roy et al. (1997), Primack et al. (2000), and Canham et al. (2004) for details

WATERSHED DELINEATION AND FLOWPATHS: We use GIS software and a 10 m resolution digital elevation model (based on USGS data) to delineate the watershed of each study lake, using the outlet of the lake as the "pour point". The same approach is used to calculate flowpath lengths from each source area in the watershed to the lake. This process is the most problematic step in the analysis, since (1) it assumes that surface topography defines the boundaries of the watershed and the flowpaths within it, and (2) that the DEM is accurate enough to identify watershed boundaries. These assumptions are problematic in areas with little relief and large wetland complexes. We omit lakes from the analysis where we do not have confidence in watershed delineation based on the surface DEM.

WETLANDS: Wetlands have been mapped and classified by the APA using visual interpretation of color infra-red imagery, with classification of the dominant and subordinate vegetation strata, with modifiers for flow regime, based on National Wetlands Inventory protocols. The APA classification recognizes hundreds of combinations of dominant and subordinate vegetation and flow regimes, but for the purposes of parsimony we typically lump these into < 12 major wetland types. See Canham et al. (2004) for details.

UPLANDS: Upland vegetation has been mapped and classified by APA using LANDSAT 5 Thematic Mapper imagery. Forests were classified into 3 major cover types (deciduous, coniferous, mixed deciduous/coniferous forests), and 2 major non-forest cover types, which correspond to (a) logged or disturbed forests, and (b) residential and developed areas.

ROADS: We use a road GIS data layer and assign a width to each road category (local, secondary, and primary highways). Roads are assumed to have no export. While they can have significant impacts on hydrologic flow paths, many of the watersheds we study are in roadless wilderness areas, and the Adirondack Park is remarkably free of roads in general, so we do not attempt to incorporate the effects of roads on flow paths.

STREAM NETWORKS: The process of watershed delineation requires calculation of flow direction and accumulation datalayers. We use these to create a stream network based on flow accumulation thresholds, but have found that a single threshold does not generate networks that correspond to maps of perennial streams from USGS maps. To get better agreement with those maps, we use a hybrid approach that weights upstream areas by landscape type, heavily weighting saturated soils.

upland vegetation of the type needed, however, are becoming much more widely available. A more fundamental issue with our approach lies in the constant tension between complexity and parsimony in model development. This is more than just a matter of taste among scientists, although it seems undeniable that researchers have a broad range of strongly held preferences for the level of mechanistic detail they build into their models. The level of complexity of the model is directly correlated with the number of parameters that must be estimated, and this has unavoidable implications for the number of lakes that must be sampled. For example, the basic model in Eq. 5 requires $3 \times n + 3$ parameters, where n is the number of cover types for vegetation in the watershed. Given that we typically begin with a minimum of a dozen different cover types, the most complex models have over 40 parameters. There are no absolute rules on the minimum sample size needed for parameter estimation, but ideally 5–10 data points (lakes) per parameter is desirable. This translates to a sample of 200 – 400 lakes. While data-intensive, our models are still conceptually very simple, and ignore many of the detailed processes that hydrologists, for example, would be tempted to build into the model. Incorporating more mechanistic detail in the model – for instance, testing for the effects of the composition of the nearshore vegetation on loss along flowpaths from more distant parts of watersheds – can cause the number of parameters in the models (and the requisite number of lakes sampled) to balloon alarmingly. Thus, one of the greatest challenges in using our approach lies in finding the most parsimonious model formulations that can address the questions of interest.

We need to stress one point about the selection of lakes to include in the analysis. Our goal is not inference about the properties of a defined universe of lakes, from which we have identified a representative sample (using traditional random sampling). Rather, our goal is robust inference about the parameter estimates of the most parsimonious model for a sample of lakes. The upshot of this distinction is that our emphasis in selecting lakes is on sampling across a broad range of conditions, since pulling signal from noise in regression almost always benefits from a wide range of values of the independent variables. The downside is that the more that we depart from representative sampling of some clearly defined population of lakes, the greater the risk that the resulting model contains idiosyncracies that are unique to the particular sample of lakes included in the analysis. The best protection against idiosyncracies is to strive to include a large number of lakes.

8.4 Statistical Analyses – Likelihood as a Basis for Linking Data and Models

In the most general terms, both our modeling approach and our statistical analyses are guided by the use of the principle of likelihood (Edwards 1992, Royall 1997). While maximum likelihood methods have a long history of use for point and interval estimation in both traditional frequentist statistics and Bayesian methods, likelihood also provides a basis for inference that represents a middle road between the frequentists and the Bayesians. Hilborn and Mangel (1997) provided one of the first primers on the approach written specifically for ecologists, and Burnham and Anderson (2002) provided a more formal treatment of the methods, in the context of information theory. There has been a great deal of (often acrimonious) debate among ecologists over the relative merits of frequentist, likelihood, and Bayesian approaches in recent years. Many of the chapters in Taper and Lele (2004) contain thoughtful (but somewhat abstract) discussions of the very fundamental statistical issues involved.

Our own interest in the use of likelihood is guided by much more pragmatic reasons. We find it to be an intuitive and powerful basis for linking data and models. Classical hypothesis testing is replaced by the much more general process of model selection and comparison using likelihood and parsimony to compare the strength of evidence for competing models. A likelihood framework stresses the process of identifying and selecting among competing models (as statements of hypotheses), or in the simplest case, among competing point estimates of a parameter of a model (i.e., the maximum likelihood estimate (MLE)). This is in stark contrast to the traditional frequentist approach of rejection of a single "null" (and usually uninteresting) hypothesis. In contrast to p-values, which contain no objective measure of the support in the data for any particular alternative hypothesis, likelihoods (or more commonly, log-likelihoods) can be calculated for an entire set of alternative models and a continuum of parameter values, and provide a quantitative measure of the strength of evidence for any particular model or parameter value. Likelihood profiles and "support intervals" (Edwards 1992) provide much more intuitive alternatives to traditional confidence intervals as a means of presenting uncertainty in a particular parameter estimate.

Our analysis is, in effect, a very data-intensive spatial regression, in which lake concentration is the dependent variable, and the independent variables are: (1) lake volume and surface area, (2) lake flushing rate, and (3) the cover type and distance from the lake for each of the grid cells in the immediate watershed. We solve for the parameter estimates that maximize the likelihood of the observed lake concentrations (e.g., DOC, Fe, N, or P), using simulated annealing (Goffe et al. 1994), an iterative, global optimization procedure. In both of our previous analyses (DOC and Fe), residuals were normally distributed with a uniform variance, so we used the probability density function for a normal distribution for the likelihood function (Canham et al. 2004, Maranger et al. 2006). The simulated annealing algorithm was very effective at converging on the maximum likelihood estimates, even given the large number of parameters we needed to estimate. However, it does so at the expense of a very large number of iterations (typically $> 10^6$). Using the computer technology available to us at the time (2002–2004), this took 1–2 weeks of computation time. Application of this optimization routine required writing software specifically for the analyses, because existing statistics packages were either (1) unable to solve models containing summation terms, or (2) in the case of more flexible statistical modeling systems such as R and SPlus because the optimization routines available in those packages are too slow when the model has summation terms of the sort present in Eq. 5. Alternate forms of the statistical models (e.g., models that use different land cover classifications, or additional terms to describe in-lake losses or other processes) represent alternate hypotheses and are compared using Aikaike's Information Criterion (AIC) (Burnham and Anderson 2002). We use two complimentary methods to assess the strength of evidence (and uncertainty) in individual parameter estimates in each model. We calculate asymptotic 2-unit support intervals for each parameter estimate (Edwards 1992), and we also calculate standard errors of the parameter estimates by inverting the Hessian matrix of the likelihood surface. An n-unit support interval is simply the range of parameter estimates within which "support" (log-likelihood) is within n units of the maximum likelihood estimate. A 2-unit support interval is roughly equivalent to a 95% confidence interval. Goodness of fit of the various models is assessed by comparing observed vs. predicted lake nutrient concentrations. Comparison of observed vs. predicted lake nutrient concentrations is also used to test for bias in the model predictions (using the slope of the relationship between observed and predicted), and to quantify prediction error (using root mean squared error). In a likelihood framework, the most general evaluation of the model is simply the goodness of fit of the predicted values relative to the observed lake chemistry data. Under this approach, all of the available lake data should be used in the calculation of the maximum likelihood parameter estimates. Others prefer to reserve a subsample of the dataset for independent testing of the model. While the latter approach has intuitive appeal as a means to test the generality of the model, it does so at an inevitable cost in terms of the empirical support for individual parameter estimates and the ability of the analysis to distinguish between alternate models. Moreover, since the subset of lakes set aside for an independent test of the model is typically a random subset of the overall sample, it is hardly a critical test of the ability of the model to be used in novel settings.

8.5 Model Comparison as a Form of Hypothesis Testing and a Basis for Model Simplification

One of the most compelling benefits of adopting a model comparison approach to hypothesis testing (Johnson and Omland 2004) is that it readily accommodates the examination of multiple working hypotheses (Chamberlain 1890). In contrast to traditional frequentist tests of a null versus a single alternate hypothesis, likelihood-based model comparisons provide an objective (although relative) measure of the strength of evidence for each alternate hypothesis (i.e., each alternate candidate model). Moreover, the use of information theory and AIC (Burnham and Anderson 2002) provides an objective basis for choosing the most parsimonious model. We have used model comparison in our studies mostly in the context of examining nested, successively simpler models (Table 8.1; Canham et al. 2004, Maranger et al. 2006). For example, while our basic model allows for a decline in loading to the lake as a function of the flowpath length from the source area (Eq. 2), our parameter estimates for the distance decay terms showed that their effect on loading of both DOC and Fe was negligible for most cover types (i.e., Fig. 8.2) If there is no distance decay in the model, the model no longer needs to be spatially explicit, and the statistical analysis would be vastly simplified (i.e., Eq. 5 no longer has a summation term, simply a set of multiple regression terms in the numerator that take the total area of each cover type as input data, and estimate export as a regression coefficient). There are obvious advantages in terms of ease of use of the approach if this is a reasonable approximation. In fact, for the full set of headwater and downstream lakes combined in our analysis of DOC, and our analysis of a smaller set of lakes for Fe, a non-spatial model sacrificed

Table 8.1 Comparison of alternate models for DOC and iron. AIC_c is Aikaike's Information Criterion corrected for small sample size. ΔAIC is the difference between the AIC_c of that model and the best model (lowest AIC_c). The "full" models include the distance decay terms for loading. The "reduced" models eliminate distance decay for either some or all of the models. See Canham et al. (2004) for details about the models for DOC and Maranger et al. (2006) for the models for iron

Model	Sample size	Number of parameters	Log likelihood	AIC_c	ΔAIC	R^2
DOC: Headwater lakes	–	–	–	–	–	–
Full model	355	41	−818.7	1730.3	34.1	0.55
Full model + k = f(depth)	355	42	−814.9	1725.4	29.2	0.56
Full model + k = f(wetland loading)	355	42	−816.1	1727.8	31.5	0.55
Reduced model – distance decay for five cover types	355	22	−824.6	1696.2	0.0	0.53
Reduced model – no distance decay	355	17	−838.3	1712.3	16.1	0.50
	–	–	–	–	–	–
DOC: All lakes	–	–	–	–	–	–
Full model	428	29	−1040.2	2142.8	12.8	0.48
Reduced model – no distance decay	428	18	−1046.2	2130.0	0.0	0.46
	–	–	–	–	–	–
Iron	–	–	–	–	–	–
Full model	93	21	−362.1	779.2	20.6	0.68
Reduced model – distance decay for two cover types	93	15	−361.4	759.1	0.5	0.68
Reduced model – no distance decay	93	13	−364.0	758.6	0.0	0.67

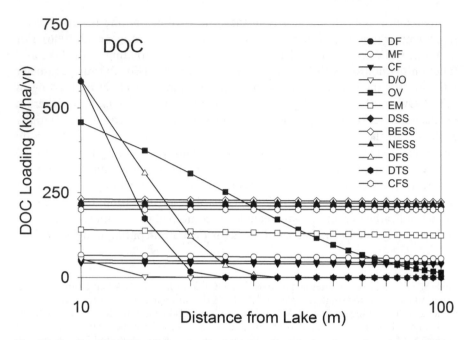

Fig. 8.2 Predicted DOC loading to headwater lakes (kg C/ha/year) as a function of distance from the lake for 12 upland and wetland cover types. Cover type codes: DF, deciduous forest; MF, mixed forest; CF, coniferous forest; DO, deciduous/open upland vegetation; OV, open upland vegetation; EM, emergent marsh; DFS, deciduous forest swamp; CFS, coniferous forest swamp; DTS, dead tree swamp; DSS, deciduous shrub swamp; BESS, broad-leaved evergreen shrub swamp; NESS, needle-leaved evergreen shrub swamp. Redrawn from Canham et al. (2004) with permission from Ecological Application

only a modest amount of predictive power for a significant drop in the number of parameters required, and was in fact the most parsimonious model (Table 8.1; Canham et al. 2004; Maranger et al. 2006).

Our other main use of model comparison has been to examine alternate (and more complex) formulations of the in-lake "decay" terms (i.e., Eqs. 4, 4a, 4b, and 4c). In this context, the likelihood of the model provides a measure of the strength of evidence (relative to other models) for a particular formulation, and AIC provides a means to balance the expected increase in likelihood due to incorporation of additional terms with the cost in terms of added complexity. For DOC, models that incorporated the effects of (1) lake depth, (2) the proportion of loading originating from wetlands, and (3) ANC of the lake all performed better (higher likelihood and lower AIC) than models that omitted those terms. For iron, models that incorporated lake color and depth performed better than models that ignored their effect on the in-lake loss term (Maranger et al. 2006).

8.6 Goodness of Fit: How "Predictive" Are the Models, and Why Does This Matter?

Statisticians appear to devote relatively little effort to the issue of model evaluation, at least compared to the scientists who use their statistics but who are mostly concerned about how "good" the model is. There is little consensus over terminology on the issue of model evaluation (Oreskes et al. 1994, Canham et al. 2003), but in the context of regression models such as the ones we use, traditional measures of goodness of fit (R^2), bias, and prediction error are widely used, although each has limitations. Our models are unbiased (i.e., do not consistently over- or under-predict lake chemistry), and explain a respectable proportion of the variance in the concentration of DOC and Fe (46 – 69%, depending on the form of the model and the nutrient). The models, however, still have relatively large root mean squared errors (RMSE): $\sim$ 2.5 mg/l for DOC (which varied from near zero to >20 mg/l among the lakes) and ~114 µg/l for Fe (which varied from near zero to $\sim$ 1000 µg/l in the sample of lakes). This limits the predictive power of the models. The uncertainty in our predictions is likely caused by many factors, including measurement error, but there is an almost inevitable cost – in the more mundane terms of time and money and the more fundamental terms of model complexity – associated with increasing the predictive power of the model. If prediction is a specific goal of the analysis, particularly in a management context, this cost may be acceptable, but if the real motivation of the research is toward basic understanding of watershed controls on lake nutrient concentrations, then there are diminishing returns in seeking to reduce uncertainty further, simply for the sake of increasing predictive power.

8.7 Benefits of the Approach

8.7.1 Robust Empirical Estimates of Export from Different Source Areas in the Watershed

In the context of nutrient loading to lakes, "export" from an upland forest within the watershed is typically defined in terms of either (1) the movement of a nutrient below the rooting zone of the ecosystem, and into hydrologic flow-paths that eventually lead to the lake, or (2) overland flow from source areas immediately adjacent to the lake. Direct measurement of the former is normally attempted through the use of lysimeters buried at the bottom of the rooting zone (to measure nutrient concentration in soil water), coupled with soil water budgets to estimate the volume of water flux. Both parts of that effort are costly and time consuming, and as a result are typically done in very limited numbers of locations, and only within otherwise intensively studied watersheds. Flux to a lake via overland flow is even harder to quantify. Our approach inverts the problem of directly measuring these fluxes and asks "what would the export have to be to explain the observed concentration in a large sample of lakes?" Our approach also quantifies uncertainty in the estimates.

Our analyses of Adirondack lakes have both confirmed and challenged conventional wisdom about export of nutrients from upland and wetland ecosystems (Canham et al. 2004, Maranger et al. 2006). DOC export estimated from different cover types is remarkably consistent with studies that calculate total watershed export of DOC (Aitkenhead and McDowell 2000, Dillon and Molot 1997). Those studies typically estimate whole watershed losses that are intermediate between our estimates for upland forests and our estimates for wetlands, and presumably reflect the weighted average loading from different cover types within the watershed. As expected, wetlands export far more DOC and Fe, on average, than forests (Fig. 8.3). There was, however, a wide range of variation among wetland types in export rates, and there was considerable uncertainty in our estimates for wetland export, suggesting that vegetation structure was not a definitive predictor of nutrient export (Fig. 8.3). Surprisingly, there was almost no variation in the predicted export of DOC from deciduous vs. mixed vs. coniferous forests. We had expected that the conifer forests might be a larger net exporter of DOC, because of generally slower rates of litter decomposition and higher levels of soil organic matter

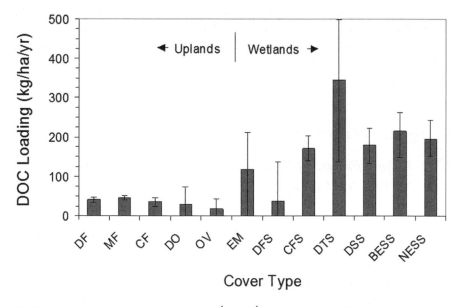

Fig. 8.3 Predicted DOC loading (kg C ha^{-1} year^{-1}) in a sample of 428 lakes from the non-spatial model in which loading was independent of distance from the source area to the lake (DOC – All Lakes, "reduced model" in Table 8.1). Error bars are 1.96-unit support intervals (analogous to 95% confidence intervals) on the mean loading. See the legend for Fig. 8.2 for description of the cover type codes. Redrawn from Canham et al. (2004). Used with permission

(D'Arcy and Carignan 1997). Soil solution concentrations of DOC in B horizons of coniferous forest plots in the Adirondacks are almost twice as high as in hardwood and mixed forest plots (8.9 mg C/l for conifer plots vs. 5.5 and 6.8 mg C/l for hardwood and mixed forest plots) (Cronon and Aiken 1985). Our results suggest that these differences disappear by the time the soil solution mixes into groundwater. Moreover, the support intervals on our estimates of export of DOC from the three different forest types are extremely narrow. The lack of uncertainty suggests that there is little room for other factors such as land-use or disturbance history to play a large role in influencing DOC export.

We have generally focused on vegetation structure and composition as the basis for classifying watersheds into functionally distinct source areas. There are many reasons for this ranging from a very academic interest in comparing the functional effects of different ecosystem types to the more pragmatic issue that relatively high quality vegetation mapping (of both wetlands and uplands) based on remote sensing is becoming much more widely available. In addition to classifying wetlands of the Adirondacks based on vegetation structure, the Park Agency recorded visual evidence of the hydrologic flow regime within each wetland mapping unit. For our analysis of regional variation in lake iron concentrations, we were interested in whether the degree and seasonal duration of inundation affected the net export of Fe to lakes (Maranger et al. 2006), because the degree and extent of anaerobic conditions should affect the sequestration and mobilization of iron within a wetland (Kaelke and Dawson 2003). We used the mapped flow regime modifiers to distinguish between "wet" (permanently or semi-permanently flooded) and "dry" (only occasionally flooded, but still generally saturated soil) variants of two of the most common wetland types in the region: deciduous shrub swamps dominated by speckled alder (*Alnus incana* sp., *rugosa*) and willows (*Salix* spp.), and evergreen shrub swamps dominated by ericaceous shrubs and stunted bog conifers.

The maximum likelihood estimates of iron export were strikingly different between the "wet" and "dry" variants of these two wetland types. Maranger et al. (2006) discuss the biogeochemical processes that might be responsible for much lower net export of iron from "wet" alder swamps but higher export from "wet" evergreen shrub swamps. Our current analyses of phosphorus loading show a similar importance of hydrologic flow regimes to net export from wetlands, but in this case the pattern is more consistent, with the permanently inundated wetlands having much lower export to lakes. While the reasons for this may still be partly a function of effects of hydrology on P biogeochemistry, there may be a more direct link to hydrology that is not explicitly incorporated in our model. It may simply be that the wetlands that are permanently inundated are that way because they are local "sinks" in the watershed, and do not drain readily to the lakes. We consider the absence of detail about hydrologic flow within the watershed a strength rather than a weakness of our model, since requiring accurate characterization of variable hydrologic source areas within the watershed (sensu Beven and Wood 1983) would add enormous complexity (in terms of numbers of parameters and resulting sample size of lakes) as well as extraordinarily detailed watershed-specific data requirements to the model. However, this issue highlights the inherent compromises that occur in the development of simple, parsimonious, statistical models for hundreds of ecosystems.

8.7.2 Quantifying Loss of Nutrients Along Flowpaths to a Lake

There are many physical and biological processes by which nutrients exported from a forest or wetland distant from a lake can be lost before reaching the lake. Thus, many studies of nutrient input to lakes implicitly or explicitly assume that the bulk of loading to a lake originates from regions of the watershed near the lake. There are a number of approaches to direct measurement of loss along flowpaths, typically using tracers or stable isotopes (i.e., Schiff et al. 1997). Again, the direct measurements are typically costly and time-consuming, and again, our method inverts the question and using the principles of mass-balance, asks "what pattern of loss as a function of distance along a flowpath is consistent with the observed lake concentration?"

For this question, with a few notable exceptions, there is little evidence of loss as a function of the length of the flowpath to the lake for either DOC or iron (Table 8.1 and Fig. 8.2). In effect, these Adirondack watersheds are very well "plumbed". A source area of forest or wetland of a given type has essentially the same net loading to the lake when it is 5 km away from the lake as when it is 50 m from the lake (Fig. 8.2). The main exceptions for upland vegetation types are disturbed areas. Those areas have much higher loading to the lake when they are immediately adjacent to the lake (Canham et al. 2004, Maranger et al. 2006).

In retrospect, the lack of a distance effect for many cover types seems reasonable. In the case of DOC, the material that is exported beneath the rooting zone of a forest is assumed fairly recalcitrant. It is also well-known that breakdown of DOC slows down dramatically in the absence of light (Gron et al. 1992). DOC adsorbs readily onto soil particles (McDowell and Wood 1984), but that adsorptive capacity must be finite. Moreover, the parent materials for the most common soils in Adirondack watersheds are coarse glacial tills over shallow bedrock. Thus, our assumption is that much of the transport of DOC from upper reaches of the watershed happens via macropores that have saturated their adsorptive capacity, and in which the rates of degradation of DOC (through microbial respiration) are too slow relative to transit time through the coarse soils to be significant.

The lack of distance decay of inputs from most wetland types (Fig. 8.2) may be caused by a different set of factors. Many of the wetlands either fringe lakes or are distributed along stream channels feeding into lakes. Thus, while the area-weighted mean flowpath distance for the wetland cover types was not generally shorter than for upland forests (Canham et al. 2004), stream water represented a much greater percentage of the flowpath length for the wetlands. The relatively high rates of decay once DOC reaches the lake reflect the importance of sunlight in the degradation process (Moran and Zepp 1997). This degradation takes place in streams as well as in the lake, but the transit time for DOC in streams is rapid and there is less time for solar-driven decay, particularly in small streams where forest canopy cover limits light penetration.

It is critical to note that our analyses to date have ignored testing specifically for loss that happens when the very end of the flowpath is a riparian wetland. There is an enormous literature on the role of riparian wetlands in filtering nutrients out before they reach surface waters (reviewed in Lowrance 1998). Much of that literature has focused on the role of saturated soils in denitrification (e.g., Gold et al. 2001). We are currently working to develop modifications to our models that can quantify these effects on the scale of whole watersheds. The modeling challenge in this case is that there are so many permutations of essentially unique combinations of different flowpaths that pass through the spatially complex patterns of wetlands adjacent to lakes and the streams that feed them. The computational needs for addressing this type of spatial variation are enormous.

8.7.3 Giving Forests Their Due – the Power of Mass Balance

While our results confirm the conventional wisdom that wetlands typically have much greater rates of export of nutrients than forests *per unit area* (e.g., Fig. 8.3), wetlands occupy less than 15% of the Adirondack landscape. Our approach, based on the principles of mass balance, allows us to calculate the predicted loading of a nutrient to the lake from any given source area within the watershed. When we sum up the loading from wetlands for either DOC (Canham et al. 2004) or Fe (Maranger et al. 2006), it is clear that they have a disproportionate rate of loading, relative to the percent of the watershed in wetlands. For example, wetlands occupied 12.3% of the watershed area for a set of 355 headwater lakes we analyzed, but contributed on average 30.4% of the total DOC loading to those lakes (Canham et al. 2004). Thus, as many studies have shown, there is typically a positive correlation between the proportion of a watershed occupied by wetlands and the concentration of DOC (and many other nutrients) (Engstrom 1987, Kortelainen 1993, Watras et al. 1995). However, these correlative studies essentially explain the variation around the mean, not the mean itself. Our models, based on mass balance, allow us to partition the total loading among source areas. On average, 70 and 75% of the total loading of DOC and Fe, respectively, to Adirondack lakes originates from upland forests, rather than from wetlands (Canham et al. 2004, Maranger et al. 2006). We know of no other current approach that can address this issue on such a broad scale.

8.7.4 Assessing the Relative Importance of Direct Lake Inputs and in-Lake Processing

Our approach cannot distinguish between direct inputs of a nutrient to the surface of a lake (in either wet or dry deposition) versus autochthonous "production" within a lake (with the latter term including the effects of processes that mobilize nutrients return to the water column from sediments). On a per unit area basis, our analyses indicate that even both of these processes combined do not generate significant "loading" to the lake ecosystem relative to either upland or wetland source areas (Canham et al. 2004, Maranger et al. 2006). Moreover, since lake surface area is typically a small proportion of an Adirondack watershed, the proportion of total loading attributed to the lake (rather than the surrounding watershed) is generally negligible. There are exceptions, however: the most notable are headwater lakes that are large relative to a small watershed in mountainous terrain containing very little wetland area. These lakes are typically among the most oligotrophic in the region, and also among the most susceptible to acidification by acid rain.

Besides in-lake "production", the only other terms in Eq. 5 that depend on features of the lake itself are the lake volume, the flushing rate (which is itself a function of lake volume and watershed runoff), and the in-lake decay term. Of these three parameters, variation in the two physical features of the lake (volume and flushing rate) has much greater impact on regional variation in lake nutrient concentration than the biological process of in-lake decay. Many of the headwater lakes have large watersheds relative to lake volume and have correspondingly high flushing rates (median $= 4.34$ year^{-1} for the 355 headwater lakes in our study of DOC; Canham et al. 2004). Given these relatively high flushing rates, examination of Eq. 5 suggests that losses due to lake discharge have a much larger effect on lake nutrient concentrations than the in-lake decay coefficient. Note that in our model, "decay" incorporates a broad suite of biotic and abiotic processes that result in loss of a nutrient from the water column, including sedimentation. Depending on the nutrient in question, the exact nature of the processes will vary, and the in-lake decay coefficient estimated by our analysis will often vary as a function of other attributes of a lake such as lake depth and alkalinity (Canham et al. 2004, Maranger et al. 2006).

8.8 Refining the Approach

8.8.1 Incorporating the Effects of Nearshore and Riparian Zones on Watershed Loading

One of the strengths of our analysis is that we can test hypotheses related to the effects on lake chemistry of the spatial configuration of a given watershed, including the effects of the spatial distribution, structure and composition of wetlands, and the effects of the composition, management, and land-use history of uplands. Our current research on loading of N and P to Adirondack lakes is focused on expanding our modeling approach to address the importance of nearshore and riparian areas. Specifically, prior versions of our analyses did not explicitly test for modification of loading from more distant parts of watersheds as a function of the composition of nearshore vegetation and riparian wetlands through (or under) which a flow-path passed on its way to the lake. It is almost axiomatic that riparian zones act as "buffers" between uplands and aquatic systems (e.g., Lowrance 1998, Broadmeadow and Nisbet 2004, Anbumozhi et al. 2005) by acting as either sinks for nutrients or sites of accelerated loss (i.e., denitrification). In effect, our first analyses envisioned a watershed as a grid of source areas, with each source "cell" conceptually connected to the lake by a pipe (consisting of both groundwater and surface water flow). Our analyses empirically tested for evidence that loading to the lake declined as a function of the length of the pipe, but did not examine whether loss along the flowpath depended on the nearshore or riparian vegetation through which the flowpath passed. The relatively high goodness of fit of the models provided strong empirical evidence that loading did not decline as a function of the distance of the source area from the lake. This result was suggestive that loadings of DOC and Fe are not strongly modified by nearshore (and near-stream) vegetation. There are ample reasons, however, to suspect that the spatial signature of inputs of P and N may be quite different as a consequence of biogeochemical transformations such as denitrification, which occurs at high rates in riparian zones (Gold et al. 2001, Groffman and Crawford 2003).

8.8.2 Quantifying the Effects of Regional Variation in N Deposition on Loading to Lakes

Our models to date have treated the spatial variation in vegetation cover type (and to a lesser degree, flow regime in wetlands) as the only basis for spatial variation in the sources of nutrients that reach the lake from within a watershed. In particular, we ignore factors such as spatial variation in soils within watersheds, partly due to very limited data on soils. For our current research on nitrogen loading to lakes, however, there is a form of regional variation that we cannot ignore: the well known and documented regional gradient in N deposition from air pollution (e.g., Ollinger et al. 1993). Moreover, there is a great deal of academic and public policy interest in the question of how much of the anthropogenic loading to the landscape is retained in uplands and wetlands, rather than passed through to surface waters. Weathers et al. (2006) have recently developed a new model of atmospheric deposition in mountainous landscapes that maps the total (wet + dry + cloud) deposition of N across the Adirondacks. This new information will allow us to modify the input terms of our model (Eq. 2) to estimate export from a given source area as a function of the mapped vegetation and N deposition, with explicit empirical estimation of the "retention coefficient" for each cover type (i.e., the proportion of N deposition that is not exported). Given the growing body of evidence for N saturation in northeastern forests (Aber et al. 2003) and evidence that the degree of N saturation (and hence retention) by forests is influenced by forest history (particularly in oldgrowth forests), we will explicitly compare models that test for differences in N export as a function of both forest type and land-use history (logged vs. un-logged).

8.9 Summary

Many of the most pressing challenges in ecology have moved beyond issues of causality – after all, ecological systems are rife with causal connections. Virtually anything that can be plausibly hypothesized as causal can be shown to be "significant" under an experimental design with sufficient statistical power. Manipulative experiments are uniquely effective at demonstrating that X causes Y, but precisely because of their often tightly controlled conditions, are far less effective at answering questions of "where", "when", and most importantly, "how much" X affects Y. One alternative approach actively pursued in limnology has been to seek predictive models, regardless of their mechanistic basis (Rigler 1982, Peters 1986). We share an interest in simple, predictive regression models, but it is clear that even simple models can be mechanistically based, and that there are many benefits when they are. While data-intensive, our analyses are obviously just a form of regression. By formulating a regression model that has a mechanistic basis, however, the estimated regression coefficients have clearly defined operational definitions (i.e., nutrient export rates, in-lake decay rates). Ecologists and hydrologists have devoted enormous effort to the direct measurement – typically in one or a few intensively studied watersheds – of many of the processes embodied in our analyses. We would argue that there are compelling benefits to inverting the question and asking "what would the rate of the process have to be" to generate the much more readily measured lake chemistry, across a very large sample of lakes, rather than attempting to directly measure the rates of the many processes controlling lake chemistry at an appropriate spatial and temporal scale. This inverse approach is widely used in other disciplines (e.g., Tarantola 2005) and has been applied in ecology particularly in the analysis of flows through food webs (e.g., Vezina and Pace 1994, Vezina and Pahlow 2003). The development of model comparison methods based on likelihood and information theory is also clearly at the heart of our approach. Understanding where, when, and how much X affects Y is fundamentally the challenge of finding a suitable model for the relationship between X and Y, and then the rest is simply but critically an issue of parameter estimation.

Acknowledgments Our work has been supported by grants from the U.S. Environmental Protection Agency, the U.S. National Science Foundation, and the Mellon Foundation. The work would not have been possible without the extraordinary efforts of both the Adirondack Lakes Survey Corporation and the Adirondack Park Agency. We would like to thank our many collaborators, particularly Michael Papaik and Roxane Maranger.

References

Aber, J.D., Goodale, C.L., Ollinger, S.V., Smith, M.L., Magill, A.H., Martin, M.E., Hallett, R.A., and Stoddard, J.L. 2003. Is nitrogen deposition altering the nitrogen status of northeastern forests? Bioscience **53**:375–389.

Aitkenhead, J.A., and W.H. McDowell. 2000. Soil C:N as a predictor of annual riverine DOC flux at local and global scales. Global Biogeochemical Cycles **14**:127–138.

Anbumozhi, V., J. Radhakrishnan, and E. Yamaji. 2005. Impact of riparian buffer zones on water quality. Ecological Engineering **24**:517–523.

Baker, J.P., S.A. Gherini, S.W. Christensen, C.T. Driscoll, J. Gallagher, R.K. Munson, R.M. Newton, K.H. Reckhow, and C.L. Schofield. 1990. Adirondack lakes survey: An interpretive analysis of fish communities and water chemistry, 1984–87. Adirondack Lakes Survey Corporation, Ray Brook, NY.

Beven K., and E.F. Wood. 1983. Catchment geomorphology and the dynamics of runoff contributing areas. Journal of Hydrology **65**:139–158.

Boyer, E.W., G.M. Hornberger, K.E. Bencala, and D. McKnight. 1996. Overview of a simple model describing variation of dissolved organic carbon in an upland catchment. Ecological Modelling **86**:183–188.

Broadmeadow, S., and T.R. Nisbet. 2004. The effects of riparian forest management on the freshwater environment: a literature review of best management practice. Hydrology and Earth System Sciences **8**:286–305

Burnham, K.P., and D.R. Anderson. 2002. Model Selection and Multimodel Inference: A Practical Information-Theoretic Approach. Second Edition. Springer-Verlag, New York.

Canham, C.D., J.S. Cole, and W.K. Lauenroth. 2003. The role of models in ecosystem science. pp 1–12 in Canham, C.D., J.S. Cole, and W.K. Lauenroth (eds.). Models in Ecosystem Science. Princeton University Press.

Canham, C.D., M.L. Pace, M.J. Papaik, A.G.B. Primack, K.M. Roy, R.J. Maranger, R.P. Curran, and D.M. Spada. 2004. A spatially-explicit watershed-scale analysis of dissolved organic carbon in Adirondack lakes. Ecological Applications **14**:839–854.

Chamberlain, T.C. 1890. The method of multiple working hypotheses. Science **15**:92–96.

Cronon, C.S., and G.R. Aiken. 1985. Chemistry and transport of soluble humic substances in forested watersheds of the Adirondack Park, New York. Geochimica et Cosmochimica Acta **49**:1697–1705.

Curran, R.P. 1990. Biological Resources and Diversity of the Adirondack Park. Technical Report 17. In: The Adirondack Park in the Twenty First Century. Tech. Rep Volume One, Albany, NY. pp. 414–461.

D'Arcy, P., and R. Carignan. 1997. Influence of catchment topography on water chemistry in southeastern Quebec Shield lakes. Canadian Journal of Fisheries and Aquatic Sciences **54**:2215–2227.

Dillon, P.J., and L.A. Molot. 1997. Dissolved organic and inorganic carbon mass balance in central Ontario lakes. Biogeochemistry **36**:29–42.

Driscoll, C.T., M.D. Lehtinen, and T.J. Sullivan. 1994. Modeling the acid-base chemistry of organic solutes in Adirondack, New York, lakes. Water Resources Research **30**:297–306.

Edwards, A.W.F. 1992. Likelihood – Expanded Edition. Johns Hopkins University Press, Baltimore.

Engstrom, D.R. 1987. Influence of vegetation and hydrology on the humus budgets of Labrador lakes. Canadian Journal of Fisheries and Aquatic Sciences **44**:1306–1314.

Gergel, S.E., M.G. Turner, and T.K. Kratz. 1999. Dissolved organic carbon as an indicator of the scale of watershed influence on lakes and rivers. Ecological Applications **9**:1377–1390.

Goffe, W.L., Ferrier, G.D., and J. Rogers. 1994. Global optimization of statistical functions with simulated annealing. Journal of Econometrics **60**:65–99.

Gold, A.J., P.M. Groffman, K. Addy, D.Q. Kellogg, M. Stolt, and A.E. Rosenblatt. 2001. Landscape attributes as controls on ground water nitrate removal capacity of riparian zones. Journal of the American Water Resources Association **37**:1457–1464.

Goodale, C.L., and J.D. Aber. 2001. The long-term effects of land-use history on nitrogen cycling in northern hardwood forests. Ecological Applications **11**:253–267.

Groffman, P.M., and M.K. Crawford. 2003. Denitrification potential in urban riparian zones. Journal of Environmental Quality **32**:1144–1149.

Gron, C., J. Torslov, H.J. Albrechtsen, and H.M. Jensen. 1992. Biodegradability of dissolved organic-carbon in groundwater from an unconfined aquifer. Science of the Total Environment **118**:241–251.

Hilborn, R., and M. Mangel. 1997. The Ecological Detective: Confronting Models with Data. Princeton University Press, Princeton.

Houle, D., R. Carignan, and M. Lachance. 1995. Dissolved organic carbon and sulfur in southwestern Quebec lakes: relationship with catchment and lake properties. Limnology and Oceanography **40**:710–717.

Johnson, J.B., and K.S. Omland. 2004. Model selection in ecology and evolution. Trends in Ecology & Evolution **19**:101–108.

Kaelke, C.M., and J.O. Dawson. 2003. Seasonal flooding regimes influence survival, nitrogen fixation, and the partitioning of nitrogen and biomass in *Alnus incana ssp rugosa*. Plant Soil **254**:167–177.

Kortelainen, P. 1993. Content of total organic carbon in Finnish lakes and its relationship to catchment characteristics. Canadian Journal of Fisheries and Aquatic Sciences **50**:1477–1483.

Kretser, W.A., J.Gallagher, and J. Nicholette. 1989. Adirondack Lakes Study 1984–87: An evaluation of fish communities and water chemistry. Adirondack Lakes Survey Corporation, Ray Brook, NY.

Linthurst, R.A., D.H.Landers, J.M. Eilers, D.F.Brakke, W.S. Overton, E.P. Meier, and R.E. Crowe. 1986. Characteristics of Lakes in the Eastern United States. Volume 1. Population Descriptions and Physico-Chemical Relationships. EPA/600/4-86/007a. U.S. Environmental Protection Agency, Washington, D.C. 136 pp.

Lowrance, R. 1998. Riparian forest ecosystems as filters for nonpoint-source pollution. In M.L. Pace. and P. Groffman (eds.) Successes, Limitations and Frontiers in Ecosystem Science. Springer Verlag, NY.

Maranger, R., C.D. Canham, M.L. Pace, and M.J. Papaik. 2006. A spatially explicit model of iron loading to lakes. Limnology and Oceanography **51**:247–256.

McDowell, W.H., and T. Wood. 1984. Podzolization – soil processes control dissolved organic carbon concentrations in stream water. Soil Science **137**:23–32.

Molot, L.A., and P.J. Dillon. 1997. Photolytic regulation of dissolved organic carbon in northern lakes. Global Biogeochemical Cycles **11**:357–365.

Moran, M.A., and R.G. Zepp. 1997. Role of photoreactions in the formation of biologically labile compounds from dissolved organic matter. Limnology and Oceanography **42**:1307–1316.

Ollinger, S.V., J.D. Aber, G.M. Lovett, S.E. Milliham, R.G. Lathrop, and J.M. Ellis. 1993. A spatial model of atmospheric deposition for the Northeastern United States. Ecological Application **3**:459–472.

Oreskes, N., K. Shrader-Frechette, and K. Belitz. 1994. Verification, validation, and confirmation of numerical models in the earth sciences. Science **263**:641–646.

Pace, M.L., and J.J. Cole. 2002. Synchronous variation of dissolved organic carbon in lakes. Limnology and Oceanography **47**:333–342.

Peters, R.H. 1986. The role of prediction in limnology. Limnology and Oceanography 31:1143–1159.

Primack, A.G.B., D.M. Spada, R.P. Curran, K.M. Roy, J.W. Barge, B.F. Grisi, D.J. Bogucki, E.B. Allen, W.A. Kretser, and C.C. Cheeseman. 2000. Watershed Scale Protection for Adirondack Wetlands: Implementing a Procedure to Assess Cumulative Effects and Predict Cumulative Impacts From Development Activities to Wetlands and Watersheds in the Oswegatchie, Black and Greater Upper Hudson River Watersheds of the Adirondack Park, New York State, USA, Part I. Resource Mapping and Data Collection, Part II. Resource Data Analysis, Cumulative Effects Assessment, and Determination of Cumulative Impacts. New York State Adirondack Park Agency, Ray Brook, New York

Rasmussen, J.B., L. Godbout, and M. Schallenberg. 1989. The humic content of lake water and its relationship to watershed and lake morphometry. Limnology and Oceanography 34:1336–1343.

Reche, I., M.L. Pace, and J.J. Cole. 1999. Relationship of trophic and chemical conditions to photobleaching of dissolved organic matter in lake ecosystems. Biogeochemistry 44:259–280.

Reckhow, K.H, and S.C. Chapra. 1983. Engineering approaches for lake management. Vol 1: Data Analysis and Empirical Modeling. Butterworth, Boston, MA.

Reckhow, K.H., and J.T. Simpson. 1980. A procedure using modeling and error analysis for the prediction of lake phosphorus concentration from land use information. Canadian Journal of Fisheries and Aquatic Sciences 37:1439–1448.

Rigler, F.H. 1982. Recognition of the possible: an advantage of empiricism in ecology. Canadian Journal of Fisheries and Aquatic Sciences 39:1323–1331.

Roy, K.M., E.B. Allen, J.W. Barge, J.A. Ross, R.P. Curran, D.J. Bogucki, D.A. Franz, W.A. Kretser, M.M. Frank, D.M. Spada, and J.S. Banta. 1997. Influences on wetlands and lakes in the Adirondack Park of New York State: A catalog of existing and new GIS datalayers for the 400,000 hectare Oswegatchie/Black River watershed. New York State Adirondack Park Agency, Ray Brook, NY.

Royall, R.M. 1997. Statistical Evidence: A Likelihood Paradigm. London: Chapman and Hall.

Schiff, S.L., R. Avena, S.E. Trumbore, M.J. Hinton, R. Elgood, and P.J. Dillon. 1997. Export of DOC from forested catchments on the Precambrian Shield of Central Ontario: clues from 13C and 14C. Biogeochemistry 36:43–65.

Soranno, P.A., S.L. Hubler, S.R. Carpenter, and R.C. Lathrop. 1996. Phosphorus loads to surface waters: a simple model to account for spatial pattern of land use. Ecological Applications 6:865–878.

Taper, M.L., and S.R. Lele. 2004. The Nature of Scientific Evidence: Empirical, Statistical, and Philosophical Considerations. Chicago: University of Chicago Press.

Tarantola, A. 2005. Inverse Problem Theory and Methods for Parameter Estimation. Society for Industrial and Applied Mathematics, Philadelphia.

Vézina, A.F., and M.L. Pace. 1994. An inverse model analysis of planktonic food webs in experimental lakes. Canadian Journal of Fisheries and Aquatic Sciences 51:2034–2044.

Vézina, A.F., and M. Pahlow. 2003. Reconstruction of ecosystem flows using inverse methods: how well do they work? Journal of Marine Systems 40–41:55–77.

Vollenweider, R.A. 1976. Advances in defining critical loading levels for phosphorus in lake eutrophication. Memorie dell'Istituto Italiano di Idrobiologia. 33:53–83.

Watras, C.J., K.A. Morrison, J.S. Host, and N.S. Bloom. 1995. Concentration of mercury species in relationship to other site-specific factors in the surface waters of a northern Wisconsin lake. Limnology and Oceanography 40:556–565

Weathers, K.C., S.M. Simkin, G.M. Lovett, and S.E. Lindberg. 2006. Empirical modeling of atmospheric deposition in mountainous landscapes. Ecological Applications 16: 1590–1607.

Wetzel, R.G. 2001. Limnology: Lake and River Ecosystems. Academic Press, San Diego, California.

Chapter 9
Forecasting and Assessing the Large-Scale and Long-Term Impacts of Global Environmental Change on Terrestrial Ecosystems in the United States and China

Hanqin Tian, Xiaofeng Xu, Chi Zhang, Wei Ren, Guangsheng Chen, Mingliang Liu, Dengsheng Lu, and Shufen Pan

Abbreviations: AVHRR – Advanced Very High Resolution Radiometer, CLC – Scenario combined Climate, Land Use, and CO_2 effects, CLM – Climate Change only, DGVM – Dynamic Global Vegetation Model, DLEM – Dynamic Land Ecosystem Model, ETM – Enhanced Thematic Mapper, fPAR – Fraction of Photosynthetically Active Radiation, GCM – General Circulation Model, GEOMOD – Geographical Modeling, GIS – Geographic Information System, GISS – Goddard Institute for Space Studies, GPP – Gross Primary Production, IPCC – Intergovernmental Panel on Climate Change, LAI – Leaf Area Index, LUCC – Land Use Cover and Change, LULC – Land Use and/or Land Cover, MATCH – Multi-scale Atmospheric Transport and Chemistry, MIT IGSM – Massachusetts Institute of Technology's Integrated Global System Model, MODIS – Moderate-resolution Imaging Spectroradiometer, NARR – North American Regional Reanalysis, NCE – Net Carbon Exchange, NCEP – National Center for Environmental Prediction, NEP – Net Ecosystem Production, NPP – Net Primary Production, NOAH – NCEP, Oregon State University, Air Force, and Hydrologic Research Lab, OCLC – Scenario combined O_3, Climate, Land Use, and CO_2 effects, PFT – Plant Functional Type, RCM – Regional Climate Model, RegEM – Regional Ecosystem Model, RISE – Regional Integration System for Earth's ecosystem, SEUS – Southeastern U.S., SOC – Soil Organic Carbon, TEM – Terrestrial Ecosystem Model, TOTEC – Total Terrestrial Carbon Storage

H. Tian (✉)
Ecosystem Science and Regional Analysis Laboratory, School of Forestry and Wildlife Sciences, Auburn University, AL 36849, USA
e-mail: tianhan@auburn.edu

S. Miao et al. (eds.), *Real World Ecology*, DOI 10.1007/978-0-387-77942-3_9,
© Springer Science+Business Media, LLC 2009

9.1 Introduction

Since the Industrial Revolution, the structure and function of Earth's terrestrial ecosystems have been altered significantly by human-induced changes in climate, atmospheric composition, and land-use/land-cover (LULC) (Vitousek et al. 1997, Millennium Ecosystem Assessment 2005). These changes, which occur on large spatial-temporal scales and interactively affect individual organisms and ecological systems, are too complex to be studied using traditional organism-focused ecological experiments (Ehleringer and Field 1993). There is no well-developed approach for understanding how these large-scale environmental changes have affected or will affect our life support system – the Earth's ecosystems. Developing such a methodology will require extrapolating the growth of plants, animals, or ecosystems into the future when climate, CO_2, and other factors may be different, and extrapolating individual plant or site studies onto a regional or global scale. System behaviors may emerge in a regional ecosystem through complex feedbacks among the sub-regional components (e.g., interactions between forests and grasslands). It is nearly impossible to reveal and capture these system behaviors at a higher hierarchical level by directly scaling up from site level results.

Developing and choosing an appropriate scaling approach is important for the regional extrapolation of knowledge or conclusion gained from site level studies. Because of the heterogeneity of terrestrial ecosystems, it is usually inappropriate to scale up by directly multiplying a measurement made at a small scale to a broader scale (King 1991, Tian et al. 1998a). To account for the spatial variability, an additive approach is frequently adopted in assessing environmental resources. This approach identifies discrete land/ecosystem categories and the values of which are calculated using a simple multiplicative approach. The regional pattern or behavior is then estimated as the sum of all the subsystem elements. In a study in the Amazon basin (Tian et al. 1998a), it was found that the traditional multiplicative approach may have overestimated the regional net ecosystem productivity by 0.5 to 1.5 P g C/ year. In fact, the calculation based on the assumption of homogeneous Amazon basin will lead to wrong conclusions about large regional carbon sinks in El Nino years, when the Amazon ecosystem is stressed by severe drought (Fan et al. 1990). When scaling across different hierarchical levels, new ecosystem behavior may emerge as a consequence of the complex interactions within a system. These ecosystem dynamics on a higher level can only be understood by investigating or simulating lower-level processes (O'Neill et al. 1986). Although numerical models have been used for summarizing, organizing, and synthesizing our knowledge of ecology since the early twentieth century (Lauenroth et al. 2003), these modeling methods use only empirical equations to regress the variation of a target variable along within single or multiple climatic and/or environmental factors.

These approaches are usually at site level due to their inability to deal with the heterogeneity of ecological variables. Therefore, in order to forecast and assess large-scale and long-term impacts of global change, it is necessary to explore an innovative approach that exceeds the limitations of both site-specific ecological studies and empirically based quantitative methods.

Built upon improved knowledge of the fundamental mechanisms of molecular systems to the planetary ecosystem, and supported by rapidly developing technology from high speed computer systems to high resolution global remote sensing sources, Regional Ecosystem Models (RegEM) now play an increasingly important role in solving these complex large-scale environmental problems (Tian 2006a). Combining RegEM with multiple sources data and models of land use and regional climate, the Regional Integration System for Earth's ecosystem (RISE) provides a powerful tool in addressing large-scale and long-term ecological issues. By using multiple sources data and a terrestrial ecosystem model (TEM), for example, Tian and his colleagues examined regional carbon dynamics in Monsoon Asia and United States (Tian et al. 1999, 2003). More specifically, the RISE can be used to examine interactive effects of multiple environmental stresses. In Ren et al.'s (2007b) study, the interactive effects of ozone with environmental factors such as elevated CO_2, climate change, and land use on the net primary production of China's terrestrial ecosystems, was assessed using a highly integrated ecosystem model. The RISE holds a predictive ability, which exceeds the traditional field studies. RISE's predictive ability also exceeds the traditional model methods in its accuracy and precision because RISE incorporates information from multiple data sources. In the first case study of this chapter, we established the predictive ability of the RISE in examining ecosystem properties such as carbon storages and fluxes. Recently developed databases utilizing field experiments, forest inventories, remote sensing observations, and climate records, provide detailed information for model parameterization and validation, and can provide comprehensive inputs that make regional/global integrated simulations feasible. By integrating ecosystem models, land use, and regional climate data, the RISE has the capacity of dealing with the complex interactions among ecosystems, climate, and humans. As regional and global environmental problems become an important ecological topic and the demands for regional assessments of present and future environmental qualities increase, the RISE will play a central role in large-scale terrestrial ecosystem studies.

In this chapter, we summarize the RISE used to integrate models of terrestrial ecosystem, land use/land cover, and regional climate with multiple sources of data to improve our understanding of large-scale ecological processes and address present and future regional and global environmental issues. We used two case studies to illustrate the overall merits and applications of the RISE in terrestrial ecosystem research.

9.2 Overview of the Regional Integration System for Earth's Ecosystem (RISE)

The RISE presented here is based on the development of integrated process-based ecosystem models, which synthesizes our current knowledge of major ecosystem components and important ecological processes. The model was parameterized and validated with intensively studied site data. After model uncertainty is assessed, simulation schemes are designed according to the research objectives. The model is driven by a comprehensive collection of spatial datasets gathered from multiple sources, usually across a wide range of scales, which include field and observational data, remotely sensed data, and simulated data from models such as a land use model and a regional climate model (Fig. 9.1). Model outputs include variables from ecosystem models that are of investigative interests. These outputs are usually first compared to the field observations for validation purposes before being used for regional assessments and predictions. The integrative study involves multiple disciplines such as information technology, statistics, and ecology. It demands knowledge of multiple technical fields such as software engineering, geographical information system, remote sensing, ecological experiments, and comprehensive knowledge of biogeochemical and hydrological cycles.

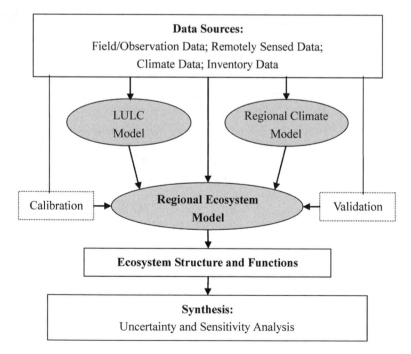

Fig. 9.1 Diagram of the Regional Integration System for Earth's Ecosystem (RISE) assimilating multiple sources data and regional ecosystem model

9.2.1 Development of Spatially Explicit Ecosystem Model

9.2.1.1 Conceptualization and Formulation

Conceptualization defines the purpose and boundaries of the model, identifies key variables, describes the behaviors of those key variables, and diagrams primary processes and feedback loops of the system. Model formulation is the process of converting feedback diagrams to level and rate equations so that the theory, or hypothesis, can be tested. The power of an ecosystem model rests on its ability to simulate the actual ecosystem processes, and to reflect the complex feedbacks in the ecosystem under changed/changing environments. As ecological processes are better understood, ecosystem models have become and will continue to become more process-based than empirically based. Recent ecosystem models such as Hybrid III (Friend et al. 1997) and LPJ (Sitch et al. 2003) are more sophisticated and reflect much more detailed physiological mechanisms than previous models like TEM (Raich et al. 1991). Our increased understanding of complex regional terrestrial ecosystems allows ecosystem models to become increasingly integrated. The first generation of process-based models included only water and carbon cycles. Gradually, nitrogen and phosphorus cycles (e.g., CENTURY, Parton et al. 1988), effects of LULC change (Tian et al. 2003), and ozone stress (Felzer et al. 2004) were included. In a recently developed Dynamic Land Ecosystem Model (DLEM, see the first case study for details), the major biogeochemical cycles, hydrological cycle, disturbances (LULC change and ozone stress), and trace gas (CH_4 and N_xO_y) emissions have been united. It should be noted that model development involves continuing modification and evaluation.

9.2.1.2 Model Parameterization and Calibration

Parameterization is the process of identifying model parameter values ideally collected from field measurements and/or manipulated experiments. When parameters are hard to obtain experimentally, or represent intermediate values in the model that do not exist in nature, the parameters can be determined through model calibration, an empirical process of estimating unknown coefficients by fitting the model outputs to the observations.

Parameter values are generally set up according to the purposes for which the model is used (Jackson et al. 2000). If the model is developed to explore the consequences of different parameter values, the model runs will use a wide range of different parameters without reference to particular ecological systems. However, if the model is developed to predict behavior in a specific system, a more limited value (if a single value system is studied it achieves single-equilibrium, several values if the studied system is in equilibrium in several conditions) is chosen for each parameter (Jackson et al. 2000). Two methods are used to evaluate the fitness of model results to observations while seeking parameters that lead to the best fit in the process of parameterization and calibration

(Hilborn and Mangel 1997). The least squares criteria is the most commonly used method for linear correlation between model results and field measurements. When model parameters are nonlinear, maximum likelihood algorithms are used to determine parameter values (Hilborn and Mangel 1997).

9.2.1.3 Model Validation and Evaluation

Once the model is assembled, parameterized, and calibrated, it must be tested to ensure that it functions properly (i.e., the process of model validation). Field measurements and remote sensing derived data sets are often used to validate model runs under different scenarios. As mentioned above, regional inventory data and outputs from remote sensing data may be compared to the regional assessments by ecosystem models for the purposes of model validation. Validation of regional ecosystem models, field-based net primary production (NPP) or net ecosystem production (NEP) measurements reflect strongly the local weather and soil characteristics at the time of measurement, and do not test the large-scale aggregated flux. To use these data for evaluating the results of an ecosystem model on a regional scale, we need to extrapolate information from the stand level to regional scale. Due to the great spatial heterogeneity of vegetation, soil, and climate, the exchange of CO_2 over a region could be different from that over a forest stand (Tian et al. 1998a). The real limitation on practical application of spatial extrapolation from stand to regional, or to global scales, may be due to our lack of knowledge and data about the relationship between small and large scales.

Regional ecosystem models cannot be fully validated with currently available data (Oreskes et al. 1994, Rastetter 1996), but the validation process forces us to examine the interaction among data, model structure, parameter sets, and predictive uncertainty. The real strength comes from the interplay of models and empiricism, whereby our intuition about the relations among variables is formalized in models and used to guide the next round of field or bench work.

9.2.2 Development of Time Series Spatial Data Sets

9.2.2.1 Data Sources

The RISE usually involves preparation of comprehensive spatial datasets for input into models (see the following case studies for examples). These data mainly come from four sources: (1) field experimental data, (2) inventory data or census records, (3) long-term climate/atmospheric records, and (4) remotely sensed data. The field data from intensively studied sites are useful for model calibration and point scale validation. Inventory data can be used for validation, as well as for spatial data development for model input (Zhang et al. 2007a, b). Climate and atmospheric records are interpolated into spatial maps which are used for model input (Raich et al. 1991, Aber and Federer 1992, Parton et al. 1993, Tian et al. 1998a, Running et al. 1999, Li et al. 2001). Remotely sensed data can be used for regional validation as well as model input (e.g., vegetation and land-cover maps). With the advancement of remote sensing technology, remote sensing becomes an increasingly important data source for ecosystem models.

An important utilization of remote sensing data for models include: LULC change, forest structure attributes, soil moisture, fire mapping, leaf area index (LAI), fraction of Photosynthetically Active Radiation (fPAR), biomass, foliar chemistry, water content, species composition, temperature, precipitation, solar radiation, humidity, fractional vegetation coverage, broadband albedo, incident solar radiation, downward thermal radiation, emissivity, and all-sky all-wave net radiation (Nightingale et al. 2004, Liang 2007). In particular, LAI and fPAR may be the most important ecological parameters used in ecosystem models. Many techniques have been developed for estimating LAI and fPAR from remote sensing data (Bacour et al. 2006, Gobron et al. 2006) and now both parameters are routinely estimated from Moderate-resolution Imaging Spectroradiometer (MODIS) data.

An additional contribution of remote-sensing data to modeling is its spatial estimations of NPP and evapotranspiration. Satellite data have been used to describe spatial and temporal variations in gross primary production (GPP), and NPP using the MODIS17 algorithm (Running et al. 1999). Since 2000, NPP and GPP are routinely estimated from MODIS data and other ancillary data such as climate on regional and global scales. NPP/GPP estimations are based on light use efficiency and respond only to current vegetation conditions (Running et al. 2004, Zhao et al. 2006), but do not relate to ecological and biogeochemical processes (Turner et al. 2004). In addition, Advanced Very High Resolution Radiometer (AVHRR) data have been used to estimate spatial variations in GPP/NPP. Ecosystem model results can be compared with corresponding regional satellite-derived estimates to identify the causes of these discrepancies, which may lead to improved GPP and NPP estimates by both approaches in the future.

9.2.2.2 Scaling Algorithms in the RISE

Scaling is one of the greatest challenges in ecological modeling (Levin 1992), and it usually involves both data preparation and model development. Most of the spatially explicit ecosystem models require a fixed spatial resolution during the simulation, while their input data may come from sources of various scales. For instance, field data and inventory records are collected at site level while remote sensing data reflect landscape and regional levels. Reorganizing data to make them consistent to the study scale while preventing information loss and distortion presents a big challenge to ecologists. Two algorithms, up-scaling and down-scaling, are used to transform data. Up-scaling is the process of extrapolating data from a smaller scale to a larger scale while down-scaling is the reverse. Many spatial interpolation methods are available for both algorithms. Selection of a suitable interpolation method that matches the properties of the data source and the study objectives is important in maintaining the quality of input data. Sometimes, a more complex process-based model (as presented in case study 1) is used to rescale the input data so that the developed dataset is more process-based than the empirically interpolated dataset.

Ecosystem models can be powerful tools to upscale knowledge gained from smaller scales (e.g., site study) to larger scales (e.g., regional assessments). Usually there are four approaches for up-scaling (King 1991): (1) lumping, (2) direct extrapolation, (3) extrapolating by expected value, and (4) explicit integration. Lumping involves retaining the original mathematical model by selecting new parameter values applicable to the larger scale. An example is the "big leaf" model which treats the whole canopy as one big leaf. In direct extrapolation, the model's inherent spatial unit is replicated many times to encompass the larger spatial scale with some information and material flow between the units. This approach may be impractical because it involves enormous computations. Extrapolating by expected value is to scale local output to a wider region by multiplying the area of the large region by the expected local output. One problem with this approach is defining which of the local outputs to use or how to combine them into an aggregated variable. Another problem is that scientists must estimate the probability distribution given incomplete and uncertain knowledge. Explicit integration is an analytical solution that requires mathematical integration of the local function over two or three dimensional spaces. This approach is usually impractical because of the complex and non-linear models (King 1991, Haefner 2005).

9.2.2.3 Application of Land Use and Climate Models in Generating Time Series Spatial Data Sets

Time series gridded data of land use history and future scenarios: To reconstruct land use and land cover (LULC) history before satellite period and to project LULC changes in the future, we need to use LULC models. In our previous study, we used GEOMOD, a spatially explicit dynamic land use simulation model, which predicts the rate and spatial pattern of land conversion, particularly for those anthropogenically derived (Hall, et al. 1995, Pontius 2001). GEOMOD is extraordinarily effective in helping scientists understand the dynamics of land use change, identify where most at risk of fragmentation and development are, visualize future conditions, and plan strategic approaches to the mitigation of harmful trends. Knowing how, where, and why these changes are likely to occur can be a powerful tool for conservation organizations, community leaders, and citizens. In partnership with local and regional experts, we gathered commonly available geographic, economic and social data on factors believed to influence land use change. We statistically analyzed these drivers to find out which are the best predictors of land use change in the study areas. We then used GEOMOD to analyze changes in land use classification data, determining a background rate of conversion from forest cover to non-forest cover. That rate was validated with data from current satellite imagery and aerial photography (e.g., 2005) and was used to predict future land development. The rate of change is derived by comparing the area of forest found in a land cover/use map at one point in time to that found in another at a different (either earlier or later) point in time. The model can be run either forward or backward and simulated results can be tested against the actual landscape as derived from satellite imagery and/or aerial photography. Future trends can also be established using regression analysis of various independent variables that may influence deforestation such as population growth, economic activity, employment history, infrastructure establishment, etc. The final products of LULC changes in the past, present, and future will be time series gridded data sets used for input of regional ecosystem models.

Downscale future climate change scenarios from GCM: Most General Circulation Models (GCMs) are run at relatively coarse spatial resolutions generally greater than 2.0° for both latitude and longitude. Their performance in reproducing regional and local climatic details is poor, which does not address climate impact assessment on a regional scale. Downscaled global climate information can give insights to processes that occur on the regional and local scales. Two common methods for downscaling future climate scenarios are statistical downscaling and regional climate modeling. To do statistical downscaling, we need to develop quantitative relationships between large-scale atmospheric variables (predictors) and local surface variables (prediction). There are different statistical downscaling methods such as weather classification schemes, regression models, and weather generators. Regional climate model (RCM) downscaling can significantly reduce GCM biases in simulating the present climate and future projection of regional climate change. The RCM downscaling has been widely used in different regions of the world. In first case study of this chapter, for example, the regional climate scenario for the year of 2050 was derived from the Penn State/NCAR Meso-scale model (MM5) downscaling. The resulting products from the statistical or RCM downscaling will be gridded climate data sets with fine spatial resolution, which is used as input to a regional ecosystem model.

9.2.3 Uncertainty Analysis

Uncertainty analysis is a method for understanding behavior, structure, and mechanisms of a model. As summarized by O'Neill and Gardner (1979), uncertainties in a model can come from three sources: hypotheses and mathematical formulation, parameter values, and natural variation. The first source results from our lack of understanding of the correct biological processes involved; the second one could be traced to our lack of knowledge regarding the mean and variance of the population from which parameter estimates are drawn; and the third one is caused by our inability to capture all components that must be treated as stochastic. All approaches for formal assessment of uncertainty in complex models include several steps: (1) identification of the output(s) of interest, (2) identification of a limited set of input parameters to which outputs are most sensitive and may vary depending on the output of interest, (3) development of the distribution for inputs and their correlation structures, (4) development of an efficient sampling technique when computer processing time is a constraint, and (5) design and evaluation of the Monte Carlo experiment (Haefner 2005).

Uncertainty analysis of input data is also important. Uncertainties exist in field measurement and data processing. As described above, the preparation of model input data usually involves interpolation, rescaling, projection, transformation, and format conversion. Errors may occur in each of these steps. Therefore, consequences of uncertainties from model inputs should be examined in the synthesis stage of simulation (Ramankutty and Foley 1999, McGuire et al. 2001). Sometimes, simulation results from input datasets of different sources are compared to biased assessment caused by different input data; thus provide a guide to improve the data quality.

9.3 Case Studies

In this section, we use two case studies to elucidate the RISE by coupling multiple sources data and dynamic ecosystem models. The model used here is the DLEM (Tian et al. 2005), and the data used include eddy flux, climate data, forest biomass and MODIS data.

9.3.1 Case Study 1: Responses of Terrestrial Ecosystem of Southeastern U.S. to Future Climate Change

In this case study, we addressed the potential impact of future environmental changes, including elevated CO_2, increased N deposition, and climate change, on carbon storage and carbon flux in terrestrial ecosystems of the southeastern U.S. (SEUS). The predictive capacity of the integrative approach was emphasized. The study region includes thirteen southeastern states: Alabama, Arkansas, Florida, Georgia, Kentucky, Louisiana, Mississippi, North Carolina, Oklahoma, South Carolina, Tennessee, Texas, and Virginia. The DLEM (Tian et al. 2005, Chen et al. 2006, Ren et al. 2007a, b, Zhang et al. 2007a, b) was used to assess responses of ecosystem productivity and total carbon storage to environmental changes from 2002 to 2050. The spatial resolution of this study was 32 km × 32 km. Five simulations were conducted to study the impacts of climate and atmospheric changes on future carbon dynamics in SEUS: 2002 baseline (2002_BASELINE), 2050 climate change only (2050_CLM), 2050 CO_2 change only (2050_CO$_2$), 2050 N deposition change only (2050_NDEP), and 2050 combination (2050_COMBINE) (Table 9.1).

Table 9.1 Design of simulation experiments for the terrestrial ecosystems of the southeastern U.S.

Scenario	Climate	CO_2	Nitrogen deposition
2002_Baseline	2002*	2002	2002
2050_Climate	2050	2002	2002
2050_CO₂	2002	2050	2002
2050_Nitrogen deposition	2002	2002	2050
2050_Combined	2050	2002	2050

2002*: the input data of year 2002 is used in the simulation; 2050: the predicted value of year 2050 is used in the simulation. The baseline is the initial equilibrium state as driven by 2002 dataset. 2005_Climate is the scenario in which only the 2005 climate was considered with other keep unvaried with 2002 condition. It is same for 2050_CO₂ and 2050_Nitrogen deposition. 2050_Combined is the overall scenario in which all the driving input data of 2050 were used

9.3.1.1 Model Description

The DLEM used in this study couples major biogeochemical cycles, hydrologic cycles, and vegetation dynamics to estimate daily spatially explicit estimates of water, carbon (CO_2, CH_4) and nitrogen fluxes (N_2O), and pool sizes (C and N) in terrestrial ecosystems (Tian et al. 2005, Chen et al. 2006, Ren et al. 2007a, b, Zhang et al. 2007a, b). DLEM is built on the experience and heritage of the existing TEM (Raich et al. 1991, Melillo et al. 1993, McGuire et al. 2001, Tian et al. 1998a, 1999, 2003, Felzer et al. 2004, 2005) and Biome-BGC (BioGeochemical Cycle) (Running et al. 1999) DLEM includes five core components: (1) biophysics, (2) plant physiology, (3) soil biogeochemistry, (4) dynamic vegetation, and (5) land use and management (Fig. 9.2). DLEM also integrates algorithms of N_2O emission from DNDC (DeNitrification-De-Composition) (Li 2000) and CH_4 emissions from other previous studies (Huang et al. 1998, 2004, Zhuang et al. 2004). The dynamic vegetation component in DLEM simulates two kinds of processes: the biogeography redistribution when climate changes and post-disturbance plant competition and succession during vegetation recovery. Like most Dynamic Global Vegetation Models (DGVM), DLEM is built on the concept of plant functional types (PFT) to describe vegetation distributions. DLEM has also emphasized the modeling and simulation of managed ecosystems including agricultural ecosystems, plantation forests, and pastures. DLEM version 1.0 has been used to simulate the effects of climate variability and change, atmospheric CO_2, tropospheric ozone, land use change, nitrogen deposition, and disturbances (e.g., fire, harvest, hurricanes) on terrestrial carbon storage and fluxes (Tian et al. in prep). This model has been fully calibrated against field data of forests, grassland, and croplands. The simulation results have been compared with independent field data and satellite products.

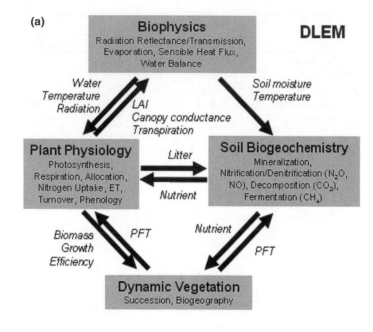

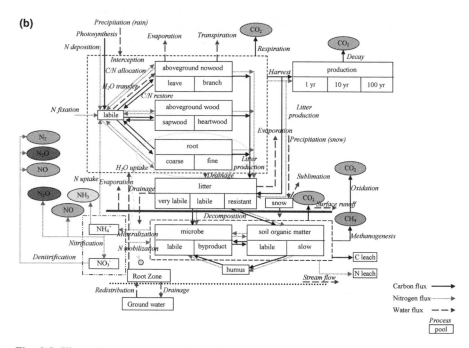

Fig. 9.2 The major components (**a**) and conceptual structure (**b**) of the Dynamic Land Ecosystem Model (DLEM)

9.3.1.2 Model Validation

Forest carbon pools comprise the largest fraction of carbon pools in the SEUS. Therefore, accurate simulation of forest carbon dynamics is very important in assessing regional carbon balances. We used environmental measurements from two intensively studied southern forests: the Duke loblolly pine forest in North Carolina (latitude: 35°58′; Longitude: 79°5′) and the deciduous broadleaf forest in Walker Branch Watershed, Tennessee (Latitude: 35° 57′; Longitude: 84°17′) as model inputs, and compared model outputs to measured carbon flux (public.ornl.gov/ameriflux) (Fig. 9.3).

9.3.1.3 Data Acquisition

Input data required by DLEM included transient climate, atmospheric CO_2 concentration, tropospheric ozone stress (AOT40 index), nitrogen (N) deposition, and ten base maps (Table 9.2). All these maps were prepared or rescaled to 32×32 km^2 using geographic information system (GIS) software.

The 2002 climate data sets (daily precipitation, average, maximum, and minimum temperatures) were generated by the North American Regional Reanalysis (NARR) project (http://www.emc.ncep.noaa.gov/mmb/rreanl/). NARR is a dynamically consistent, high-resolution (32-km), high-frequency (daily), atmospheric and land surface hydrology dataset for the North American domain. It uses the National Center for Environmental Prediction (NCEP) Eta Model and the NCEP Data Assimilation System, and introduces a new version of the N̲CEP, O̲regon State University, A̲ir Force, and H̲ydrologic Research Lab (NOAH) land surface model to reconstruct a long-term (from 1979 to present) climate dataset. It is perhaps the best high resolution (both temporal and spatial) climate dataset of the study region currently available.

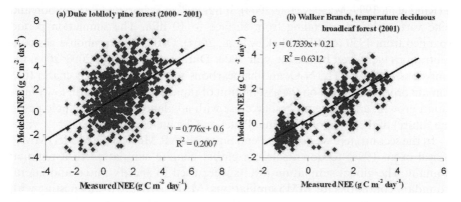

Fig. 9.3 Comparison of the simulated net ecosystem exchange (NEE) and measured site fluxes in the (**a**) Duke forest and the (**b**) Walker Branch watershed to validate the simulations

Table 9.2 Data types and sources for model input in the simulation of the southeastern U.S.

Data types	Unit	Source
Vegetation	6 categories	USGS National Land Cover Datasets (http://edc.usgs.gov/products/landcover.html)
Soil clay content	%	General soil association map
Soil-sand content	%	(STATSGO) by United States
Soil silt content	%	Department of Agriculture (USDA)
Soil depth	M	
Soil acidity	pH	
Soil bulk density	g/cm^3	
Elevation map	M	USGS National Elevation Dataset
Aspect map	Degree	(http://edcnts12.cr.usgs.gov/ned/
Slope map	Degree	ned.html)
Precipitation	mm /year	For year 2002: North American Regional
Maximum temperature	Celsius	Reanalysis (NARR). For year 2050:
Minimum temperature	Celsius	model simulation.
Average temperature	Celsius	
CO_2	ppmv	Year 2002: 375 ppmv; Year 2050: 480 ppmv*.
Ozone concentration, AOT40@	ppb-hr	Spatial interpolation based on records from 181 ozone stations (Felzer et al.2004).
NH_x deposition	$mgN/ (m^2 \cdot year)$	Atmospheric chemical model simulation
NO_y deposition	$mgN/ (m^2 \cdot year)$	by Dentener (2006)

* CO_2 concentration in 2002 is based on standard Intergovernmental Panel on Climate Change (IPCC, 2001); CO_2 concentration in 2050 is based on IPCC A1B scenario (http://www.ipcc.ch/).
@ AOT40 is the accumulated dose over a threshold of 40 ppb during daylight hours.

The 2050 climate dataset was generated in two steps. First, future climate changes were projected by the Goddard Institute for Space Studies (GISS) GCM 2 (Rind et al. 1999, Mickley et al. 2004). It has a $4° \times 5°$ horizontal resolution and nine vertical layers extending from surface to 10 hPa. The simulation period spanned from 1950 to 2055 (Mickley et al. 2004). Observed greenhouse gas concentrations were used before the year 2000. During the period of 2000–2055, the emissions of CO_2, CH_4, N_2O, and halocarbons followed the A1B scenario (i.e., climate projection based on the assumption of rapid economic growth, low population growth, and moderate resource use with a balanced use of technologies in the future) of the Intergovernmental Panel on Climate Change (IPCC).

In the second step, we used the Penn State/NCAR Meso-scale Model (MM5) (Grell et al. 1994) to downscale the global climate simulation to a regional domain. The global simulation results were used to specify initial and lateral boundary conditions for MM5 simulations. MM5 used a two-way nesting with grid distances of 108 km for the outer domain and 36 km for the inner domain. Hourly meteorological fields such as wind, pressure, and temperature were generated for the inner domain. Finally, bilinear interpolation was used to transform the 36-km resolution climate outputs into the 32-km resolution climate dataset required by this regional study.

From 2002 to 2050, the regional annual precipitation was predicted to decrease about 10% (114.97 mm) for the whole SEUS, while annual average, maximum, and minimum air temperatures were predicted to increase 0.52, 0.66 and 1.90 °C, respectively (Fig. 9.3A). The average air temperature was projected to decrease in the eastern part of the SEUS, while increasing in the western and central parts (Fig. 9.4A). The largest temperature increase will take place in the western part of the SEUS. Precipitation is predicted to increase in the eastern, lower central and upper western parts of the SEUS, but decrease in the other

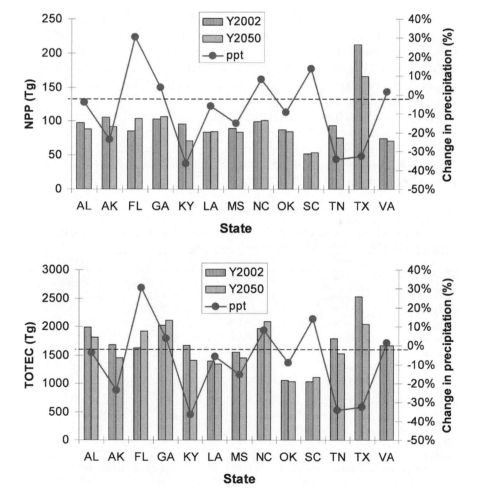

Fig. 9.4 Rate of changes (%) in productivity (**a**) and carbon storage (**b**) from 2002 to 2050 as simulated by DLEM (Dynamic Land Ecosystem Model)

parts (Fig. 9.4A). According to our climate projection, from 2002 to 2050, the biggest decrease (over 50%) in precipitation would occur in Texas, which is an arid and semiarid region, thereby increasing droughts. In contrast, precipitation will increase over 50% in most areas of Florida.

9.3.1.4 Terrestrial Ecosystem Productivity and Carbon Storage in Southeastern U.S.

Considering all factors, the total terrestrial ecosystem NPP of the southeast is estimated to be 1.28 Pg C/year in 2002 (Table 9.3). Texas has the highest total NPP (0.21 Pg C/year) (Fig. 9.5A). South Carolina has the lowest total NPP of 0.05 Pg C/year. The total terrestrial carbon storage (TOTEC) in the southeast is estimated to be 21.94 Pg C (Table 9.3). According to our simulation results, Texas has the highest carbon storage of 2.5 Pg C, followed by Georgia which is estimated to have carbon storage of 2.0 Pg C in 2002 (Fig. 9.5B). Both South Carolina and Oklahoma have low carbon storage of about 1.0 Pg. Soil organic carbon (SOC) is estimated to account for more than 61% of the total carbon storage in the southeast.

The high productivity and total carbon storage of Texas is mainly due to its large area. The average carbon density of Texas, however, is the lowest (3,669 g C/m^2) according to our model estimation (Fig. 9.6). Our model results show that in 2002 Tennessee had the highest average carbon density of 16,360 g C/m^2 followed by Kentucky (16,267 g C/m^2).

Table 9.3 Impacts of climate change, CO_2 and N deposition on productivity and carbon storage of SEUS (Southeastern U.S.) (1Pg = 10^{15}g)

	NPP (Pg C/year)	VEGC(Pg C)	SOC(Pg C)	TOTEC(Pg C)
2002(BASELINE)	1.28	8.54	13.40	21.94
2050_COMBINE	1.18	8.42	12.55	20.97
change@	−0.09	−0.12	−0.85	−0.97
%change@	−7%	−1%	−6%	−4%
2050_CLM	0.99	7.56	11.17	18.73
Change	−0.29	−0.98	−2.23	−3.21
%change	−23%	−11%	−17%	−15%
2050_CO2	1.48	9.44	14.79	24.22
Change	0.20	0.90	1.39	2.28
%change	16%	11%	10%	10%
2050_NDEP	1.31	8.64	13.48	22.12
Change	0.03	0.10	0.08	0.18
%change	3%	1%	1%	1%
Interaction effect*	0.04	0.14	0.09	0.23
% Effect@	3%	2%	1%	1%

@ Relative to 2002 baseline.
*Interaction effect = (2050_CLM + 2050_CO$_2$ + 2050_NDEP) − 3 × 2050_COMBINE.

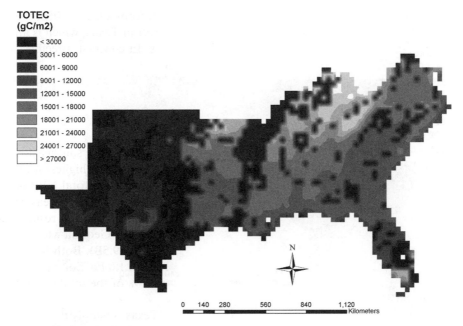

TOTEC (gC/m2)

- ◼ < 3000
- ◼ 3001 - 6000
- ◼ 6001 - 9000
- ◼ 9001 - 12000
- ◼ 12001 - 15000
- ◼ 15001 - 18000
- ◼ 18001 - 21000
- ◼ 21001 - 24000
- ◼ 24001 - 27000
- ☐ > 27000

N

0   140   280   560   840   1,120
                                              Kilometers

Fig. 9.5 Comparisons of (**a**) the productivity and (**b**) carbon storage of southeastern states between 2002 and 2050

9.3.1.5 Response of Ecosystem Productivity to Climate Change from 2002 to 2050

The regional NPP and TOTEC are predicted to decrease by 7% and 4%, respectively, due to the climate change from 2002 and 2050, despite the positive effects of the elevated CO_2 (480 ppmv in 2050 in compare to the 375 ppmv in 2002) and elevated nitrogen deposition on plant growth (Table 9.3). The interaction of all environmental factors could result in positive effects of about 3% and 1% on productivity and total carbon storage, respectively.

Climate change is one of the major factors that control carbon dynamics in terrestrial ecosystems. Figure 9.5 indicates that future climate change may have different impacts on different regions in the SEUS. In general, precipitation is the major factor that controls the productivity and carbon storage in the Southeast, while temperature change may also alter the magnitude of future regional carbon dynamics. Precipitation in Kentucky, Texas, and Tennessee is predicted to decrease by more than 30% in 2050 compared to 2002. From the late spring to the autumn in 2050, these states are expected to experience severe drought. Consequently, the NPPs of these three states are predicted to decrease by more than 20%, according to model simulations. The largest decline in NPP (26%) will occur in Kentucky, which is expected to experience severe drought in 2050 (37% decrease in precipitation). Total terrestrial ecosystem carbon storage in Texas is predicted to decline more significantly (19%) than in Kentucky

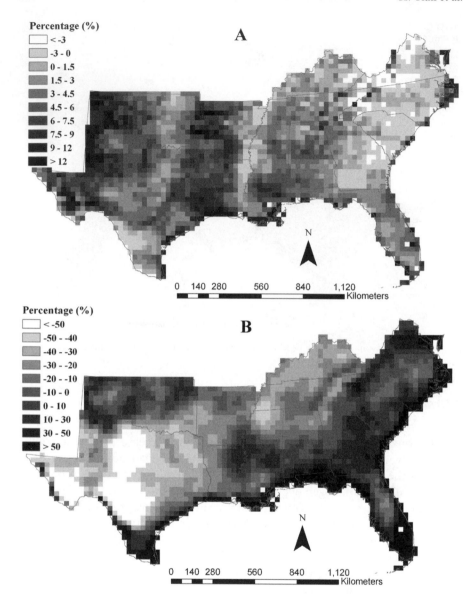

Fig. 9.6 Spatial pattern of carbon density in the southeastern U.S. in 2002 as simulated by the DLEM (Dynamic Land Ecosystem Model)

(16%) because Texas will experience significant warming (1°C) as well as drought in the future (Fig. 9.3). Increased temperature could enhance ecosystem respiration and therefore decrease carbon storage in Texas. In contrast to Kentucky and Texas, Florida's precipitation was predicted to rise over 30%, increasing NPP by 21% and ecosystem carbon storage by 18% by 2050.

The spatial distribution and intensity of carbon sinks and sources in the southeast is controlled by the spatial pattern of climate change and the carbon density of the terrestrial ecosystem. Our simulation results (Figs. 9.7A and 9.7B) show that the northern (Kentucky and Tennessee) and western (Texas and Arkansas) areas of the study region will become significant carbon sources, while the coastlines and the southern region will become carbon sinks during

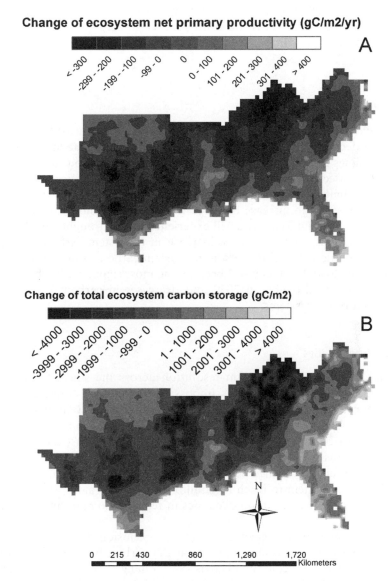

Fig. 9.7 Rate of changes in productivity (**a**) and carbon storage (**b**) from 2002 to 2050 as simulated by DLEM (Dynamic Land Ecosystem Model)

the study period. This pattern generally matches precipitation changes (Fig. 9.3B). It should be noted that although the most severe droughts are predicted to take place in Texas, high carbon losses will occur in Kentucky and Arkansas because of the original high carbon densities in these two states (Figs. 9.6 and 9.7B). The average productivity of deciduous broadleaf forests in Kentucky will decrease by 243 g $C/m^2\cdot$year, while the average productivity of Texas (composed primarily of shrub land and grasslands) will decrease by only 68 g $C/m^2\cdot$year by 2050 (Fig. 9.7A). As a result of increased precipitation, the productivity and the total carbon storage of coastal regions are predicted to increase up to 420 gC/m^2/year and 5107 gC/m^2, respectively, by 2050 (Figs. 9.6 and 9.7).

9.3.1.6 Conclusions

Climate change will dominate carbon dynamics of the southeast from 2002 to 2050. The decrease of annual precipitation and increase in temperature may lead to declines in regional ecosystem productivity and carbon storage despite the positive effects of elevated CO_2 and increased N deposition. The spatial pattern of climate change will be highly heterogeneous. While precipitation in Florida is expected to increase by more than 30%, the model indicates that Texas, Kentucky, and Tennessee will experience severe droughts by 2050. The model simulations predict an 18% increase in ecosystem carbon storage in Florida, and a decrease of 16–19% in ecosystem carbon storage in Kentucky and Texas by 2050. Kentucky will become the most significant carbon source due to the severe droughts in 2050 and decreases in carbon density of its deciduous forest ecosystems.

9.3.1.7 Uncertainties

Application of MM5 at a 32-km resolution across the whole Southeast is computationally intensive, so it is impossible to produce transient climate change for each year from 2002 to 2050 due to limitations of computer capacity. Due to lack of continuous climate data from 2002 to 2050, we run the model to equilibrium in 2050 and compared the model outputs against the outputs of 2002 simulations, also run to equilibrium. This approach could overestimate the impacts of climate change because in reality it usually takes more than 50 years for the ecosystem to reach its equilibrium. Furthermore, this approach cannot reflect ecosystem carbon dynamics in response to inter-annual climate fluctuations.

Uncertainties may also emerge due to our exclusion of land use change. Human activity has important impacts on the carbon balance of the southeastern ecosystems (Delcourt and Harris 1980, Zhang et al. 2007b), which have experienced rapid land use changes in the past 200 years. Rapid land use change is expected to continue in the Southeast (Wear and Bolstad 1998); therefore, the actual future ecosystem productivity and carbon storage could differ from our prediction as a result of land use change during the study period.

9.3.2   Case Study 2: Impacts of Tropospheric Ozone Pollution on Productivity and Carbon Storage of China's Terrestrial Ecosystems from 1961 to 2000

In this study, the potential effects of elevated tropospheric ozone (O_3) and its combination with changing climate, increasing CO_2, and land use on NPP and carbon storage in China's terrestrial ecosystems for the period 1961–2000 were investigated using DLEM in conjunction with historical spatial data of tropospheric ozone and other environmental factors. Six experiments were designed to analyze the effects of O_3 on NPP, Net Carbon Exchange (NCE), and carbon storage in terrestrial ecosystems of China (Table 9.4).

Unlike the first case study, which runs the model to an equilibrium state, we conducted transient simulations (i.e., climate varies from year to year during the study period). We first ran the model to an equilibrium state using average climate datasets to determine the baseline. By using detrended climate data, the model ran for an initial period of 150 years to eliminate the noise caused by time series fluctuation. Finally, we ran the model using transient data from 1961 to 2000.

9.3.2.1   Description of Input Data

Input datasets used include: (1) elevation, slope, and aspect maps derived from digital elevation model with 1 km spatial resolution for China (http://www.wdc.cn/wdcdrre), (2) soil datasets derived from the 1:1,000,000 soil map based on the second national soil survey of China (pH, bulk density, depth to bedrock, soil texture represented as the percentage content of loam, sand and silt) (Wang et al. 2003, Shi et al., 2004, Tian et al. 2006b), (3) a vegetation map (or land cover map) from the 2000 land use map of China (LUCC_2000) which was developed from Landsat Enhanced Thematic Mapper (ETM) imagery (Liu et al. 2005a), (4) a potential vegetation map, which was constructed by replacing the croplands of LUCC 2000 with potential vegetation defined by global potential vegetation maps developed by Ramankutty and Foley (1998), (5) a standard IPCC (2001) historical CO_2 concentration dataset, (6) the AOT40 dataset (see below for detailed information), (7) the long-term land use history (cropland and urban distribution of China from 1961–2000) which was developed based on three recent (1990, 1995 and 2000) land-cover maps (Liu et al. 2005a, b) and historic land use records of China (Xu 1983, Ge et al. 2003), and (8) a historical daily climate data (maximum, minimum, and average temperatures, precipitation, and relative humidity). Seven hundred forty six climate stations in China and 29 stations from surrounding countries were used to

Table 9.4 Design of simulation experiments for terrestrial ecosystems in China

Scenarios	O_3	Climate	CO_2	Land use	Fertilizer	Irrigation
O_3 only	Historical	Constant	Constant	Constant	No	No
$O_3\_CO_2$ (OC)	Historical	Constant	Historical	Constant	No	No
$O_3\_$Climate (OCL)	Historical	Historical	Constant	Constant	No	No
Climate_LUCC_CO_2 (CLC)	Constant	Historical	Historical	Historical	Historical	Historical
$O_3\_$Climate_LUCC_CO_2 (OCLC)	Historical	Historical	Historical	Historical	Historical	Historical
$O_3\_$Climate_LUCC_$CO_2\_$N	Historical	Historical	Historical	Historical	No	No

produce daily climate data for the time period from 1961 to 2000, using an interpolation method similar to that used by Thornton et al. (1997). To account for cropland management, we used data from the National Bureau of Statistics of China, which recorded annual irrigation areas and fertilizer amounts in each province from 1978 to 2000. All datasets had a spatial resolution of $0.5° \times 0.5°$, and climate and AOT40 datasets were developed for a daily time step while CO_2 and land use datasets were developed for a yearly time step.

The methods used for monitoring ozone vary among the limited ground ozone monitoring sites in China (Chameides et al. 1999). Therefore, it is difficult to develop a spatial historical AOT40 dataset based on the interpolation of site-level data like Felzer et al. (2004) developed for the U.S. In this study, the AOT40 dataset was derived from the global historic AOT40 datasets constructed by Felzer et al (2005). This AOT40 index was calculated by combining geographic data from the MATCH model (Multiscale Atmospheric Transport and Chemistry) (Lawrence and Crutzen 1999, Rasch et al. 1997) with hourly zonal ozone data from the Massachusetts Institute of Technology's Integrated Global Systems Model (MIT IGSM). The average monthly boundary layer MATCH ozone values for 1998 are scaled according to hourly values interpolated zonal average ozone from the Integrated Global Systems Model (IGSM), which had 3-hourly values, to the zonal ozone from the monthly MATCH to maintain the zonal ozone values from the IGSM (Wang et al. 1998, Mayer et al. 2000).

9.3.2.2 Impact of Ozone on Carbon Storage and Flux in Terrestrial Ecosystems of China

In the simulation experiments, O_3 produced negative effects on total average NPP and carbon storage during the study period. Average annual NPP and total C storage from the 1960s to the 1990s in China increased by 0.66% and 0.06%, respectively, under the full factorial (O_3, climate, land use, and CO_2 were changed, hereafter referred to as OCLC), while they increased by 7.77% and 1.63%, respectively, under the scenario without O_3 (hereafter CLC). This difference indicates that O_3 decreased NPP (about 1.64% in the 1960s and 8.11% in the 1990s) and total C storage (about 0.06% in the 1960s and 1.61% in the 1990s) in China's terrestrial ecosystems. Although NPP and total C storage in both scenarios increased over time, the soil and litter C storage decreased (–0.18% and –0.67%, respectively) under the OCLC scenario, while they increased by 0.30% and 1.63%, respectively, under the CLC scenario. Therefore O_3 reduced soil and litter C storage by 0.03 and 0.16% in the 1960s and by 0.52 and 1.84% in the 1990s, respectively, in China. The model results showed that NPP and carbon storage, including vegetation carbon, soil carbon, and litter carbon, decreased with O_3 exposure, and the reduced NPP was greater than the decrease in carbon storage. The changing rates in the 1960s and 1990s indicated that increasing O_3 concentrations could lower NPP and decrease carbon storage, which further implied that under the influence of O_3 alone,

Table 9.5 Overall changes in carbon fluxes and pools during 1961–2000 as estimated by the Dynamic Land Ecosystem Model (DLEM)

Scenarios	O_3_Climate_LUCC_CO_2			Climate_ LUCC_CO_2			Difference with O_3 and without O_3	
	1960s (Pg)	1990s (Pg)	Net exchange (%)	1960s (Pg)	1990s (Pg)	Net exchange (%)	1960s (%)	1990s (%)
NPP	3.04	3.06	0.66	3.09	3.33	7.77	−1.64	−8.11
Vegetation C	26.02	26.32	1.15	26.03	27.38	5.19	−0.04	−3.87
Soil C	59.79	59.68	−0.18	59.81	59.99	0.30	−0.03	−0.52
Litter C	19.32	19.19	−0.67	19.35	19.55	1.03	−0.16	−1.84
Total C	105.14	105.2	0.06	105.2	106.92	1.63	−0.06	−1.61

Note: O_3_Climate_LUCC_CO_2 means the scenario considering the effects of O_3, Climate variability; land use and land cover change (LUCC), and CO_2 increase.
Climate_LUCC_CO_2 means the scenario considering the effects of climate variability, LUCC, and CO_2 increase.

China's soil ecosystem would be a net C source, while without O_3, it might be a net C sink (Table 9.5).

The results of accumulated NCE across China under three simulation experiments, including O_3-only, combined effects without O_3 and with O_3 from 1961–2000, indicate that O_3 effects may cause carbon release from the terrestrial ecosystem to the atmosphere. The accumulated NCE was –919.1 TgC under the influence of O_3 only, 1,177.4 TgC under the combined influence of changed climate, CO_2 and land use (CLC), and 677.3 TgC under the full factorial (OCLC). These values imply that China would be a CO_2 source when influenced only by O_3, but would be a sink under the influence of both the CLC and OCLC scenarios. The accumulated NCE influenced by OCLC was 620.4 TgC less than that influenced by CLC, implying that interactions between O_3 and other factors (CO_2, climate, or land use change) were very strong and decreased CO_2 emissions from terrestrial ecosystems. For the 1990s, a decade with rapid atmospheric O_3 change, our simulations showed that the cumulative NCE under the full factorial (OCLC) throughout China decreased, compared to the results without O_3 influence (CLC) (Table 9.5 and Fig. 9.8). Parts of central-eastern China and northeastern China released 150 g /m^2 more C into the atmosphere under O_3 influences during the 1990s. In our study, average annual NPP decreased 0.01PgC/year and accumulated NCE was −0.92 PgC (Fig. 9.8) from 1961 to 2000 under the effect of O_3 only. Similar to most field experiments in the U.S., Europe, and China, and to other model results (e.g., Heagle 1989), our results show that ozone's negative effects on terrestrial ecosystem production are caused by direct ozone-induced reductions in photosynthesis.

The combined effects of O_3, CO_2, climate, and land use show very different results. Ozone may compensate for CO_2 fertilization, resulting in NPP losses in different plant types over time across China (Fig. 9.8). Climate variability increased the ozone-induced reduction of carbon storage and led to substantial

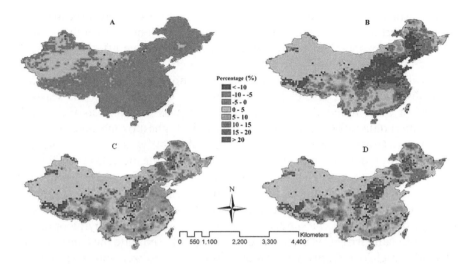

Fig. 9.8 Rate of change in mean annual NPP from the 1960s to the 1990s under different scenarios: (**a**) O_3 only; (**b**) O_3 and Climate; (**c**) O_3, climate variability, and LUCC; (**d**) climate variability, LUCC, and CO_2

year-to-year variations in carbon fluxes (NPP and NCE). Similarly, the contribution of land use change to the terrestrial carbon budget varied over time and by ecosystem type (e.g., Houghton and Hackler 2003). In our study, we accounted for the combined effects of land use change with historic ozone concentrations, atmospheric CO_2 concentrations, and climate variability. The other two aspects are the sensitivity of different biomes to ozone exposure and agriculture management. The former results in different carbon loss rates; however, the latter's effects combined with ozone pollution on carbon storage are related to changing soil environment such as water and nitrogen conditions. In contrast to land use change in the eastern U.S. (Felzer et al. 2004), it is necessary to consider how to manage irrigation in arid areas, an important issue in China. Reasonable irrigation management can both enhance the water use efficiency and mitigate ozone damage.

9.3.2.3 Uncertainties

An integrated assessment of O_3 impacts on terrestrial ecosystem production on a large scale requires the following information: (1) O_3 data that reflect air quality in the study area and other environmental data, including climate (temperature, light, and precipitation), plant (types, distribution, and parameters), and soil information(texture, moisture, etc.), (2) mechanisms of O_3 impacts on ecosystem processes, which describe relationships between air pollutant dose and eco-physiological processes such as photosynthesis, respiration, allocation, toleration, and competition, and (3) an integrated process-based model, which is able to quantify ozone damage to ecosystem processes. To reduce uncertainty in our current work, future research needs to address ozone data quality, ozone-plant physiological effects, and ozone affects on agricultural crops. First, we used one set of O_3 data derived from a combination of global atmospheric chemistry models which had not been well validated against field observations because of limited field data. We found that the seasonal pattern of our simulated ozone data (AOT40) was the same as the limited observation data sets (Wang et al. 2007) with the highest ozone concentrations in summer. However, we still need observed AOT40 values to calibrate and validate our O_3 data set in the future. Second, the ozone module in our DLEM focuses on the direct effects on photosynthesis and indirect effects on other processes, such as stomata conductance, carbon allocation, and plant growth. The quantitative relationship between O_3 and these processes remains untested by field studies. Third, we simulated land use change accompanying optimum fertilization and irrigation management. It is hard to separate the contributions of land use change from their combination with fertilizer and irrigation. In particular, according to the studies of Felzer et al. (2004, 2005), ozone effects combined with fertilizer management could increase the damage to ecosystem production. However, irrigation management in dry lands could reduce the negative effect of ozone (Ollinger et al. 1997). In our model, crops were classified as dry farmland and paddy farmland which improved the crops' simulation compared to previous process-based models. Different crop types, such as spring wheat, winter wheat, corn, rice, and soybean, vary in their sensitivities to ozone and could produce different ecosystem production changes due to the effects of ozone (e.g., Wang et al. 2007).

9.4 Summary and Perspectives

The RISE was summarized with description of each step, and two case studies, China and southeastern U.S., were used for elucidating the large-scale ecological issues and understanding the large-scale patterns and processes of terrestrial ecosystems in a changing global environment. In contrast to the traditional approach, the RISE is a modern ecological method designed to address large-scale spatial or temporal ecological issues, especially in combination with the important development of long-term ecological networks and a great amount of field experiments and satellite techniques. To reach that point, a spatially explicit model and multiple sources datasets for driving that model need to be developed using an appropriate scaling algorithm. After the model is assembled, it must be parameterized, calibrated and validated before it is applied for specific scientific questions.

The spatial model is the key component, and the multiple sources of data are the premise of the RISE. Data sources provide the knowledge for model development, parameters for model parameterization and calibration, and data for model validation. Models play a central role in addressing target questions because they are built toward specific ecological issues. The interaction between field ecologist and modeler is the key in the development and application of the integrative regional method. Data sharing and model sharing are two primary practice options for the near future.

9.5 The Way Forward

To better address large-scale and long-term impacts of global changes, it is of critical importance to fully couple biogeochemical and biophysical processes with the climate system across spatial and temporal scales and develop innovative technologies for data acquisition and processing. To improve the structure and abilities of dynamic ecosystem models, Tian et al (1998b) summarizes five aspects that must be addressed: (1) modeling functional dynamics of terrestrial ecosystems, (2) linking vegetation dynamics with biogeochemical processes, (3) linking biogeographic processes with biogeochemical processes, (4) linking land use changes to biogeochemical processes, and (5) modeling interactions among energy, water and biogeochemical cycles. In the past decade, significant progress in ecosystem modeling has been made, but how to fully couple energy, water, biogeochemical cycles, and vegetation dynamics still remains a challenge. To accurately assess how human activities have affected or will affect the Earth's ecosystem, we need to better address the integration of socioeconomic processes and ecosystem dynamics in our integrated system model.

Besides the model, there is an urgent need to develop accurate spatial data by using innovative methods in data acquisition and processing. The shorter time scales and small areas are the main drawbacks to current ecological field experiments (Rastetter 1996). Long-term and large-scale experiments are needed for better understanding and validating a model to address large-scale ecological issues.

For large-scale studies, satellite images become an important data source, but converting spectral information to ecological parameters remains a challenge (Running et al. 2004, Lu 2006). Extraction of categorical (e.g., LULC classes) and continuous (e.g., LAI, biomass) variables from remote sensing data on regional and global scales is still very difficult because of the complex landscape, limitations of remote sensing techniques, and limitations of human labor and facilities (Lu and Weng 2007). Most research focusing on remote sensing application is still on a landscape scale. High to medium spatial resolution images are frequently used. Some previous literature has already discussed the potential variables that can be derived from these remote sensing data (Cohen and Goward 2004, Wulder et al. 2004). Higher spatial resolution data has the potential to derive more accurate and detailed biophysical variables that can be used for model validation and evaluation.

Acknowledgment This research has been supported by NASA Interdisciplinary Science Program (NNG04GM39C), NASA Land Use and Land Cover change Program, DOE NICCR Program, US EPA 2004-STAR-L1 (RD-83227601) and AAES Program. We thank Drs. Yuhang Wang, Tao Zeng, and L. Ruby Leung for providing the regional climate data of the southeastern U.S., Dr. Felzer for providing the troposheric ozone data set for China, and Drs. Martha K. Nungesser and ShiLi Miao for providing valuable comments.

References

Aber, J. D., and C.A. Federer. 1992. A Generalized, Lumped-Parameter Model of Photosynthesis, Evapotranspiration, and Net Primary Production in Temperate and Boreal Forest Ecosystems. Oecologia **92**: 463–474.

Bacour, C., F. Baret, D. Beal, M. Weiss., and K. Pavageau. 2006. Neural network estimation of LAI, fAPAR, fCover and LAIxC(ab), from top of canopy MERIS reflectance data: principles and validation. Remote Sensing of Environment **105**: 313–325.

Chameides, W., X. Li, X. Zhou, C. Luo, C.S. Kiang, J.S. John, R.D. Saylor, S.C. Liu., and K.S. Lam 1999. Is ozone pollution affecting crop yield in China? Geophysical Research Letter **26**: 867–870.

Chen, G., H. Tian, M. Liu, W. Ren, C. Zhang., and S. Pan. 2006. Climate Impacts on China's Terrestrial Carbon Cycle: An Assessment with the Dynamic Land Ecosystem Model. In: Tian, H.Q. editor. Environmental Modeling and Simulation, ACTA Press, Anaheim/Calgary/Zurich, pp. 56–70.

Cohen, W.B., and S.N. Goward. 2004. Landsat's role in ecological applications of remote sensing. BioScience **54**: 535–545.

Delcourt, H., and W. Harris. 1980. Carbon budget of the Southeastern U.S. biota: analysis of historical change in trend from source to sink. Science, **210**: 321–323.

Ehleringer J.R., and C. Field. 1993. Scaling Physiological Processes: Leaf to Globe. Academic Press, San Diego, CA, pp. 141–158.

Fan S.M., S.C., Wofsy, P.S. Bakwin, and D.J. Jacob. 1990. Atmosphere-biosphere exchange of CO_2 and O_3 in the central Amazon forest. Journal of Geophysical Research **95**: 16851–16864.

Felzer, B., J.M. Reilly, D. Kicklighter, M. Sarofim, C. Wang, R.G. Prinn, and Q. Zhuang (2005), Future effects of ozone on carbon sequestration and climate change policy using a global biochemistry model. Climatic Change **73**: 195–425.

Felzer, B., D. Kicklighter, J. Melillo, C. Wang, Q. Zhuang, and R. Prinn. 2004. Effects of ozone on net primary production and carbon sequestration in the conterminous United States using a biogeochemistry model. Tellus **56B**: 230–248.

Friend, A.D., A.K. Stevens, R.G. Knox, and M.G.R. Cannell. 1997. A process-based, biogeochemical, terrestrial biosphere model of ecosystem dynamics (Hybrid v3.0). Ecological Modeling **95**: 249–287.

Ge, Q., J. Dai, F. He, J. Zheng, Z. Man, and Y. Zhao. 2003. Spatiotemporal dynamics of reclamation and cultivation and its driving factors in parts of China during the last three centuries. Progress in Natural Science **14(7)**: 605–613.

Gobron, N., B. Pinty, O. Aussedat, J. Chen, W. Cohen, R. Fensholt, V. Gond, K.F. Husmmrich, T. Lavergne, F. Melin, J.L. privette, I. Sandholt, M. Taberner, D.P. Turner, M.M Verstraete, and J.L. Widlowske. 2006 Evaluation of fraction of absorbed photosynthetically active radiation products for different canopy radiation transfer regimes: methodology and results using Joint Research Center products derived from Sea WiFS against ground-based estimation. Journal of Geophysical Research **111**: Art. No.D1311o.

Grell, G.A., J. Dudhia, and D.R. Stauffer. 1994. A description of the fifth-generation Penn State / NCAR Mesoscale Model (MM5). NCAR Technical Note NCAR / TN3981STR, 119 pp.

Haefner, W. 2005. Modeling biological systems: principles and applications. Springer, New York.

Hall C.A.S., H. Tian, Y. Qi., G. Pontius, J. Cornell, and J. Uhlig. 1995. Modeling spatial and temporal patterns of tropical land use change. Journal of Biogeochemistry **22**: 753–757.

Heagle, A.S. 1989. Ozone and crop yield. Annual Review of Phytopathology **27**: 397–423.

Hilborn, R., and M. Mangel. 1997. The Ecological Detective: Confronting Models with Data. Princeton (NJ): Princeton University Press.

Houghton, R. A., and J. L. Hackler 2003. Sources and sinks of carbon from land-use change in China. Global Biogeochemical Cycles **17(2)**: 1029–1034.

Huang, Y., W. Zhang, X. Zheng, J. Li, and Y. Yu. 2004. Modeling methane emission from rice paddies with various agricultural practices. Journal of Geophysical Research **doi:** 10.1029/2003JD004401.

Huang, Y., R.L. Sass, and F.M. Fisher. 1998. A semi-empirical model of methane emission from flooded rice paddy soils. Global Change Biology **4**: 247–268.

IPCC. 2001. Climate change 2001: The scientific basis. Contribution of working group I to the third assessment report of the Intergovernmental Panel on Climate Change [Houghton, J. T.Y. Ding, D.J. Griggs, M. Noguer, P.J. van der Linden, X. Dai, K. Maskell, and C.A. Johnson (eds.)]. Cambridge University Press, Cambridge, United Kingdom and New York, NY, USA, 881 pp.

Jackson, L.J., A.S. Trebitz, and K.L. Cottingham. 2000. An introduction to the practice of ecological modelling. BioScience, **50(8)**: 694–706.

King A.W. 1991 Translating models across scales in the landscape. pp 479–517 *in* M.Turner and R. Gardner, editors. Quantitative Methods in Landscape Ecology: the Analysis and Interpretation of Landscape Heterogeneity. Springer-Verlag, NY.

Lawrence, M.G., and P.J. Crutzen. (1999), Influence of NO_x emissions from ships on tropospheric photochemistry and climate. Nature **402**: 167–170.

Lauenroth, W.K., I.C. Burke, and J.K. Berry. 2003. The state of dynamic quantitative modelling in ecology, *in* Canham C.D., J.J. Cole, and W.K. Lauenroth. editors. Models in Ecosystem Science. Princeton University Press, Princeton, NJ.

Levin, S.A. 1992. The problem of pattern and scale in ecology. Ecology **73(6)**: 1943–1967.

Li C.S., Y. Zhuang, M. Cao, P. Crill, Z. Dai, S. Frolking, B. Moore, W. Salas, W. Song and X. Wang. 2001. Comparing a process-based agro-ecosystem model to the IPCC methodology for developing a national inventory of N_2O emissions from arable lands in China. Nutrient Cycling in Agroecosystems, **60**: 159–175.

Li C. S. 2000. Modeling trace gas emissions from agricultural ecosystems. Nutrient Cycling in Agroecosystems **58**: 259–276.

Liang, S. 2007. Recent developments in estimating land surface biophysical variables from optical remote sensing. Progress in Physical Geography **31**: 501–516.

Liu, J., H. Tian, M. Liu, D. Zhuang, J.M. Melillo, and Z. Zhang. 2005a. China's changing landscape during the 1990s: Large-scale land transformation estimated with satellite data. Geophysical Research Letters **32**: L02405, doi: 10.1029/2004GL021649.

Liu, J., M. Liu, H. Tian, D. Zhuang, Z. Zhang, W. Zhang, X. Tang, and X. Deng. 2005b. Spatial and temporal patterns of China's cropland during 1990–2000: An analysis based on Landsat TM data. Remote Sensing of Environment **98**: 442–456.

Lu, D., and Q. Weng. 2007. A survey of image classification methods and techniques for improving classification performance. International Journal of Remote Sensing **28(5)**: 823–870.

Lu, D. 2006. The potential and challenge of remote sensing–based biomass estimation. International Journal of Remote Sensing **27(7)**: 1297–1328.

Mayer, M., C. Wang, M. Webster., and R.G. Prinn. 2000. Linking local air pollution to global Chemistry and climate. Journal of Geophysical Research **105(D18)**: 22,869–22,896.

McGuire, A.D., S. Sitch, J.S. Clein, R. Dargaville, G. Esser, J. Foley, M. Heimann, F. Joos, J. Kaplan, D.W. Kicklighter, R.A. Meier, J.M. Melillo, B. Moore III, I.C. Prentice, N. Ramankutty, T. Reichenau, A. Schloss, H. Tian, L.J. Williams, and U. Wittenberg. 2001 Carbon balance of the terrestrial biosphere in the twentieth century: analyses of CO_2, climate and land-use effects with four process-based ecosystem models. Global Biogeochemical Cycles **15(1)**: 183–206.

Melillo, J.M, A.D. McGuire, D.W. Kicklighter, B. Moore, C.J. Vorosmarty, and A.L. Schloss. 1993. Global climate change and terrestrial net primary production. Nature **363**: 234–240.

Mickley, L.J., D.J. Jacob, B.D. Field, and D. Rind. 2004. Climate response to the increase in tropospheric ozone since preindustrial times: A comparison between ozone and equivalent CO_2 forcing. Journal of Geophysical Research **109(D05106)**: doi: 10.1029/2003JD003653.

Millennium Ecosystem Assessment. 2005. Ecosystems and Human Well-being: Synthesis. Island Press, Washington DC.

Nightingale, J.M., S.R. Phinn, and A.A. Held. 2004. Ecosystem process models at multiple scales for mapping tropical forest productivity. Progress in Physical Geography **28**: 241–281.

Ollinger, S.V., J.D. Aber, and P.B. Reich. 1997. Simulating ozone effects on forest productivity: interactions among leaf-canopy and stand-level processes. Ecological Applications **7(4)**: 1237–1251.

O'Neill, R.V., and R.H. Gardner. 1979. Source of uncertainty in ecological models. *In* Zeigler, B.P., M.S. Elzas, G.J. Klir, and T.I. Oren. editors, Methodology in Systems Modeling and Simulation. North Holland, Amsterdam. 447–463.

O'Neill, R.V., D.L. Deangelis, J.B. Waide, and T.F.H. Allen. 1986. A Hierarchical Concept of Ecosystems. Princeton University Press, Princeton, New Jersey.

Oreskes N., K. Shrader-Frechette, and K. Belitz. 1994. Verification, validation, and confirmation of numerical models in the earth sciences. Science, **263**: 641–646.

Parton, W.J., W.K. Lauenroth, D.L Urban., D.P. Coffin, H.H. Shugart, T.B. Kirchner, and T.M. Smith. 1993. Modelling vegetation structure-ecosystem process interactions across sites and ecosystems. Ecological Modeling **67(1)**: 49–80.

Parton, W.J., J.W.B. Stewart, and C.V. Cole. 1988. Dynamics of C, N, P and S in grassland soils: a model. Biogeochemistry **5(1)**: 109–131.

Pontius R.G., 2001. Quantification error versus location error in comparison of categorical maps. Photogrammetric Engineering and Remote Sensing **66(8)**: 1011–1016.

Raich, J.W., E.B. Rastetter, J.M. Melillo, D.W. Kicklighter, P.A. Steudler, B.J. Peterson, A.L. Grace, B. Moore III, and C.J. Vorosmarty. 1991. Potential net primary productivity in South America: application of a global model. Ecological Applications **1**: 399–429.

Ramankutty, N., and J. Foley. 1999. Estimating historical changes in land cover: North American croplands from 1850 to 1992. Global Ecology and Biogeography **8**: 381–396.

Ramankutty, N., and J.A. Foley. 1998. Characterizing patterns of global land use: An analysis of global croplands data. Global Biogeochemical Cycles **12(4)**: 667–685.

Rasch, D.A.M.K., E.M.T. Hendrix, and E.P.J. Boer. 1997. Replication-free optimal designs in regression analysis. Computer Statistic **12**: 19–52.

Rastetter, E.B., 1996. Validating models of ecosystem response to global change. BioScience, **46(3)**: 190–198

Ren, W., H. Tian, M. Liu, C. Zhang, G. Chen, S. Pan, B. Felzer, and X. Xu. 2007a. Effects of tropospheric ozone pollution on net primary productivity and carbon storage in terrestrial ecosystems of China. Journal of Geophysical Research **112**: D22S09, doi: 10.1029/2007JD008521.

Ren, W., H. Tian, G. Chen, M. Liu, C. Zhang, A. Chappelka, and S. Pan. 2007b. Influence of ozone pollution and climate variability on grassland ecosystem productivity across China. Environmental Pollution **149**: 327–335.

Rind, D., J. Lerner, K. Shah, and R. Suozzo. 1999. Use of on-line tracers as a diagnostic tool in general circulation model development: 2. Transport between the troposphere and stratosphere. Journal of Geophysical Research **104**: 9151–9167.

Running, S.W., R.R. Nemani, F.A. Heinsch, M. Zhao, M. Reeves, and H. Hashimoto. 2004. A continuous satellite-derived measure of global terrestrial primary productivity: Future science and applications. Bioscience **56**: 547–560.

Running, S., D. Baldocchi, D. Turner, S. Gower, P. Bakwin, and K. Hibbard. 1999. A global terrestrial monitoring network integrating tower fluxes, flask sampling, ecosystem modeling and EOS satellite data. Remote Sensing of Environment **70**: 108–127.

Shi, X., D. Yu, X. Pan, W. Sun, H. Wang, and Z. Gong. 2004. 1:1,000,000 soil database of China and its application. *In*: Proceedings of 10th National Congress of Soil Science of China. Science Press, Beijing, China, 142–145. (in Chinese).

Sitch S., B. Smith., I.C. Prentice, A. Arneth, A. Bondeau, W. Cramer, J.O. Kaplan, S. Levis, W. Lucht, M.T. Sykes, K. Thonicke, and S. Venevsky. 2003. Evaluation of ecosystem dynamics, plant geography and terrestrial carbon cycling in the LPJ dynamic global vegetation model. Global Change Biology, **9**: 161–185.

Thornton, P.E., S.W. Running., and M.A. White. 1997. Generating surfaces of daily meteorological variables over large regions of complex terrain. Journal of Hydrology **190**: 241–251.

Tian, H. (Eds.) 2006a. Environmental Modeling and Simulation, ACTA Press. New York.

Tian, H., S. Wang, J. Liu, S. Pan, H. Chen, C. Zhang, and X. Shi. 2006b. Patterns of soil nitrogen storage in China. Global Biogeochemical Cycles, **20**, GB1001, doi:10.1029/2005GB002464.

Tian, H., M. Liu, C. Zhang, W. Ren, G. Chen, X. Xu., and C. Lv. 2005. DLEM-The Dynamic Land Ecosystem Model, User Manual. The ESRA (Ecosystem Science and Regional Analysis) Laboratory, Auburn University, Auburn, AL.

Tian, H., J.M. Melillo, D.W. Kicklighter, S. Pan, J. Liu, A.D. McGuire, and B. Moore III. 2003. Regional carbon dynamics in monsoon Asia and its implications to the global carbon cycle. Global and Planetary Change **37**: 201–217

Tian, H., J.M. Melillo, D.W. Kicklighter, A.D. McGuire, and J. Helfrich. 1999. The sensitivity of terrestrial carbon storage to historical atmospheric CO_2 and climate variability in the United States. Tellus **51B**: 414–452.

Tian, H., J.M. Melillo, D.W. Kicklighter, A.D. McGuire, J.V.K. Helfrich, B. Moore, and C.J. Vörösmarty. 1998a. Effect of interannual climate variability on carbon storage in Amazonian ecosystems. Nature **396**: 664–667.

Tian, H., C.A. Hall, and Y. Qi. 1998b. Modeling primary productivity of the terrestrial biosphere in changing environments: toward a dynamic biosphere model. Critical Reviews in Plant Sciences **15(5)**: 541–557.

Turner, D.P., S.V. Ollinger, and J.S. Kimball. 2004. Integrating remote sensing and ecosystem process models for landscape- to regional-scale analysis of the carbon cycle. BioScience **54**: 573–584.

Vitousek, P.M., H.A. Mooney, J. Lubchenco, and J.M. Melillo. 1997. Human domination of Earth's ecosystems. Science **277**: 494–499.

Wang, S., H. Tian, J. Liu, and S. Pan. 2003. Pattern and change in soil organic carbon storage in China: 1960s–1980s. Tellus **55B**: 416–427.

Wang, C., R.G. Prinn, and A. Sokolov. 1998. A global interactive chemistry and climate model: formulation and testing. Journal of Geophysical Research **103(D3)**: 3399–3418.

Wang, X., W. Manning, Z. Feng., and Y.G. Zhu. 2007. Ground-level ozone in China: Distribution and effects on crop yields. Environmental Pollution **147(2)**: 394–400.

Wear, D., and P. Bolstad. 1998. Land-use changes in southern Appalachian landscapes: Spatial analysis and forecast evaluation. Ecosystems **1**: 575–594.

Wulder, M.A., R.J. Hall, N.C. Coops, and S.E. Franklin. 2004. High spatial resolution remotely sensed data for ecosystem characterization. Bioscience **54**: 511–521.

Xu, D.F. 1983. Statistical Data of Agricultural Production and Trade in Modern China. Shanghai: Shanghai People's Press.

Zhang, C., H. Tian, A. Chappelka, W. Ren, H. Chen, S. Pan, M. Liu, D. Styers, G. Chen, and Y. Wang. 2007a. Impacts of climatic and atmospheric changes on carbon dynamics in the Great Smoky Mountains. Environmental Pollution **149**: 336–347.

Zhang, C., H. Tian, S. Pan, G. Lockaby, E. Schilling, and J. Stanturf. 2007b. Effects of forest regrowth and urbanization on ecosystem carbon storage in a rural-urban gradient in the southeast US. Ecosystems, doi: 10.1007/s10021-006-0126-x.

Zhao, M., S.W. Running, and R. Nemani. 2006. Sensitivity of Moderate Resolution Imaging Spectroradiometer (MODIS) terrestrial primary production to the accuracy of meteorological re-analyses. Journal of Geophysical Research **111**: G01002, doi:10.029/2004JG000004.

Zhuang, Q, J.M. Melillo, D.W. Kicklighter, R. Prinn, A.D. McGuire, P.A. Steudler, B. Felzer., and S. Hu. 2004. Methane Fluxes Between Terrestrial Ecosystems And the Atmosphere at Northern High Latitudes During the Past Century: A Retrospective Analysis with a Process-based Biogeochemistry Model. Global Biogeochemical Cycles **18**: GB3010, doi:10.1029/2004GB002239.

Chapter 10
Gradual Global Environmental Change in the Real World and Step Manipulative Experiments in Laboratory and Field: The Necessity of Inverse Analysis

Yiqi Luo and Dafeng Hui

10.1 Introduction

Due to land-use change and fossil fuel combustion, the atmospheric CO_2 concentration has gradually increased from 280 ppm in pre-industrial times (Neftel et al. 1982, Friedli et al. 1986) to ~379 ppm in 2005 and is expected to exceed 700 ppm in 2100 (IPCC 2007). As a consequence of rising CO_2 and other greenhouse gases, the Earth's surface temperature has increased by 0.74°C in the twentieth century and is expected to increase by another $1.8 \sim 4.0$°C by the end of this century (IPCC 2007). Anthropogenic climate change likely leads to increasingly altered precipitation regimes. Precipitation is anticipated to increase by about 0.5–1% per decade over most of the middle- and high-latitude land areas in the northern hemisphere in this century (IPCC 2007).

Changes in multiple global environmental factors potentially trigger complex influences on ecosystem structure and function with feedbacks to climate change. Carbon dioxide, for example, is a substrate of plant photosynthesis. Rising atmospheric CO_2 concentration can directly affect photosynthetic rates (Farquhar et al. 1980), plant growth, and ecosystem productivity (Luo et al. 2006). Almost all physical, chemical, and biological processes are sensitive to temperature (Shaver et al. 2000). Climate warming can affect ecosystem function and community structure (Rustad et al. 2001, Luo 2007). Plant growth is also regulated by precipitation (Fay et al. 2003, Weltzin et al. 2003, Huxman et al. 2004), nitrogen fertilization (Ägren 1985, Mack et al. 2004), and ozone concentration (Ashmore 2005, Sitch et al. 2007). Thus, it is imperative to assess impacts of global environmental change on ecosystem processes.

Since it is impossible to duplicate the gradual global environmental change under experimental conditions or to meaningfully detect ecosystem responses to such a gradual change in the real world, ecologists usually manipulate one or a few variables while other factors are maintained at the

Y. Luo (✉)
Department of Botany and Microbiology, University of Oklahoma, 770 Van Vleet Oval, Norman, OK 73019, USA
e-mail: yluo@ou.edu

S. Miao et al. (eds.), *Real World Ecology*, DOI 10.1007/978-0-387-77942-3_10,
© Springer Science+Business Media, LLC 2009

ambient or control levels. To generate large enough effects on ecosystems within an abbreviated time period so that they can be detected, the manipulative experiments usually implement a step change in the factor(s) under study. So far, numerous manipulative experiments have been conducted using step changes in the laboratory and under field conditions (Allen et al. 1992, Hendrey et al. 1999, Hui et al. 2002, Körner et al. 2005, Rustad 2006). While step increases in global change factors may be an effective way to gain insights into ecosystem behavior, one fundamental question about these experiments has to be carefully addressed: Can observed experimental effects of a step change in global environmental factors (such as atmospheric CO_2) be directly extrapolated to infer responses of ecosystems to a gradual change in the real world?

This chapter discusses how results from manipulative experiments can be analyzed to improve our predictive understanding of ecosystems responses to gradual global change in the real world. We first describe gradual changes in several of the global environmental variables and corresponding manipulative experiments. Then we review a modeling study by Luo and Reynolds (1999) of differential responses of ecosystems to gradual vs. step changes in CO_2 concentration. We also review results from several experiments to verify that ecosystem responses to step CO_2 increases are different from those to gradual changes. Finally, we introduce a framework of analytical techniques – inverse analysis – that extract information from experimental data toward predictive understanding in ecological research.

10.2 Gradual Increases in Global Environmental Variables in the Real World and Step Changes in Experiments

Most global change variables evolve gradually with small annual increments over a long time span in the real world. For example, the annual mean increase in atmospheric CO_2 concentration is approximately 1.4 ppm per year from 1960 to 2005 and 1.9 ppm per year from 1995 to 2005 (IPCC 2007). Projected CO_2 concentration is also gradually increasing at 2–4 ppm per year in the next 100 years. Global surface temperature increased by an average of about 0.74°C over the past 100 years and at a rate of 0.13°C per decade in the past 50 years (IPCC 2007). The surface temperature is expected to increase by another 1.8–4°C in this century. Precipitation, on the other hand, shows large inter-annual fluctuation and large regional variation that may increase or decrease in the future (Fay et al. 2003, Weltzin et al. 2003). Many other global change variables, such as methane, ozone, and nitrogen decomposition are also gradually increasing although not all of them substantially influence ecosystem processes.

In spite of the fact that most global environmental variables evolve in a gradual fashion, experimentalists have no way to duplicate such long-term, gradual changes in experimental facilities. Nor can ecosystem responses to such gradual changes in any of those global environmental factors be meaningfully measured. Most studies use a perturbation approach that creates treatment levels (i.e., magnitudes in changes of treatment factors) that are large enough to generate detectable ecosystem responses. For example, ecologists usually double CO_2 concentration, increase it by 200 ppm or set a specific level at the onset of experiments conducted in greenhouses, growth chambers, open-top chambers, and Free Air CO_2 Enrichment (FACE) facilities (Allen et al. 1992, Hendrey et al. 1999, Miglietta et al. 2001). Scientists then measure responses of plants and soil processes to such a step increase in CO_2 concentration through time.

Similar to elevated CO_2 studies, to examine effects of climate warming on ecosystem processes and community structure, researchers usually raise soil or air temperature by 1–4°C in treatment plots instantaneously above that in the control. The methods used to manipulate temperature include soil warming (Melillo et al. 2002, Hartley et al. 2007), infrared heaters (Harte and Shaw 1995, Wan et al. 2002, Kimball 2005), passive heating (Klanderud and Totland 2007), and open-top chambers with heated air (Norby et al. 1997, Dermody et al. 2007). Most of the experiments with infrared heaters also have constant energy input to the ecosystem, resulting in increases in soil surface temperature being higher at night than during daytime (Wan et al. 2002).

Although other global change factors, such as precipitation and nitrogen deposition, may not evolve gradually in a regular fashion over time and/or uniformly over space as atmospheric CO_2 concentration does, experimental studies mostly use the step-change approach as well. Precipitation amounts were doubled in a grassland experiment in central US Great Plains (Sherry et al. 2007) and reduced by 30, 55 and 80% in Argentina (Yahdjian and Sala 2002). Precipitation amounts were changed together with timing (Fay et al. 2000). To understand ecosystem responses to nitrogen deposition, a certain amount of nitrogen was usually applied once at the start of an experiment or once a year for the duration of a study (Mack et al. 2004, Verburg et al. 2004).

The contrast between the gradual increases of global change factors in the real world and step changes in manipulative experiments lead to a question: Can results from step-change experiments be directly extrapolated to infer responses of ecosystems to the gradual global environmental change in the real world?

10.3 Modeled Ecosystem Responses to Gradual Versus Step Changes in CO_2

Although there is a clear difference in global change factors between most experimental setting and the real world conditions, a few studies have attempted to address this issue. Luo and Reynolds (1999) conducted a modeling study to examine ecosystem responses to gradual and step changes in CO_2 concentration. They used a terrestrial ecosystem carbon (C) sequestration (TCS) model, which consists of a carbon submodel and a nitrogen submodel. The carbon submodel simulates photosynthesis and carbon partitioning among 12 pools such as leaves, woody tissues, fine roots, and litter pools, and respiratory release from plants, detritus, and soil organic matter (SOM). Photosynthesis is determined by two components: sensitivity and acclimation (Luo et al. 1994). Photosynthetic sensitivity to CO_2 change is simulated by the Farquhar model (Farquhar et al. 1980). Photosynthetic acclimation is strongly correlated with CO_2-induced changes in area-based leaf nitrogen concentration (Luo et al. 1994) and the nitrogen–photosynthesis relationship is simulated according to Harley et al. (1992). Photosynthetically fixed carbon is allocated to leaves, woody tissues, and fine roots. Leaf and root litter are partitioned to metabolic and structural pools using a lignin/nitrogen ratio (Parton et al. 1987). Decomposition of litter and SOM is calculated according to the first-order kinetics model (Olson 1963, Wieder and Lang 1982). The nitrogen submodel has a mineral nitrogen (N) pool in addition to the carbon pools in the carbon submodel. Nitrogen and carbon are coupled stoichiometrically with flexible C/N ratios in each pool.

Luo and Reynolds (1999) used the IPCC business-as-usual scenario (IPCC 1992) to represent a gradual increase in atmospheric CO_2 concentration (Fig. 10.1A) and mimicked a step CO_2 increase from 350 to 700 ppm in 1987, with a constant CO_2 concentration at 700 ppm until 2100 (Fig. 10.1D). In response to a gradual CO_2 increase, both canopy photosynthesis and ecosystem respiration increased, but with a delay in carbon efflux or loss due to respiration (Fig. 10.1B). The difference between photosynthesis and respiration is the rate of carbon (C) sequestration, which is $27 \text{ g C m}^{-2} \text{y}^{-1}$ in 1987 at 350 ppm and is predicted to be $58 \text{ g C m}^{-2} \text{y}^{-1}$ in 2085 at 700 ppm (Fig. 10.1C), in a modeled forest ecosystem with gross primary productivity of $1200 \text{ g C m}^{-2} \text{y}^{-1}$ in pre-industrial times.

In response to a step increase in CO_2 concentration from 350 to 700 ppm in 1987, ecosystem carbon influx as a result of photosynthesis increased to 154% of the pre-industrial level (Fig. 10.1E). Ecosystem carbon efflux via plant and microbial respiration gradually increases, approaching the level of carbon influx during 113 years of the modeled CO_2 treatment (Fig. 10.1E). The difference between carbon influx and efflux (i.e., carbon sequestration) abruptly increased from 27 to $263 \text{ g C m}^{-2} \text{y}^{-1}$ immediately after the step CO_2 increase and then gradually declines (Fig. 10.1F). The modeled carbon sequestration in the first year of a step CO_2 increase to 700 ppm in an experiment is fivefold higher than that when CO_2 concentration is gradually increased to 700 ppm. In addition, ecosystem carbon sequestration at any point in time after the initiation of a step CO_2 increase is not equivalent to that in response to a gradual increase. Carbon sequestration declines with time in response to a step increase but slowly increases in response to a gradual increase in CO_2.

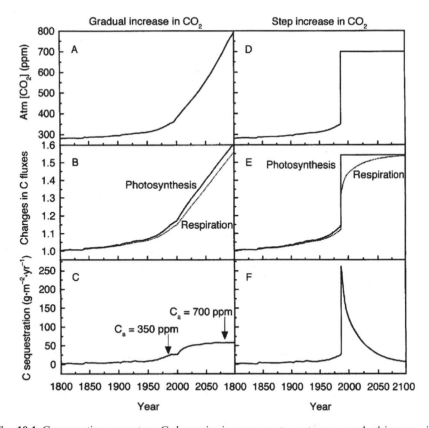

Fig. 10.1 Comparative ecosystem C dynamics in response to a step vs. gradual increase in atmospheric CO_2 (C_a) (Luo and Reynolds 1999 with permission from Ecology). (**a**) C_a gradually increases from 280 ppm in 1800 to 793 ppm in 2100 in the natural world. (**b**) In response to the gradual increase in C_a, both photosynthesis and respiration increased gradually. (**c**) The rate of ecosystem C_{seq} gradually increased. (**d**) C_a gradually increased to 350 ppm in 1987 and then was elevated to 700 ppm to mimic CO_2 experiments using FACE or OTC facilities. A constant CO_2 level was assumed at 700 ppm for 113 years for the purpose of illustrating long-term C and N dynamics. (**e**) In response to the step CO_2 increase, photosynthetic C fixation increased immediately but respiratory C release increased slowly. The constant photosynthesis after the CO_2 treatment resulted from assumed constant LAI and no environmental disturbance. (**f**) The rate of C_{seq} was high right after the step increase in CO_2 and then declined

Additional carbon sequestration requires additional nitrogen supply and sequestration (Luo et al. 2004), which also differs between step and gradual increases in atmospheric CO_2 concentration. With a step CO_2 increase, ecosystem carbon sequestration of 263 g C m^{-2} y^{-1} (Fig. 10.1F) requires an additional nitrogen supply of 4.1 g N m^{-2} y^{-1} (Fig. 10.2). As atmospheric CO_2 concentration gradually increases to 350 and 700 ppm in 1987 and 2085, ecosystem carbon sequestration can be balanced by additional nitrogen supply of 0.6 and 1.7 g N m^{-2} y^{-1}, respectively (Fig. 10.2).

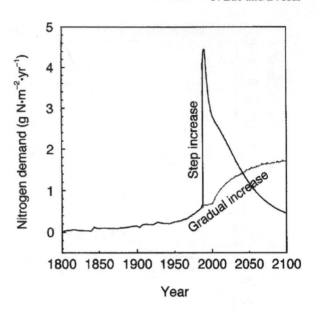

Fig. 10.2 N demand to balance extra C storage in an ecosystem in response to a gradual vs. step increase in CO_2. In response to the gradual increase in CO_2, the N demand gradually increased to 0.6 g N m^{-2} y^{-1} in 1987 and 1.7 g N m^{-2} y^{-1} in 2085 (Luo and Reynolds 1999 with permission from Ecology). In response to the step increase in CO_2, the N demand was 4.1 g N m^{-2} y^{-1} in the first year, slightly increased in year 3 and 4, and then declined

Luo and Reynolds (1999) also examined responses of four types of ecosystems (i.e., forest with high productivity, grassland with high productivity, forest with low productivity, and grassland with low productivity) (Fig. 10.3) to gradual vs. step increases in CO_2 concentration. In all four cases, modeled C sequestration reaches the highest level in the first year of a step CO_2 increase and then declines. In response to a gradual CO_2 increase, carbon sequestration gradually increases at a much lower rate than that in response to the step change in the four ecosystems. Other modeling studies at regional and global scales have also showed slow increases in net primary production and carbon sequestration (Cramer et al. 2001, McGuire et al. 2001). The pulse responses to the step change in CO_2 concentration have been demonstrated in other studies (e.g., Comins and McMurtrie 1993, Rastetter et al. 1997). Modeled net primary productivity, for example, increased from 450 to 750 g C m^{-2} y^{-1} in the first year of a step CO_2 increase and then rapidly declined to a range from 450 to 550 g C m^{-2} y^{-1} within 10 years, depending on nitrogen regulation (Rastetter et al. 1997). Similar transient responses have been reported for other environmental driving variables. Reynolds and Leadley (1992) examined effects of changes in season length on arctic plants by imposing step (single-year) vs. gradual (over a 50-year period) changes. Modeled net gas exchange increased by 19% in response to a step change in season length from 90 to 110 days, but by only 1% in response to a gradual increase.

The comparison study with four types of ecosystems also suggests that additional carbon influx and carbon residence time are two key parameters in determining magnitudes of ecosystem carbon sequestration. In response to either step or gradual CO_2 increases, the ecosystems with high productivity

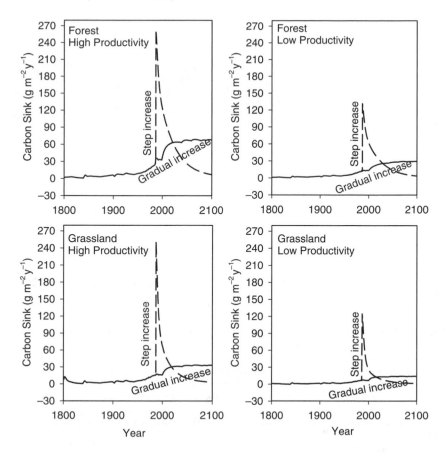

Fig. 10.3 Comparative simulation studies between grasslands and forests were used to illustrate that C relaxation time and additional photosynthetic C influx in response to an increase in CO_2 determined the capacity of an ecosystem to sequester C (Luo and Reynolds 1999 with permission from Ecology). The modeling study assumes that residence times in grasslands are shorter then that in forests

generally have more additional carbon influx than the ecosystems with low productivity. As a result, the ecosystems of grasslands or forests with high productivity have higher carbon sequestration rates in responses to either step or gradual CO_2 increases than their counterparts with low productivity (Fig. 10.3). Modeled carbon residence time is higher in forests than grasslands based on an assumption that woody tissues and woody litter elongate carbon residence time in the forests. Carbon sequestration rates in forests are higher than in grasslands for both gradual and step CO_2 increases with a given productivity.

Overall, comparison of ecosystem responses to a step vs. gradual increase in CO_2 (Figs. 10.2, 10.3, and 10.4) suggest that carbon sequestration obtained in a step CO_2 experiment cannot be considered a direct estimate of the potential of

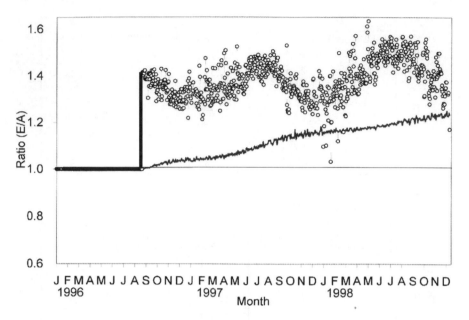

Fig. 10.4 Results from the Duke Forest FACE experiment demonstrate contrasting patterns of ecosystem photosynthetic (*open circles that*) vs respiratory (*line*) responses to the step CO_2 increase. The ratio of modeled daily canopy C fluxes at elevated relative to that at ambient CO_2 (E/A) from August 1996 to December 1998 displays a step CO_2 stimulation of C influx (Luo et al. 2001a). However, the step increase in the C influx did not lead to a step increase in respiratory C release. Rather the ratio of modeled soil respiration shows little change in the first year but a gradual increase in the second and third years under the elevated CO_2 treatment compared to that under the ambient CO_2 treatment (Luo et al. 2001b)

ecosystems to sequester carbon when atmospheric CO_2 concentration gradually increases to the level in the experiment. Similarly, nitrogen limitation revealed in a step CO_2 change experiment may differ from that in response to a gradual CO_2 increase in the real world. However, ecosystem responses to the gradual and step CO_2 increases are all regulated by additional carbon influx and carbon residence time, which are the two key parameters for a predictive understanding of ecosystem response to global change (Luo et al. 2003). In section 10.6 below, we will address how to derive these key parameters from manipulative experiments using inverse analysis approaches.

10.4 Experimental Evidence of Different Responses to Gradual vs. Step Changes

Since results of the modeling study by Luo and Reynolds (1999) have profound implications for manipulative experiments of CO_2 concentration, in particular, and other global change factors, in general, it is necessary to test the modeling results experimentally. Luo (2001) gathered experimental evidence from the Duke Forest Free Air CO_2 Enrichment (FACE) experiment to demonstrate that a step change in CO_2 resulted in an immediate increase in ecosystem carbon influx and a gradual increase in ecosystem C efflux via respiration. The Duke Forest FACE experiment started in August 1996 and has been well described in terms of experimental design and other environmental conditions (Hendrey et al. 1999). In the Duke FACE loblolly pine forest, leaf-level photosynthesis increased by approximately 40–60% in response to the step increase of CO_2 concentration by 200 ppm (Herrick and Thomas 2001, Ellsworth 2000). To quantify whole ecosystem C influx, Luo et al. (2001a) used a canopy photosynthesis model MAESTRA to simulate photosynthetic CO_2 assimilation of the loblolly pine canopy at the FACE site. Elevated CO_2 concentration increased average ecosystem C influx by approximately 40% after the start of CO_2 fumigation in August 1996. The CO_2 enhancement in canopy carbon influx exhibits a step function after the CO_2 fumigation (Fig. 10.4). The step increase in carbon influx into the elevated CO_2 plots did not result in a step increase in respiratory carbon release. Monthly measurements of soil respiration at the Duke Forest FACE site show that elevation of CO_2 concentration did not statistically change soil respiration in 1996 but caused significant increases of 23.2% and 35.5% in 1997 and 1998, respectively (Andrews and Schlesinger 2001, Hui and Luo 2004). The observed gradual increase in soil respiration is qualitatively consistent with the modeling results (Luo and Reynolds 1999).

The respiratory carbon release from the soil surface is the convolution of root respiration and microbial decomposition of litter and soil organic matter from various pools with different time constants (Luo et al. 2001b). As additional carbon in the elevated CO_2 plots flows to various pools, pool sizes gradually increase. Amounts of carbon released by microbial respiration are usually proportional to pool sizes. Respiratory carbon release is slower from long residence time pools than short residence time pools, delaying the conversion of organic matter from photosynthesis to CO_2 release by respiration. Thus, the step increase in photosynthetic C influx is transformed into a gradual increase in respiratory C release.

Hui et al. (2002) designed a specific experiment to examine effects of a gradual vs. step increases in CO_2 concentration on plant photosynthesis and growth at two nitrogen levels. They grew *Plantago lanceolata* for 80 days before treating plants with ambient CO_2 (as the control), gradual CO_2 increase, and step CO_2 increase at low and high N additions for 70 days. CO_2 concentration was kept at a constant 350 ppm for the control and 700 ppm for the step CO_2 treatments. It was raised by 5 ppm per day from 350 ppm in the beginning of the treatment to 700 ppm by the end of experiment for the gradual CO_2 treatment. The step CO_2 treatment immediately resulted in an approximate 50% increase in leaf photosynthesis at both the low and high N additions (Fig. 10.5) and a 20–24% decrease in leaf N concentration in week 3 after the CO_2 treatment. The gradual CO_2 increase caused much less stimulation of photosynthesis and less decrease in leaf N concentration than did the step CO_2 increase. Specific leaf

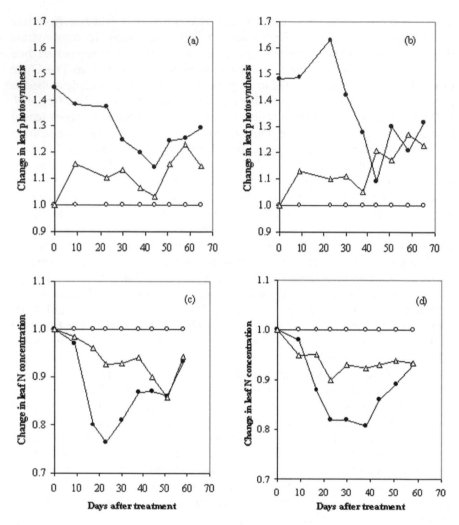

Fig. 10.5 Change of leaf photosynthesis (**a, b**) and leaf N concentration (**c, d**) of plants grown under the gradual CO_2 increase and the step CO_2 increase treatments at the low (**a, c**) and high (**b, d**) N levels compared with the ambient CO_2 treatment. Open circle, open triangle and filled circle represent the ambient CO_2, graduate CO_2 and the step CO_2 treatments. (Hui et al.2002, with permission from Elsevier)

area and leaf N concentration decreased but plant biomass increased in both the gradual and step CO_2 treatments in comparison with the ambient CO_2 treatment (Hui et al. 2002). Magnitudes of changes in all physiological and growth traits measured were larger under the step than under the gradual CO_2 treatments.

Klironomos et al. (2005) examined responses of a mycorrhizal fungal community to a single-step CO_2 increase and a gradual CO_2 increase over many generations. They conducted an experiment with *Bromus inermis* and its associated mycorrhizal community over a period of 6 years during which CO_2 concentration either increased abruptly as is typical of most CO_2 experiments, or gradually over 21 generations. They found that plant photosynthesis did not differentially respond to step vs. gradual CO_2 increases. Belowground plant production was higher in the step than the gradual CO_2 treatments. The abrupt CO_2 increase resulted in an immediate decline in fungal species richness and a significant change in mycorrhizal function whereas a gradual CO_2 increase did not significantly cause changes in fungal diversity and function from those at the ambient CO_2 treatment (Fig. 10.6).

Overall, evidence from the Duke Forest FACE project and results from two manipulative experiments with both step and gradual CO_2 increases all suggest that ecosystem responses to step CO_2 changes as in most manipulative

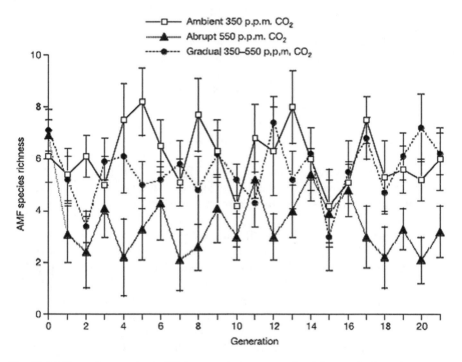

Fig. 10.6 Effect of atmospheric CO_2 concentration on species richness of arbuscular mycorrhizal fungi (AMF). Overall, mean richness was significantly higher in the 'ambient' compared with the 'abrupt' treatment. Richness in the gradual CO_2 and the ambient CO_2 treatments did not significantly differ from each other but did differ from the abrupt CO_2 treatment. AMF species richness at generation 21 in the ambient and gradual treatments remained similar and contained twice as many AMF taxa as in the abrupt treatment. (Kilronomos et al., 2005, with permission from Macmillan Publishers Ltd: [Nature])

experiments are different from those in response to a gradual CO_2 increase that occurs in the real world. Thus, it is imperative to evaluate possibilities to either develop new experimental approaches that can approximate gradual changes, or develop data-analysis methods that can extract useful information from the step change experiments. These new approaches will improve predictive understanding of ecosystem responses to gradual global change in the real world.

10.5 Experimental Approaches to Approximate Gradual Change

Given experimental verification of the modeling results as shown above that ecosystem responses to step CO_2 increases in manipulations are different from responses to a gradual CO_2 increase in the real world, we now ask: are there experimental approaches that we can improve or develop to approximate gradual changes in CO_2 concentration?

First, it is not feasible to conduct an experiment that lasts hundreds of years as in the real world to mimic gradual changes in CO_2. It is not possible, either, to detect changes in ecosystem processes in response to the long-term, gradual climate change in the real world. When atmospheric CO_2 concentration increases at 1.5 ppm per year as happened in the past four decades, this CO_2 increase is estimated to stimulate a 0.17–0.37% increase in ecosystem C influx (Luo and Mooney 1996), equivalent to an average of 3.3 g C $m^{-2}y^{-1}$ in a productive grassland or temperate forest ecosystem with a gross primary productivity of 1200 g C $m^{-2}y^{-1}$ This increment in carbon influx is far below detectable levels by any measurement techniques against strong variability and random errors in natural ecosystems.

Second, greenhouse and growth chamber experiments as conducted by Hui et al. (2002) and ctgeq(1999 et al. (2005) are effective in demonstrating differential responses of plants and microorganisms to step vs gradual changes in atmospheric CO_2concentration. These experiments illustrated that plant and microbial responses to step CO_2 treatments are usually stronger than responses to gradual CO_2 treatments. Thus, step CO_2 experiments may overestimate ecosystem responses compared to responses to gradual CO_2 changes in the real world, as shown in the modeling study by Luo and Reynolds (1999). Although it may be technically feasible to conduct similar gradual CO_2 experiments in the field, it would not be easy to extract useful information from these gradual CO_2 experiments to enhance our predictive understanding of ecosystem responses to global change, as the time scales of climate change factor used in these studies are quite different from the real world.

Third, ecologists have used natural CO_2 springs to examine long-term plant and ecosystem responses to rising atmospheric CO_2 concentration

(Koch, 1993, Raschi et al., 1997). Natural CO_2 springs have the potential to provide long-term CO_2 enrichment of vegetation and soils. Studies using natural CO_2 springs can take advantage of the natural CO_2 gradients generated from vents to the surrounding areas. Plants and ecosystems exposed to the perimeter of a CO_2 spring have had enough time for adaptation and acclimation and thus are considered in an equilibrium state with different CO_2 levels. Thus, natural CO_2 springs are potentially useful to address several major long-term issues such as: (1) regulation of physiological acclimation through slow plant and soil processes, (2) genetic adaptation to a gradual increase in atmospheric CO_2, (3) changes in population structure and species composition of plant communities, and (4) dynamics of soil organic matter and nutrient supply. However, CO_2 concentration strongly fluctuates in all types of CO_2-enriched springs due to wind speed and direction changes, whereas atmospheric CO_2 concentration is steadily increasing with annual variation only up to ten parts per million. Short-term fluctuations likely induce changes in physiological processes. In addition, some of the CO_2 springs emit sulfuric chemicals, methane, saturated hydrocarbon compounds, and other geochemical materials. Therefore, it is difficult to locate control sites with comparable vegetation, soil type, and environmental characteristics.

Fourth, ecologists have used multiple CO_2 levels (Körner, 1995, Luo et al., 1998) or a CO_2 tunnel to create CO_2 gradients (Polley et al. 2003, 2006) to study ecosystem responses to different levels of CO_2 enrichment. The CO_2 tunnel, for example, used elongated field chambers to expose a C_3/C_4 grassland in central Texas, USA to a continuous gradient in CO_2 from 200 to 560 ppm. Those experiments with multiple CO_2 levels or continuous gradients have been very effective to reveal nonlinearity of plant and soil processes in response to global change (Luo et al. 1998, Gill et al. 2002). But at a given level of CO_2 treatment, plant and soil are still exposed to a step change in CO_2 concentration. Interpretation of results from the multi-level and gradient CO_2 treatments is still subject to similar limitations for two-level CO_2 experiments. In addition, other environmental factors in these studies, such as temperature and relative humidity, may be difficult to control and could compound the CO_2 treatment effects.

10.6 New Approaches to Analysis of Data from Step Change Experiments

While several experimental approaches have been proposed to approximate the gradual change, all of them have certain drawbacks as discussed in Section 10.5. The common manipulative experimental method used today to study ecosystem responses to global change is still the step change experiment. This type of experiment generates a perturbation to an ecosystem, primarily by altering input of C flux in a CO_2 study, which is large enough so that we can trace responses in a reasonable time frame. Such pulse signals are routinely employed in systems study in physics and chemistry (Laidler and Meiser 1982). The question we need to answer is how can we generate useful information from the perturbation experiments in such a way that we can characterize ecosystem structure and function to predict future ecosystem changes?

To develop effective approaches to data analysis, we first have to understand fundamental properties of ecosystems. The structure of ecosystem carbon processes can be conceptually characterized by compartmentalization, donor pool-controlled transfer, and sequential linearity (Luo and Reynolds 1999). The carbon processes are highly compartmentalized due to the fact that photosynthetically fixed C goes to distinctive compartments, such as plant litter and SOM. Donor pool-controlled transfer is reflected by the fact that C release from each compartment through plant and microbial respiration is controlled by size of donor pools and to a much lesser extent by products of respiration. In addition, the majority of photosynthetically fixed C sequentially transfers from one compartment to another, following a first-order linear function (Wieder and Lang 1982). Only a small fraction of C can be recycled between soil microbes and organic matter. The three properties have been incorporated in virtually all pool-based biogeochemical models (Jenkinson and Rayner 1977, Parton et al. 1987, Comins and McMurtrie 1993, Rastetter et al. 1997, Thompson and Randerson 1999). Thus, model structure of ecosystem carbon dynamics is highly robust. To improve model prediction of ecosystem responses to gradually rising atmospheric CO_2 concentrations in the real world, we need to estimate key parameters, which are: ecosystem carbon influx, carbon partitioning coefficients among pools, and transfer coefficients from donor pools (Luo et al. 2003).

Ecosystem-scale C influx is usually estimated from measured leaf photosynthesis or net ecosystem exchange with assistance of canopy models. Carbon partitioning and transfer coefficients can be estimated by inverse and/or deconvolution analysis. Deconvolution partitions the response into different processes based on their different rates. Deconvolution analysis is applied to a measurable quantity that is a convolution of several processes with distinguishable characteristics (Luo et al. 2001b). For example, soil respiration is a composite of CO_2 production from multiple processes, including root exudation, root respiration, root turnover, and decomposition of litter and soil organic matter. Those processes have distinctive response times to C perturbation, which are related to carbon residence times – the time of carbon remaining in an ecosystem from entrance via photosynthesis to exit via respiration (Thompson and Randerson 1999, Luo and Reynolds 1999). For example, belowground carbon cycling through the pathway of root exudation takes only a few weeks from photosynthesis to respiratory release (Cheng et al. 1994, Rouhier et al. 1996). In contrast, carbon cycling through the pathway of wood growth, death, and decomposition takes several decades from photosynthesis to respiratory release. In response to either an increase in carbon influx (e.g., in an elevated

CO_2 experiment) or a decrease in substrate supply (e.g., in a tree girdling experiment), root exudation and root respiration change first while soil organic matter changes slowly. Using the distinctive response times of various carbon processes, Luo et al (2001b) applied the deconvolution approach to partitioning of soil respiration observed in the Duke Forest FACE experiment. The deconvolution analysis suggests that the increases in root exudation and root respiration may be of minor importance in carbon transfer to the rhizosphere whereas root turnover and aboveground litterfall are the major processes delivering carbon to soil.

The inverse analysis is an approach that fundamentally focuses on data analysis for estimation of parameters and their variability (Fig. 10.7). It also can be used to evaluate model structure and information content of data. In brief, inverse modeling usually starts with data and asks what the observed responses to a perturbation can tell us about the system in question. By combining prior knowledge about the system, we try to identify processes underlying the observations, which are incorporated into a model for an inverse analysis. The latter is implemented with optimization algorithms to adjust parameter values to the extent that differences between model predictions and observations (i.e., a cost function; Raupach et al. 2005) are minimized. Those parameter values that satisfy the minimized cost function are considered the optimized parameter estimates given the observations and model structure. The optimized parameter values can be used in forward analysis, which is usually implemented using simulation models. In general the forward analysis asks what a model can tell us about the ecosystems whereas the inverse analysis

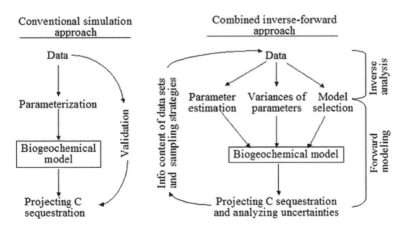

Fig. 10.7 Compared with the traditional simulation modeling, the inversion-based approaches has the potential to (1) estimate parameter and state variables, (2) quantify variances of estimated parameters and predicted C sinks, (3) reject and/or select a model from its alternatives, and (4) evaluate measurements schemes to maximize information contents of individual data sets for predicting carbon sinks

asks what the data can tell us about the same system. The combination of the two approaches allows us to probe mechanisms underlying ecosystem responses to global change.

Next, we will use an inverse analysis on estimation of the three sets of model parameters (i.e., photosynthetic parameters P, carbon turnover time τ_E, and initial carbon pool size X_0) as an example to illustrate the general procedure (see detailed description by Luo et al. 2003, Xu et al. 2006). This analysis is based on a terrestrial ecosystem (TECO) model. Eight carbon pools are used in the model include leaf, wood, root carbon and litter pools, microbial and SOM pools. The processes of the C cycling are mathematically represented in a matrix form (Luo et al. 2003):

$$\begin{cases} \dfrac{\mathrm{d}X}{\mathrm{d}t} = \xi A X + Bp \\ X(t = 0) = X_0 \end{cases} \tag{10.1}$$

where X is a vector of C pool sizes, X_0 is initial values of X, A is a matrix of transfer coefficients, ξ is an environmental scalar representing effects of soil temperature and moisture on C transfer (Luo et al. 2001b), p is the ecosystem C influx, B is a vector of partitioning coefficients of C to growth of leaf, wood, and fine roots, and t is time.

In the inverse analysis, first, a vector of parameters that are to be estimated can be defined as:

$$q = \{P, \tau_E, X_0\}, \tag{10.2}$$

where q represents all the parameters to be estimated in the inverse analysis. Assume that there are six data sets to be assimilated into the eight-pool TECO model. The six data sets are measured woody biomass, C content in the forest floor, C content in mineral soil, foliage biomass, litterfall, and soil respiration. Each of the data sets is denoted by $Q^j(t_i)$, where the superscript j represents the jth data set, t_i represents observation times of $i = 1, 2, \ldots, n_j$ and n_j is the number of data points in the jth data set. To link modeled variables with each of the observed data sets, one observation operator, $\varphi^j(q)$, is defined. The six observation operators in $\varphi^j(q)$ corresponding to the six data sets respectively are:

Woody biomass :	(0	1	0	0	0	0	0	0)
C in forest floor :	(0	0	0	0.75	0.75	0	0	0)
C in mineral soil :	(0	0	0	0	0	1	1	1)
Foliage biomass :	(1	0	0	0	0	0	0	0)
Litterfall :	$(a_{1,1}$	$f_a a_{2,2}$	0	0	0	0	0	0)
Soil respiration :	(0	$g_1 a_{2,2}$	$g_3 a_{3,3}$	$g_4 a_{4,4}$	$g_5 a_{5,5}$	$g_6 a_{6,6}$	$g_7 a_{7,7}$	$g_8 a_{8,8}$

$$(10.3)$$

where $a_{i,i}$ is diagonal element in matrix A in equation 10.1, f_a is fraction of aboveground woody biomass over the total, and g_i is fraction of C released as CO_2 during microbial decomposition of litter and SOM.

With the observation operators, modeled pool size $X(q)/(t_i)$ can be mapped to data set $Q^j(t_i)$, in a cost function $J(q)$:

$$J(q) = \sum_{j=1}^{m} v_j \left[\sum_{i=1}^{n_j} (\varphi^j(q)X(q)(t_i) - Q^j(t_i))^2 \right] + \sum_{r=1}^{R} C_r \delta(\alpha\prime_r - \alpha_r)^\kappa, \qquad (10.4)$$

where m is the number of data sets, equaling 6 in this example; v_j is a weighing factor for jth data set toward the evaluation criteria:

$$v_j = \frac{k_j}{m} \frac{1}{\mathrm{Var}(Q^j)}, \qquad (10.5)$$

where k_j is a proportional value of data set j contributing to the overall evaluation, $\mathrm{Var}(Q^j)$ is the variance of the observation data set Q^j. The second term in equation 10.4 is a penalty function derived from prior information on parameter α_r, $r = 1, 2 \ldots, R$; δ is penalty parameter; κ and C_r are weighing factors.

To evaluate the cost function, $J(q)$, a variety of methods can be used, such as genetic algorithms (Barrett 2002), simulated annealing, the Levenberg-Marquardt minimization method in combination with a quasi-Monte Carlo algorithm and Markov Chain Monte Carlo (MCMC) (Luo et al. 2003, White and Luo 2002, 2005, White et al. 2005, Xu et al. 2006). The use of MCMC method with Bayesian probabilistic inversion is quite common now as it can provide uncertainty estimation on both estimated parameters and projected C sinks or sources. A more detailed description of the Bayesian probabilistic inversion theory and the Metropolis-Hastings (M-H) algorithm for MCMC statistical computation can be found in Knorr and Kattge (2005) and Xu et al. (2006). In brief, Bayesian inversion is based on Bayes' theorem of inverse probability which states that the conditional probability of the model parameters can be obtained from the conditional probability of the measurement and the unconditional probability of the parameters. MCMC techniques such as the M-H algorithm and the Gibb's sampler can be used to handle the problem of numerical integration of high-dimensional posterior probability density functions. The basic idea for MCMC is to design a Markov chain with the posterior probability distribution as the stationary distribution. Once the chain has run for a period of time, it will follow the stationary distribution. Samples will be taken from the chain, and the statistics, such as mean value and the variation associated with the probability distribution, can be inferred from the samples. The probabilistic inversion with the MCMC method generates marginal distributions of parameters and state variables, which are expressed by probability density functions (PDF) or cumulative probability functions (CPF) (Fig. 10.8). The narrower a PDF is the lower the uncertainty is for the estimated parameter.

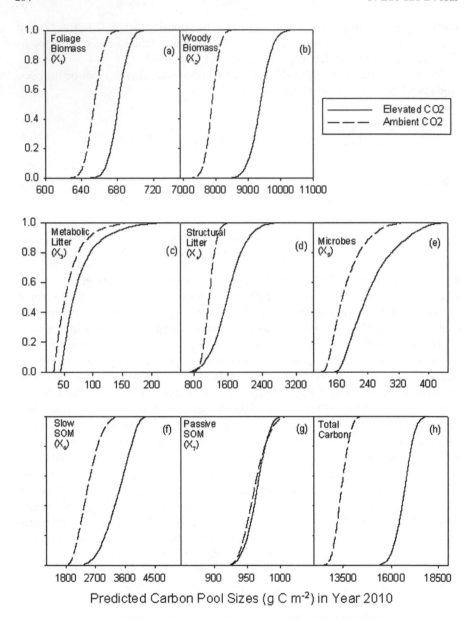

Fig. 10.8 Cumulative probability density functions (CDFs) of simulated carbon pool sizes at the Duke Forest FACE site in year 2010 (Xu et al. 2006 with permission from AGU). CDFs at ambient CO_2 was depicted by *dashed lines* and at elevated CO_2 by solid lines in non-woody (**a**), woody (**b**), metabolic litter (**c**), structural litter (**d**), microbial (**e**), slow soil organic matter (**f**), passive organic matter (**g**), and whole-ecosystem carbon (**h**) pools. Soil carbon pool sizes at elevated CO_2 are significantly higher than those at ambient CO_2 in woody plant and whole-ecosystem pools

10.7 Applications of Inverse Analysis to Manipulative Experiments

The inverse analysis has recently been applied to ecological research for parameter estimation (Schulz et al. 2001, White and Luo 2002, Hanan et al. 2002, Luo et al. 2003, Braswell et al. 2005, Xu et al. 2006, Pataki et al. 2007, Hui et al. 2008), uncertainty analysis (Knorr and Kattge 2005, Williams et al. 2005, Xu et al. 2006), and selection of alternative model structure (Sacks et al. 2006). For example, Luo et al. (2003) have applied the inverse analysis to six data sets of soil respiration, woody biomass, foliage biomass, litterfall, C content in the litter layers, C content in mineral soil measured in both ambient CO_2 (350 ppm) and elevated CO_2 (550 ppm) plots from the Duke FACE experiment from 1996 to 2000. The inverse analysis was designed to estimate carbon residence time, which quantifies the capacity for carbon storage in various plant and soil pools. The estimated residence times at elevated CO_2 decreased for plant C pools and increased for litter and soil pools in comparison to those at ambient CO_2 (Table 10.1). The estimated residence times from all the pools at elevated CO_2, however, were well correlated with those at ambient CO_2. The estimated residence times, combined with C influx, were used to simulate C sequestration rates in response to a gradual increase in atmospheric CO_2 concentration. The simulated C sequestration rate gradually increased from 69 g m^{-2} yr^{-1} in 2000 when CO_2 concentration was 378 ppm to 201 g m^{-2} yr^{-1} in 2100 when CO_2 concentration was set at 710 ppm.

Xu et al. (2006) applied the Bayesian probability inversion and a Markov Chain Monte Carlo (MCMC) technique to analyze uncertainties of estimated C transfer coefficients and simulated C pool sizes. They used a Metropolis-Hastings algorithm to construct a posterior probability density function (PPDF) of C transfer coefficients based on prior information of model parameters, model structure and the same six data sets as used by Luo et al. (2003) from the Duke Forest FACE site. According to the constructed PPDFs, the transfer coefficients from pools of non-woody biomass, woody biomass, and structural litter are well-constrained by the six data sets under both ambient and elevated CO_2. The data sets also contain moderate information to the transfer coefficient from slow soil C pool. However, the transfer coefficients from pools of metabolic litter, microbe, and passive soil C are poorly constrained. They also constructed cumulative distribution functions (CDFs) of simulated C pool sizes, which show that on average the ecosystem would store 16,616 g C m^{-2} at elevated CO_2 by the year of 2010, significantly higher than 13,426 g C m^{-2} at ambient CO_2 with 95% confidence (Fig. 10.8). Applications of inverse analysis significantly improved our estimation on model parameters as well as our understanding of carbon sequestration in forests under current and future climate conditions.

Table 10.1 Maximum likelihood estimates (MLEs), mean estimates, 95% high probability intervals (lower limit, upper limit) (Xu et al. 2006 with permission from AGU). As a comparison, the result in Luo et al., (2003) was also listed. Symbol "–" stands for "not available"

Parameters (g C g^{-1} d^{-1})		Xu et al. (2006)			Luo et al. (2003)
		MLE	Mean	95% high prob. interval	
Ambient	c_1 ($\times 10^{-3}$)	1.82	1.82	(1.72, 1.89)	1.76
	c_2 ($\times 10^{-4}$)	1.21	1.21	(0.99, 1.42)	1.00
	c_3 ($\times 10^{-2}$)	–	1.70	(0.66, 2.70)	2.15
	c_4 ($\times 10^{-3}$)	1.04	1.04	(0.80, 1.34)	0.845
	c_5 ($\times 10^{-3}$)	–	5.10	(3.10, 6.85)	8.530
	c_6 ($\times 10^{-4}$)	1.70	1.70	(0.55, 2.65)	0.898
	c_7 ($\times 10^{-6}$)	–	5.25	(1.51, 9.00)	3.1
Elevated	c_1 ($\times 10^{-3}$)	2.34	2.34	(2.25, 2.46)	2.17
	c_2 ($\times 10^{-4}$)	1.25	1.25	(1.19, 1.52)	1.41
	C_3 ($\times 10^{-2}$)	–	1.71	(0.65, 2.71)	2.268
	C_4 ($\times 10^{-3}$)	1.03	1.10	(0.50, 1.71)	0.965
	C_5 ($\times 10^{-3}$)	–	4.84	(2.90, 6.80)	2.534
	C_6 ($\times 10^{-4}$)	0.55	0.66	(0.50, 2.40)	0.558
	C_7 ($\times 10^{-6}$)	–	5.19	(1.60, 9.00)	2.700

10.8 Summary

Both modeling analysis by Luo and Reynolds (1999) and experimental evidence from studies by Hui et al. (2002) and ctgeq(1999 et al. (2005) suggest that ecosystem responses to the perturbation generated by a step CO_2 increase as in field CO_2 manipulative experiments are different from those to a gradual CO_2 increase as in the real world. Therefore, results from a step CO_2 increase cannot be directly extrapolated to predict carbon sequestration in natural ecosystems in the real world. While observations from the step experiments cannot be directly extrapolated, processes involved in the ecosystem responses are the same. These processes include photosynthetic carbon influx, carbon partitioning to various pools, and carbon residence times in each of the pools. To develop our ability to predict ecosystem responses to a gradual CO_2 increase in the real world, we need to analyze data from manipulative global change experiments to derive the mechanistic understanding using an inversion approach. The inverse analysis fundamentally focuses on data analysis for parameter estimation and evaluation of alternative model structures so as to improve our predictive understanding from both experimental observations and prior knowledge about the ecosystem processes.

Applications of inverse analysis to Duke Forest FACE experimental data demonstrated that uncertainty in both parameter estimations and carbon sequestration in forest ecosystems can be quantified to improve our understanding of ecosystem carbon responses to climate change. As more data are accumulating in many long-term manipulative experiments, inverse analysis will play a more important role in global change ecology.

Acknowledgments   We thank Dr. Miao and three anonymous reviewers for their constructive comments and suggestions that made this chapter much improved. Our work has been supported by grants from the Office of Science (BER), U.S. Department of Energy, grant DE-FG03-99ER62800 DE-FG02-006ER64319, and National Science Foundation (DEB0444518).

References

Andrews, J.A., and W.H. Schlesinger. 2001. Soil CO_2 dynamics in a temperate forest with experimental CO_2 enrichment. Global Biogeochemical Cycles **15**:149–162.

Ägren, G.I. 1985. Theory for growth of plants derived from the nitrogen productivity concept. Physiologia Plantarum **64**:17–28.

Allen, L.H., Jr., B.G. Drake, H.H. Rogers, and J.H. Shinn. 1992. Field techniques for exposure of plants and ecosystems to elevated CO_2 and other trace gases. Critical Reviews in Plant Science **11**:85–119.

Ashmore, M. R. 2005. Assessing the future global impacts of ozone on vegetation. Plant Cell Environ. **28**: 949–964.

Barrett, D.J. 2002. Steady state turnover time of carbon in the Australian terrestrial biosphere. Global Biogeochemical Cycles **16**, 1108, doi:10.1029/2002GB001860.

Braswell, B.H., W.J. Sacks, E. Linder, and D.S. Schimel. 2005. Estimating diurnal to annual ecosystem parameters by synthesis of a carbon flux model with eddy covariance net ecosystem exchange observations. Global Change Biology **11**: 335–355.

Cheng, W., D.C. Coleman, C.R. Corroll, and C.A. Hoffman. 1994. Investigating short-term carbon flows in the rhizospheres of different plant species, using isotopic trapping. Agronomy Journal **86**:782–788.

Comins, H.N., and R.E. McMurtrie. 1993. Long-term biotic response of nutrient-limited forest ecosystems to CO_2-enrichment: Equilibrium behavior of integrated plant-soil models. Ecological Applications **3**:666–681.

Cramer, W., A. Bondeau, F.I. Woodward, et al. 2001. Global response of terrestrial ecosystem structure and function to CO_2 and climate change: results from six dynamic global vegetation models. Global Change Biology **7**: 357–373.

Dermody, O., J.F. Weltzin, E.C. Engel, P. Allen, and R.J. Norby. 2007. How do elevated [CO_2], warming, and reduced precipitation interact to affect soil moisture and LAI in an old field ecosystem? Plant and Soil **301**: 255–266.

Ellsworth, D.S. 2000. Seasonal CO_2 assimilation and stomatal limitations in a Pinus taeda canopy. Tree Physiology **20**: 435–445.

Farquhar, G.D., S. von Caemmerer, and J.A. Berry. 1980. A biochemical model of photosynthetic CO_2 assimilation in leaves of C_3 species. Planta **149**:79–90.

Fay, P.A., J.D. Carlisle, A.K. Knapp, et al. 2000. Altering rainfall timing and quantity in a mesic grassland ecosystem: Design and performance of rainfall manipulation shelters. Ecosystems **3**: 308–319.

Fay, P.A., J.D. Carlisle, A.K. Knapp, et al. 2003. Productivity responses to altered rainfall patterns in a C_4-dominated grassland. Oecologia **137**: 245–251.

Friedli, H., H. Loestcher, H. Oeschger, U. Siegenthaler, and B. Stauffer. 1986. Ice core record of the $^{13}C/^{12}C$ record of atmospheric CO_2 in the past two centuries. Nature **324**:237–238.

Gill, R.A., H.W. Polley, H.B. Johnson, L.J. Anderson, H. Maherali, and R.B. Jackson. 2002. Nonlinear grassland responses to past and future atmospheric CO_2. Nature **417**: 279–282.

Hanan, N.P., G. Burba, S.B. Verma, J.A. Berry, A. Suyker, and E.A. Walter-Shea. 2002. Inversion of net ecosystem CO_2 flux measurements for estimation of canopy PAR absorption. Global Change Biology **8**: 563–574.

Harley, P.C., R.B. Thomas, J.F. Reynolds, and B.R. Strain. 1992. Modeling photosynthesis of cotton grown in elevated CO_2. Plant, Cell and Environment **15**: 271–282.

Harte, J., and R. Shaw. 1995. Shifting dominance within a montane vegetation community – results of a climate warming experiment. Science **267**: 876–880.

Hartley, I.P., A. Heinemeyer, and P. Ineson P. 2007. Effects of three years of soil warming and shading on the rate of soil respiration: substrate availability and not thermal acclimation mediates observed response. Global Change Biology **13**: 1761–1770.

Hendrey, G.R., D.S. Ellsworth, K.F. Lewin, et al. 1999. A free-air enrichment system for exposing tall forest vegetation to elevated atmospheric CO_2. Global Change Biology **5**: 293–309.

Herrick, J.D., and S.B. Thomas. 2001. No photosynthetic down-regulation in sweetgum trees (*Liquidambar styraciflua* L.) after three years of CO_2 enrichment at the Duke Forest FACE experiment. Plant Cell and Environment **24**: 53–64.

Hui, D., D.A. Sims, D.W. Johnson, W. Cheng, and Y. Luo. 2002. Effects of gradual versus step increase in carbon dioxide on Plantago photosynthesis and growth in a microcosm study. Environmental and Experimental Botany **47**:51–66.

Hui, D., and Y. Luo. 2004. Evaluation of soil CO_2 production and transport in Duke Forest using a process-based modeling approach. Global Biogeochemical Cycles **18**, GB4029, doi:10.1029/2004GB002297.

Hui, D., Y. Luo, D. Schimel, J.S. Clark, A. Hastings, K. Ogle, M. Williams. 2008. Converting raw data into ecologically meaningful products. Eos, Transactions, American Geophysical Union **89**: 5.

Huxman, T.E., M.D. Smith, P.A. Fay, et al. 2004. Convergence across biomes to a common rain-use efficiency. Nature **429**: 651–654.

Intergovernmental Panel on Climate Change (IPCC), Climate Change 1992 (IPCC, Geneva, 1992); www.ipcc.ch.

Intergovernmental Panel on Climate Change (IPCC), Climate Change 2007 (IPCC, Geneva, 2007); www.ipcc.ch.

Jenkinson, D.S., and J.H. Rayner. 1977. The turnover of soil organic matter in some of the Rothamsted classical experiments. Soil Science **123**: 298–305.

Kimball, B.A. 2005. Theory and performance of an infrared heater for ecosystem warming. Global Change Biology **11**:2041–2056.

Klanderud, K., and O. Totland. 2007. The relative role of dispersal and local interactions for alpine plant community diversity under simulated climate warming. Oikos **116**: 1279–1288.

Klironomos, J.N., M.F. Allen, M.C. Rillig, J. Piotrowski, S. Makvandi-Nejad, B.E. Wolfe, and J.R. Powell. 2005. Abrupt rise in atmospheric CO_2 overestimates community response in a model plant–soil system. Nature **433**, 621–624.

Knorr, W., and J. Kattge. 2005. Inversion of terrestrial ecosystem model parameter values against eddy covariance measurements by Monte Carlo sampling. Global Change Biology **11**: 1333–1351.

Koch, G.W. 1993. The use of natural situations of CO_2 enrichment in studies of vegetation responses to increasing atmospheric CO_2. Pages 381–391 *in* E.D. Schulze, and H.A. Mooney, editors. Design and Execution of Experiments on CO_2 Enrichment. Commission of the European Communities, Brussels.

Körner, C. 1995. Towards a better experimental basis for upscaling plant-responses to elevated CO_2 and climate warming. Plant Cell and Environment **18**: 1101–1110.

Körner, C., R. Asshoff, O. Bignucolo, S. Hättenschwiler, S.G. Keel, S. Peláez-Riedl, S. Pepin, R.T.W. Siegwolf, and G. Zotz. 2005. Carbon flux and growth in mature deciduous forest trees exposed to elevated CO_2. Science **309**: 1360–1362.

Laidler, K.J., and J.H. Meiser. 1982. Physical Chemistry. The Benjamin/Cummings Publishing Company. Menlo Park, CA.

Long, S.P., and P.R. Hutchin. 1991. Primary production in grasslands and coniferous forests with climate change – an over view. Ecological Applications **1**: 139–156.

Luo, Y. 2001. Transient ecosystem responses to free-air CO_2 enrichment: Experimental evidence and methods of analysis. New Phytologist 152: 3–8.

Luo, Y. 2007. Terrestrial Carbon-Cycle Feedback to Climate Warming. Annual Review of Ecology, Evolution, and Systematics 38: 683–712.

Luo, Y., C.B. Field, and H.A. Mooney. 1994. Predicting responses of photosynthesis and root fraction to elevated CO_2: Interactions among carbon, nitrogen, and growth. Plant, Cell and Environment 17: 1194–1205.

Luo Y., B. Medlyn, D. Hui, D. Ellsworth, J. Reynolds, and G. Katul. 2001a. Gross primary productivity in Duke Forest: Modeling synthesis of CO_2 experiment and eddy-flux data. Ecological Applications 11: 239–252.

Luo, Y., and H. A. Mooney. 1996. Stimulation of global photosynthetic carbon influx by an increase in atmospheric carbon dioxide concentration. Pages 381–397 in G. W. Koch, and H. A. Mooney, editors. Carbon Dioxide and Terrestrial Ecosystems. Academic Press, San Diego.

Luo, Y., and J.F. Reynolds. 1999. Validity of extrapolating field CO_2 experiments to predict carbon sequestration in natural ecosystems. Ecology 80: 1568–1583.

Luo, Y., D.A. Sims, K.L. Griffin. 1998. Nonlinearity of photosynthetic responses to growth in rising atmospheric CO_2: An experimental and modeling study. Global Change Biology 4: 173–183.

Luo, Y., B. Su, W.S. Currie, J.S. Dukes, A. Finzi, U. Hartwig, B. Hungate, R.E. McNurtrie, R. Oren, W.J. Parton, D.E. Pataki, M.R. Shaw, D.R. Zak, C.B. Field. 2004. Progressive nitrogen limitation of ecosystem responses to rising atmospheric carbon dioxide. BioScience 54: 731–739.

Luo, Y., L. White, J. Canadell, E. DeLucia, D. Ellsworth, A. Finzi, J. Lichter, and W. Schlesinger. 2003. Sustainability of terrestrial carbon sequestration: A case study in Duke Forest with inversion approach. Global Biogeochemical Cycles. 17, 1021, doi:10.1029/2002GB001923

Luo, Y., L. Wu, J.A. Andrews, L. White, R. Matamala, K.V.R. Schafer, and W. H. Schlesinger. 2001b. Elevated CO_2 differentiates ecosystem carbon processes: Deconvolution analysis of Duke Forest FACE data. Ecological Monographs 71: 357–376.

Luo, Y.Q., D. Hui, and D. Zhang. 2006. Elevated Carbon Dioxide Stimulates Net Accumulations of Carbon and Nitrogen in Terrestrial Ecosystems: A Meta-Analysis. Ecology 87: 53–63.

Mack, M.C., E.A.G. Schuur, M.S. Bret-Harte, et al. 2004. Ecosystem carbon storage in arctic tundra reduced by long-term nutrient fertilization. Nature 431: 440–443.

McGuire, A.D., S. Sitch, J.S. Clein, R. Dargaville, G. Esser, J. Foley, M. Heimann, F. Joos, et al. 2001. Carbon balance of the terrestrial biosphere in the twentieth century: Analyses of CO_2, climate and land use effects with four process-based ecosystem models. Global Biogeochemical Cycles 15: 183–206.

Melillo, J.M., P.A. Steudler, J.D. Aber, K. Newkirk, H. Lux, F.P. Bowles, C. Catricala, A. Magill, T. Ahrens, and S. Morrisseau. 2002. Soil warming and carbon-cycle feedbacks to the climate system. Science 298: 2173–2176.

Miglietta, F., A. Peressotti, F.P. Vaccari, A. Zaldei, P. deAngelis, and G. Scarascia-Mugnozza. 2001. Free-air CO_2 enrichment (FACE) of a poplar plantation: the POPFACE fumigation system. New Phytologist 150: 465–476.

Neftel, A., H. Oeschger, J. Schwander, B. Stauffer, and R. Zumbrunn.1982. Ice core sample measurements give atmospheric CO_2 content during the past 40,000 yr. Nature 295: 220–223.

Norby, R.J., N.T. Edwards, J.S. Riggs, C.H. Abner, S.D. Wullschleger, and C.A. Gunderson. 1997. Temperature-controlled open-top chambers for global change research Global Change Biolog 3: 259–267.

Olson, J.S. 1963. Energy-storage and balance of producers and decomposers in ecological-systems. Ecology 44: 322–331.

Parton, W.J., D.S. Schimel, C.V. Cole, and D.S. Ojima. 1987. Analysis of factors controlling soil organic matter levels in Great-Plains grasslands. Soil Science Soc. of America Journal 51: 1173–1179.

Pataki, D.E., T. Xu, Y. Luo, and J.R. Ehleringer. 2007. Inferring biogenic and anthropogenic CO_2 sources across an urban to rural gradient. Oecologia 152: 307–322.

Polley, H.W., H.B. Johnson, and J.D. Derner. 2003. Increasing CO_2 from subambient to superambient concentrations alters species composition and increases above-ground biomass in a C3/C4 grassland. New Phytologist 160: 319–327.

Polley, HW., P.C. Mielnick., W.A. Dugas, H.B. Johnson, and J. Sanabria. 2006. Increasing CO_2 from subambient to elevated concentrations increases grassland respiration per unit of net carbon fixation. Global Change Biology 12: 1390–1399.

Raschi, A., F. Milglietta, R. Tognetti, and P.R. van Gardingen, editors. 1997. Plant responses to elevated CO_2: evidence from natural springs. Cambridge University Press, New York.

Rastetter, E.B., G.I. Ågren, and G.R. Shaver. 1997. Responses of N-limited ecosystems to increased CO_2: A balanced-nutrition, coupled-element-cycles model. Ecological Applications 7: 444–460.

Raupach, M.R., P.J. Rayner, D.J. Barrett, R.S. Defries, M. Heimann, D.S. Ojima, S. Quegan, and C.C. Schmullius. 2005. Model-data synthesis in terrestrial carbon observation: methods, data requirements and data uncertainty specifications. Global Change Biology 11: 378–397.

Reynolds, J.F., and P.W. Leadley. 1992. Modeling the response of arctic plants to changing climate. Pages 413–438 in F. S. Chapin, III, R. Jefferies, J. F. Reynolds, G. Shaver and J. Svoboda, editors. Arctic Physiological Processes in a Changing Climate, San Diego, CA.

Rouhier, H., G. Billès, L. Billès, and P. Bottner. 1996. Carbon fluxes in the rhizosphere of sweet chestnut seedlings (Castanea sativa) grown under two atmospheric CO_2 concentrations: [14]C partitioning after pulse labelling. Plant and Soil 180: 101–111.

Rustad, L.E., J.L. Campbell, G.M. Marion, R.J. Norby, M.J. Mitchell, A.E. Hartley AE, and J. Gurevitch. 2001. A meta-analysis of the response of soil respiration, net nitrogen mineralization, and aboveground plant growth to experimental ecosystem warming. Oecologia 126: 543–562.

Rustad, L.E. 2006. From transient to steady-state response of ecosystems to atmospheric enrichment and global climate change: conceptual challenges and need for an integrated approach. Plant Ecology 182: 43–62.

Sacks, W.J., D.S. Schimel, R.K. Monson, and B.H. Braswell. 2006. Model-data synthesis of diurnal and seasonal CO_2 fluxes at Niwot Ridge, Colorado. Global Change Biology 12: 240–259.

Shaver, G.R., J. Canadell, F.S. Chapin, J. Gurevitch, J. Harte, G. Henry, P. Ineson, S. Jonasson, J. Melillo, L. Pitelka, L. Rustad. 2000. Global warming and terrestrial ecosystems: A conceptual framework for analysis. Bioscience 50: 871–882.

Sherry, R.A., X. Zhou, S. Gu, J.A. Arnone III, D.S. Schimel, P.S. Verburg, L.L. Wallace and Y. Luo. 2007. Divergence of Reproductive Phenology under Climate Warming. Proceedings of National Academy of Sciences USA 104: 198–202.

Schulz, K., A. Jarvis, K. Beven, and H. Soegaard. 2001. The predictive uncertainty of land surface fluxes in response to increasing ambient carbon dioxide. Journal of Climate 14: 2551–2562.

Sitch, S., P.M. Cox, W.J. Collins, and C. Huntingford. 2007. Indirect radiative forcing of climate change through ozone effects on the land-carbon sink. Nature 448: 791–794.

Thompson, M.V., and J.T. Randerson. 1999. Impulse response functions of terrestrial carbon cycle models: method and application. Global Change Biology 5: 371–394.

Verburg, P.S.J., J.A Arnone III, R.D. Evans, D. LeRoux-Swarthout, D. Obrist, D.W. Johnson, D.E. Schorran, Y. Luo, and J.S. Coleman. 2004. Net ecosystem C exchange in two model grassland ecosystems. Global Change Biology 10: 498–508.

Wan, S., Y. Luo, and L. Wallace. 2002. Changes in microclimate induced by experimental warming and clipping in tallgrass prairie. Global Change Biology 8: 754–768.

Weltzin, J.F., M.E. Loik, S. Schwinning, et al. 2003. Assessing the response of terrestrial ecosystems to potential changes in precipitation. Bioscience 53: 941–952.

White, L., and Y. Luo. 2002. Inverse analysis for estimating carbon transfer coefficients in Duke Forest. Applied Mathematics and Computation **130**: 101–120.

White, L., and Y. Luo. 2005. Model-based CO_2 data assessment for terrestrial carbon processes: Implications for sampling strategy in FACE experiments. Applied Mathematics and Computation **167**: 419–434.

White, L., Y. Luo, and T, Xu. 2005. Carbon sequestration: inversion of FACE data and prediction. Applied Mathematics and Computation **163**: 783–800.

Wieder, R.K., and G.E. Lang. 1982. A critique of the analytical methods used in examining decomposition data obtained from litter bags. Ecology **63**: 1636–1642.

Williams, M., P.A. Schwarz, B.E. Law, J. Irvine, and M.R. Kurpius. 2005. An improved analysis of forest carbon dynamics using data assimilation. Global Change Biology **11**: 89–105.

Xu, T., L. White, D. Hui, and Y. Luo. 2006. Probabilistic inversion of a terrestrial ecosystem model: Analysis of uncertainty in parameter estimation and model prediction, Global Biogeochemical Cycles **20**, GB2007, doi:10.1029/2005GB002468

Yahdjian, L., and O.E. Sala. 2002. A rainout shelter design for intercepting different amounts of rainfall. Oecologia **133**: 95–101.

Chapter 11
Ecology in the Real World: How Might We Progress?

James B. Grace, Susan Carstenn, ShiLi Miao, and Erik Sindhøj

11.1 Introduction

One might imagine some system or research problem of interest that is the ideal subject for study, where samples can be readily obtained, where controlling factors can be physically manipulated however one desires, where responses are repeatable, and where understanding emerges in direct proportion to the effort put in by the research team. In this ideal case, probabilistic statements might not even seem necessary because results are regular, findings are repeatable, and results lack uncertainty. Unfortunately, large-scale natural systems are not so easy to study.

This book is about constraints and how they impede our ability to make confident probability statements about the characteristics and behavior of ecological systems. This book is also about how we might best make scientific progress despite these constraints. As the title of this book states, we are concerned with the real world of large-scale and long-term ecological systems and the context of which the public and decision-makers are usually most concerned. As ecologists, we often feel like the data we really need to address our societal responsibilities are at odds with the conventions of sampling and experimental designs. In this context, there is a feeling of urgency about our need to move forward despite the constraints. The chapters in this book all tackle constraints, either of logistics, study design, or analytic assumptions, and demonstrate a variety of methodological approaches for various situations. In this chapter, we seek to put the ideas in this book into a general framework.

There is a long history of ecologists seeking to be self aware in their use of scientific approaches to learn about systems (e.g., Levins 1968, Pickett, Kolasa and Jones 1994). Particularly relevant to this book is the work by Holling (1978), who dealt with the question of how science might best aid the management of ecological systems. Another recent, highly relevant, and excellent effort is the book by Canham, Cole, and Laurenroth (2003) on the use of models in ecosystem science. While the objectives of these books differ somewhat from the

J.B. Grace (✉)
U.S. Geological Survey, 700 Cajundome Blvd., Lafayette, LA 70508, USA
e-mail: gracej@usgs.gov

S. Miao et al. (eds.), *Real World Ecology*, DOI 10.1007/978-0-387-77942-3_11,
© Springer Science+Business Media, LLC 2009

focus of the current book, they collectively suggest a perspective from which we might profitably discuss issues, suggest possible approaches, and focus on the contributions of alternative methodological devices to scientific progress.

11.2 A Framework for Thinking about Constraints and Alternative Approaches to the Study of Ecological Systems

Our perspective on the issue of how different approaches contribute to scientific progress can be summarized in the form of a diagram (Fig. 11.1). This diagram is inspired by a simpler one presented by Holling (1978) to classify general scientific approaches (description versus modeling versus statistics); several other derivations of the original diagram exist (Starfield and Bleloch 1986, Turner 2003). In this case, we have expanded the number of distinctions in the diagram to focus on the variety of statistical modeling methodologies, particularly what to do with the collection and analysis of data under constrained conditions.

When we tackle a new research question or topic, we are typically operating in a situation where there are little data to quantify patterns and also little understanding of the processes and mechanisms that cause those patterns. This situation corresponds to the lower left-hand corner of Fig. 11.1. Under such conditions, our science is often operating in an exploratory mode in which our focus is on accumulating observations, developing new measurement methods, and developing ideas about causal processes. As our study of a topic continues, we expect an accumulation of additional data and an increase in our understanding of underlying processes. We see two main

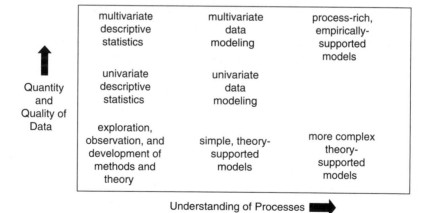

Fig. 11.1 Diagram representing the progress of science, which seeks to move from the *lower left* to *upper right* and the logical relationships among scientific methodologies relative to that progress (elaborated from Holling 1978)

avenues forward in pursuing understanding of a problem or system. The first avenue is to collect data to develop an improved description of phenomena (moving vertically in the diagram). The second avenue forward is to borrow knowledge about processes that are operating. When we say "borrow" we mean that many of our candidate models for how the system might work come from other situations or systems that seem relevant. In the early days of ecosystem science, Eugene Odum (1969) borrowed heavily from knowledge about organismal development to postulate theories about ecosystem development. Such a logical process can be described as deductive in the sense that only logical premises are involved and data are not required to develop new inferences. Such is the predominant mode for many theoretical approaches to science. We must also recognize that movement to the right in the diagram can come just as readily from inductive thinking when we infer process from data. In reality, deductive and inductive logic are intertwined in our thinking, though it may still be useful to distinguish between these two modes of learning when thinking about the scientific process.

For simplicity's sake, we presume that the scientific enterprise collectively seeks to move toward some level of information and understanding that is process-rich and empirically supported, which corresponds to the upper right-hand corner of the diagram (Fig. 11.1). Some may argue that science is more about the journey than the ultimate destination. However, reality requires accountability to managers and policy makers, funding agencies, and others for whom the destinations are most important. In either case, for studies of ecological systems, there is practical utility in striving toward such a goal, even if we are forced to recognize that such a destination may be impossible to reach for many problems. If we accept these descriptions of starting and ending points for scientific analysis, we may then view the problem of studying systems as one of moving efficiently from our diagram's lower left-hand corner (Fig. 11.1) to the upper right-hand corner. To a substantial degree, this idea meshes rather well with the concept of "theory maturation" introduced to ecology by Craig Loehle (1987) to describe the logical progression from new, unproven, and often vague ideas to clear, proven ones.

11.3 The Variety of Statistical Methodologies

Because of our interest in providing context for the question "How do we study large-scale ecosystems in the real world? " we emphasize a variety of statistical methodologies, describing when they are appropriate, what roles they play, and how they relate to our ability to make scientific progress. While finer distinctions can (and will) be drawn, our diagram (Fig. 11.1) recognizes two main classifications that fall into the traditional statistical arena, distinguishing between descriptive statistics versus "data modeling" and between univariate versus multivariate methods. Both distinctions deserve some discussion.

Descriptive Statistics versus Data Modeling – Another diagrammatic device illustrates how data and theoretical ideas contribute to statistical and modeling methodologies. Let us first consider that models that we think of as statistical derive relationships from data. They differ from dynamic simulation modeling efforts that make a number of assumptions (including assumptions based on previous data) and then proceed with a logical analysis of the implications of those assumptions. We will refer to this latter type of model analysis (one that does not directly involve data) as "theoretical". In contrast to theoretical modeling, we use the phrases "data modeling", "empirical modeling" or "statistical analysis" to refer to situations where data are being analyzed or directly incorporated into process models in a single analysis. Our representation of the range of involvement of data and theoretical ideas in models is shown in Fig. 11.2.

While not always appreciated by ecologists or other scientists, a great continuum of statistical models exists, from those that are almost purely descriptive to those that are extremely structured by theoretical ideas (Fig. 11.2). Traditionally, based on what is usually taught in statistics courses or found in introductory statistics textbooks, scientists are exposed to statistical models that have very simplistic representations of processes, though strong data requirements (Fig. 11.2A). The simplest are those that describe the distributions of observations, such as frequency histograms. Very quickly, elementary theoretical ideas begin to insert themselves as we consider abstractions such as "means" and "standard deviations" as summaries of the raw information.

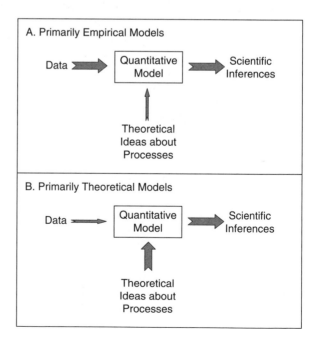

Fig. 11.2 Representation of the fact that quantitative models are informed by both the theoretical ideas about processes that are used to develop their structure and by the data that is available and appropriate to that model. Models can range widely from statistical models strongly tied to the data and with little theoretical structure (Fig. 11.2A), to models reliant primarily on theoretical knowledge (Fig. 11.2B)

When two or more variables are involved, we might consider the associations between variables and, again with some abstraction, might derive some measures of the strength of association (e.g., Spearman or Pearson correlations). When less descriptive analyses are conducted, we begin to work with models that are based on some basic premise about causal processes. Such theoretical ideas include the premise of a unidirectional influence by some variable or set of variables on a response variable. We may now begin to consider regression or other models where we describe the "effect" of factors on responses.

When we move to regression or other dependence models, we should realize that we are now relating our empirical information (data) to some model of process. A typical representation of such a situation could be

$$y_i = b \times x_i + \varepsilon, \qquad (11.1)$$

where y is some response measured over i cases, x is some causal condition also measured for those cases, and epsilon (ε) is some summarization of the irregularities in the relationship. In this model, b can be viewed as not only the coefficient that quantitatively relates x to y, it also represents the effect of the process. The nature of the process is determined by the hypothesis represented by the equation, in this case, a linear impact of x on y. Statisticians can say a great deal about equation 11.1, about the assumptions inherent in the model, and about the inferences that result. From a scientist's perspective, the representation of process by equation 11.1 is extraordinarily simplistic.

How satisfied will we be with a process model that simply says that there is some linear dependence, as in equation 1? Our understanding of the processes implied or represented by statistical models can be enhanced by the use of graphical models. Sewell Wright (1921) was among the first to emphasize the role of graphical models as a way of better representing the scientific processes associated with mathematical expressions. Recently, graphical modeling has seen a major expansion in interest (Pearl 2000, Spirtes, Glymour and Scheines 2000, Borgelt and Kruse 2002) because of its value in conveying the causal implications of equations. Ultimately in statistical models, the forms of the individual relationships and many details of the processes must be represented by the equations, while graphical representations can be used to convey structural features of hypotheses such as mediation and indirect effects (e.g., in Chapter 2 by Grace et al., Fig. 2.1). What is most important here is the idea that data can and should be integrated into models that convey processes to a much greater degree of realism.

It is worth pointing out that models based on elaborate portraits of processes are infrequently developed by statisticians simply because much theoretical information about the subject must be included, and statisticians usually are not the ones who possess such knowledge. For this reason, statisticians tend (with important exceptions) to emphasize aspects of models that subject matter experts recognize as being essential, but consider to be of secondary scientific importance (e.g., the precise formulation of error terms and standard errors in complex sampling situations). It is usually up to the quantitative ecologist to build and evaluate models having more realistic processes represented (e.g., Hilborn and Mangel 1997), though they certainly may rely heavily on statisticians for the building blocks of these models. The enterprise of data or empirical modeling as used here is meant to represent a flexible approach to developing statistical models with structures that match the major causal processes operating in the system. Such applications are very much on the increase in biology, both in relation to univariate and multivariate modeling. Ecological practitioners are, for the most part, just beginning to learn about these methods, and they represent a strong departure from the "fixed structure" statistical analyses considered "traditional" for the field. Statistical modeling, which allows development of custom models for specific situations, is capable of accommodating some of the features of data from large-scale, long-term real world studies at least when sampling is not too badly compromised beyond a certain point.

Traditionally it has been in the realm of theoretical modeling where inference has involved a heavy dose of ideas about processes and an incomplete complement of data (Fig. 11.2B). Increasingly, the gap between theoretical and statistical modeling is being bridged (Clark and Gelfand 2006), creating a greater range of intergrading possibilities.

Univariate versus Multivariate Statistical Models – In addition to distinguishing between descriptive statistics and data modeling, there are also differences between univariate and multivariate statistical models. Univariate models have been the preferred form for most hypothesis testing in biology. In univariate models, focus is placed on explaining a single response with some set of explanatory variables. It is often unappreciated that univariate models (e.g., ANOVA, multiple regression, and the general linear model) are not designed for studying systems, but instead, for analyzing individual processes. Multivariate models, in contrast, can (at least for certain types of multivariate models) represent multiprocess features of systems, and in the case of methods like structural equation modeling, permit partitioning effects among pathways and provide some evaluation of their relative importance. For these reasons, we see multivariate modeling as essential if we are to work with models that can explain system behavior.

Applications of multivariate statistical analysis in the ecological sciences have for the most part focused on description (upper left corner of Fig. 11.1). Methods such as cluster analysis, principal components, and other ordination techniques (e.g., McCune and Grace 2002) are designed to summarize and describe associations among suites of variables. Such models used alone lack the capacity to incorporate what is learned about processes from preceding studies into the structure of the model used in the next investigation. Rather, their fixed structure restricts the scientist to a series of descriptive studies in which the data from subsequent studies rarely test the conclusions about processes drawn from earlier studies (except in the most general way). Multivariate modeling, such as represented by structural equation (path) or other network models (upper middle of Fig. 11.1), has seen limited use in the biological sciences, though there have been many successful applications over the years. Interest in such applications is currently increasing, driven both by the demands on our science to provide system-relevant information and by increased exposure (e.g., Shipley 2000, Pugesek, Tomer and von Eye 2003, Grace 2006).

11.4 The Landscape of Methodologies

Even when dealing with the ideal case where replicate samples are easily obtained, factors of experimental interest can be precisely manipulated, confounding factors can be controlled and responses are repeatable, many of the often-used statistical procedures cannot, by themselves, lead to the understanding of ecological systems that we desire. We must consider, rather, that each methodology has its own domain of applicability. Here Fig. 11.1 is inadequate as a heuristic device because many specific issues may still apply for studies unconstrained by sampling or complex behavior. Models need to accommodate spatial structure, nonlinear responses, and a variety of statistical distributions while still incorporating process-relevant equations and permitting scientific interpretations of the sort that advance our understanding of processes. At this point, we emphasize the good news that much of the flexibility needed to create models that will allow us to move efficiently across the research landscape in Fig. 11.1 is achievable. A great profusion of statistical tools and procedures now exist; the chapters in this volume provide a very large number of examples. Much of this development has been recent, and it is rapid indeed. What is noticeably slower is the diffusion of knowledge about those methodologies and especially so among practicing researchers. As will be discussed below, we are currently more limited by the supply of knowledgeable, skilled data analysts than by the tools at their disposal. For this reason, continued work on the tools themselves can be only a partial solution to the problem of efficient scientific progress.

11.5 The Problem of Constraints on Sampling and Some Solutions

This book emphasizes the constraints that impair scientific progress in our understanding of ecosystems. Considered throughout the chapters in this book, and especially summarized in the Introductory chapter, is that many of the desired features of data for analysis are hard to come by in large-scale studies. In the context of Fig. 11.1, we can view this as a constraint on progress in the vertical dimension. Both the quantity and quality of data desired for rigorous estimations of parameters and probabilities are unattainable in many cases. In such studies, samples are conspicuously compromised in that they are neither random relative to the world at large nor representative in any reasonable sense. The two reasons for this lapse are the logistic infeasibility of taking random samples and the weak repeatability of many ecological phenomena. While the first of these is obvious, the second requires a little elaboration. Take for example the effects of fire on an ecosystem. Responses may vary for a great many reasons, including the characteristics of the fire itself, the environmental context in which it occurs and in the subsequent conditions that develop (e.g., rainfall events during the post-fire period). So many factors influence the outcome of the fire that scientists need both a considerable knowledge of the factors to measure and a very large and adequate sample with which to estimate all the factors.

Related to the problems of sampling mentioned here is the issue of alternative study designs. Chapter 1 (Introduction) as well as some other chapters (Chapter 4) describe in detail both the challenges of satisfying requirements for spatial replication and the various innovative approaches that have been developed to substitute temporal replication (e.g., before–after designs). The chapter by Peters et al. gives some insight into the realm of hybrid approaches where sampling designs are constructed for the purpose of estimating parameters for numerical simulations rather than for assessments of probabilities/uncertainty. Here, we can use the Bayesian concept of uncertainty to illustrate a major point emphasized in Chapters 1 and 4; namely, that when it comes to providing guidance for immediate management decisions, one sample (from unreplicated designs) is better than no samples. While unreplicated designs have difficulties in providing the information needed for quantifying uncertainty, it is undoubtedly true that the values from a single sample greatly reduce uncertainty relative to the prior state of no information. This is the real world confronted by Carpenter (1996), Miao and Carstenn (2006), as well as by others.

11.6 A Way Forward

The chapters in this book suggest an emerging theme for how to progress scientifically when our ability to collect ideal data for an ideal statistical analysis is compromised. The message that emerges from this volume can be summarized by reference to Fig. 11.2. In this figure, we illustrate how two sources of information for quantitative modeling contribute to our ability to make scientific inferences and that their relative contributions can vary. The chapters in this book illustrate a variety of contributions of the two sources of information. Relative to our diagram in Fig. 11.1, we can see that studies in which adequate data can be readily obtained tend to progress along a track from lower left to upper right that involves considerable quantity and quality of data. When access to such data are not possible, the path forward must rely more on theoretical knowledge of processes, with data filling in critical places as needed and possible. Many of our chapters illustrate very nicely the combination of these two techniques.

Historically, the gulf between data analysis and theoretical modeling has been viewed as large and disjunct. As mentioned above, considerable effort is being spent to develop statistical modeling procedures that can accommodate a greater degree of heterogeneity in data while still accommodating complex process models. Much of this work is now being done within a Bayesian modeling framework (e.g., Congdon 2003), but much of the same flexibility can be achieved using likelihood procedures as well (c.f. Lele, Dennis, and Lutscher 2007). Ultimately, though, statistical modeling methods are limited in situations where constraints on data collection and data quality are great. In such cases, theoretical modeling methods that can borrow information more freely from data sources, and at the same time forego statistical estimations of probabilities in favor of model scenarios, need to be used.

Current practice is limited to a greater degree by insufficiently distributed knowledge about modeling methods than by their actual availability. This situation implies a fundamentally important point recently considered by Urban (2003) relative to theoretical modeling, that education of both data modeling experts and general practitioners is a critical imperative if we are, as a body of scientists, to have sufficient access to and familiarity with the methods that can advance our science most effectively. We will not repeat the detailed prescription given in Urban (2003) for how this might be accomplished, but will simply lend our support to the many recommendations he makes for how we must expand our institutional capacity to incorporate quantitative advances into our scientific studies. Those with substantial data analysis expertise are keenly aware of the dauntingly small ratio of analysts to data. It is clear, in light of enterprises such as NEON (Terri and Raven 2002), as well as the general increase in our capacity to generate data, that we are well into an era in which the production of data and the capacity for advanced analyses will be greatly unbalanced. The solution to this problem must involve a concerted effort by institutions and funding agencies to create the training infrastructure needed to meet the demand. It seems unlikely that individual efforts alone will be sufficient to create the training centers and instructional materials required to meet the demand of an ecological science responsible for advising decision makers on environmental issues. This training will need to encompass the full range of statistical and theoretical methods represented in Fig. 11.1, if large-scale and long-term ecological studies, constrained as they will surely be, can be as efficiently addressed as their importance dictates.

It is our hope that the chapters in this book provide a glimpse into some of the possibilities for how we may make greater progress in the development of scientific understanding about large-scale ecosystems. Methodological advances are permitting a greater range of possibilities for dealing with quantitative data, while societal pressures are pushing us to make progress as rapidly as possible. This convergence provides the impetus for the many ideas presented in this volume and for many more yet to come.

Acknowledgment We appreciate M. Nungesser, Matt Kirwan, and D. Drum for their comments on the early draft of the manuscript.

References

Borgelt, C. and Kruse, R. 2002. Graphical Models. John Wiley & Sons, New York.
Canham, C.D., Cole, J.J., and Laurenroth, W.K. (eds.) 2003. Models in Ecosystem Science. Princeton University Press, Princeton, NJ.
Carpenter, S. 1996. Microcosm experiments have limited relevance for community and ecosystem ecology. Ecology, 77: 677
Clark, J.S. and Gelfand, A.E. 2006. A future for data and models in ecology. Trends in Ecology and Evolution. 21: 375–380.
Congdon, P. 2003. Applied Bayesian Modelling. John Wiley & Sons, New York.
Grace, J.B. 2006. Structural Equation Modeling and Natural Systems. Cambridge University Press, Cambridge UK
Hilborn, R. and Mangel, M. 1997. The Ecological Detective: Confronting Models with Data. Princeton University Press, Princeton, NJ.
Holling, C.S. 1978. Adaptive Environmental Assessment and Management. John Wiley & Sons, New York.
Lele, S.R., Dennis, B., and Lutscher, F. 2007. Data cloning: easy maximum likelihood estimation for complex ecological models using Bayesian Markov chain Monte Carlo methods. Ecology Letters, 10: 551–563.
Levins, R. 1968. Evolution in Changing Environments. Princeton University Press, Princeton, NJ.
Loehle, C. 1987. Hypothesis testing in ecology: psychological aspects and the importance of theory maturation. The Quarterly Review of Biology, 62: 397–409.
McCune, B. and Grace, J.B. 2002. Analysis of Ecological Communities. MJM, Gleneden Beach, Oregon
Miao, S. and Carstenn, S. 2006. A new direction for large-scale experimental design and analysis. Frontiers in Ecology and the Environment. 4: 227.
Odum, E.P. 1969. The strategy of ecosystem development. Science 164: 262–270.
Pearl, J. 2000. Causality. Cambridge University Press, Cambridge, UK.
Pickett, S.T.A., Kolasa, J., and Jones, C.G. 1994. Ecological Understanding. Academic Press, San Diego, California.
Pugesek, B.H., Tomer, A., and von Eye, A. 2003. Structural Equation Modeling. Cambridge University Press, Cambridge, UK.
Shipley, B. 2000. Cause and Correlation in Biology. Cambridge University Press, Cambridge, UK.
Spirtes, P., Glymour, C., and Scheiners, R. 2000. Causation, Prediction, and Search. Cambridge University Press, Cambridge, UK.
Starfield, A.M. and Bleloch, A.L. 1986. Building Models for Conservation and Wildlife Management. Macmillian Publishing, New York.
Terri, J.A. and Raven, P.H. 2002. A national ecological observatory network. Science 298: 1893.
Turner, M.G. 2003. Modeling for synthesis and integration: Forest, people, and riparian coarse woody debris. pp. 83–110, In: Canham, C.D., Cole, J.J. and Laurenroth, W.K. (eds.) 2003. Models in Ecosystem Science. Princeton University Press, Princeton, NJ.
Urban, D.L. 2003. A community-wide investment in modeling. pp. 466–470, In: Canham, C.D., Cole, J.J. and Laurenroth, W.K. (eds.) Models in Ecosystem Science. Princeton University Press, Princeton, NJ.
Wright, S. 1921. Correlation and causation. Journal of Agricultural Research, 10: 557–585.

Index

Printed in the United States of America